This publication is due for return, on or before the last date shown below.

PERMANENT LOAN

J. Fearnley

GUIDELINES FOR ENGINEERING DESIGN FOR PROCESS SAFETY

This book is one in a series of process safety guideline and concept books published by the Center for Chemical Process Safety (CCPS). Please go to *www.wiley.com/go/ccps* for a full list of titles in this series.

GUIDELINES FOR ENGINEERING DESIGN FOR PROCESS SAFETY

Second Edition

Center for Chemical Process Safety
New York, NY

A JOHN WILEY & SONS, INC., PUBLICATION

Copyright © 2012 by American Institute of Chemical Engineers, Inc.

Published by John Wiley & Sons, Inc., Hoboken, New Jersey. All rights reserved.
Published simultaneously in Canada.

No part of this publication may be reproduced, stored in a retrieval system or transmitted in any form or by any means, electronic, mechanical, photocopying, recording, scanning or otherwise, except as permitted under Section 107 or 108 of the 1976 United States Copyright Act, without either the prior written permission of the Publisher, or authorization through payment of the appropriate per-copy fee to the Copyright Clearance Center, Inc., 222 Rosewood Drive, Danvers, MA 01923, (978) 750-8400, fax (978) 750-4470, or on the web at www.copyright.com. Requests to the Publisher for permission should be addressed to the Permissions Department, John Wiley & Sons, Inc., 111 River Street, Hoboken, NJ 07030, (201) 748-6011, fax (201) 748-6008, or online at http://www.wiley.com/go/permission.

Limit of Liability/Disclaimer of Warranty: While the publisher and author have used their best efforts in preparing this book, they make no representation or warranties with respect to the accuracy or completeness of the contents of this book and specifically disclaim any implied warranties of merchantability or fitness for a particular purpose. No warranty may be created or extended by sales representatives or written sales materials. The advice and strategies contained herein may not be suitable for your situation. You should consult with a professional where appropriate. Neither the publisher nor author shall be liable for any loss of profit or any other commercial damages, including but not limited to special, incidental, consequential, or other damages.

For general information on our other products and services please contact our Customer Care Department within the United States at (800) 762-2974, outside the United States at (317) 572-3993 or fax (317) 572-4002.

Wiley also publishes its books in a variety of electronic formats. Some content that appears in print, however, may not be available in electronic formats. For more information about Wiley products, visit our web site at www.wiley.com.

Library of Congress Cataloging-in-Publication Data:

Guidelines for engineering design for process safety. — 2nd ed.
 p. cm.
Includes bibliographical references and index.
ISBN 978-0-470-76772-6 (hardback)
 1. Chemical plants—Safety measures. I. American Institute of Chemical Engineers. Center for Chemical Process Safety.
 TP155.5.G765 2012
 660'.2804—dc23   2011041436

Printed in the United States of America.

10 9 8 7 6 5 4 3

It is sincerely hoped that the information presented in this document will lead to an even more impressive safety record for the entire industry. However, the American Institute of Chemical Engineers, its consultants, the CCPS Technical Steering Committee and Subcommittee members, their employers, their employers' officers and directors, and Aon Energy Risk Engineering, and its employees do not warrant or represent, expressly or by implication, the correctness or accuracy of the content of the information presented in this document. As between (1) American Institute of Chemical Engineers, its consultants, CCPS Technical Steering Committee and Subcommittee members, their employers, their employers' officers and directors, and Aon Energy Risk Engineering, and its employees and (2) the user of this document, the user accepts any legal liability or responsibility whatsoever for the consequences of its use or misuse.

CONTENTS

Acronyms and Abbreviations	*xv*
Glossary	*xxi*
Acknowledgments	*xxxiii*
Foreward	*xxxv*
Preface	*xxxvii*

1 INTRODUCTION   1

1.1	Engineering Design for Process Safety Through the Life Cycle of the Facility	2
1.2	Regulatory Review / Impact on Process Safety	5
1.3	Who Will Benefit From These Guidelines?	7
1.4	Organization of this Book	7
1.5	Other CCPS Resources	9
1.6	References	10

2 FOUNDATIONAL CONCEPTS   13

2.1	Understanding the Hazard		14
	2.1.1	Dangerous Properties of Process Materials	14
	2.1.2	Process Conditions	19
	2.1.3	Inventory	20
2.2	Risk-Based Design		21
	2.2.1	The Concept of Risk	22
	2.2.2	Selection of Design Bases for Process Safety Systems	23
2.3	Intentional Unsteady State Condition Evaluation		27
	2.3.1	Batch Reaction Systems	29
2.4	Unintentional Unsteady State Issues		31
	2.4.1	Runaway Reactions	31
	2.4.2	Deviating from the Design Intent	32
2.5	Non-Linearity of the Design Process		33
2.6	References		36

3 BASIC PHYSICAL PROPERTIES / THERMAL STABILITY DATA — 39

- 3.1 Basic Physical Properties — 39
- 3.2 Flammability Data — 40
  - 3.2.1 Flash Point — 41
  - 3.2.2 Fire Point — 43
  - 3.2.3 Autoignition Temperature — 43
  - 3.2.4 Flammable Limits — 44
  - 3.2.5 Minimum / Limiting Oxygen Concentration — 45
  - 3.2.6 Dust Deflagration Index - K_{St} — 45
  - 3.2.7 Gas Deflagration Index - K_g — 45
- 3.3 Reactivity / Thermal Stability Data — 46
  - 3.3.1 Chemical Reactivity — 47
  - 3.3.2 Detonations and Deflagrations — 49
  - 3.3.3 Runaway Reactions — 49
  - 3.3.4 Calorimetric Data — 50
  - 3.3.5 Interaction Matrix — 53
  - 3.3.6 Testing Methods — 56
- 3.4 References — 60

4 ANALYSIS TECHNIQUES — 63

- 4.1 Hazard Identification — 63
  - 4.1.1 Process Hazards — 64
  - 4.1.2 Chemical / Material Hazards — 72
  - 4.1.3 Human Impact Data — 79
- 4.2 Hazard Analysis Techniques — 94
  - 4.2.1 A Life Cycle Approach — 94
  - 4.2.2 Qualitative — 96
  - 4.2.3 Semi-Quantitative — 100
  - 4.2.4 Quantitative — 103
  - 4.2.5 Human Factors — 104
  - 4.2.6 Selecting the Appropriate Technique — 106
- 4.3 Risk Assessment — 108
  - 4.3.1 Technical Aspects of QRA — 109
  - 4.3.2 Risk Criteria — 113
  - 4.3.3 Quantitative Risk Assessment — 117
  - 4.3.4 Risk Tolerance / Decision Making Criteria — 117
- 4.4 Reliability / Maintainability Analysis — 118
- 4.5 References — 119

5 GENERAL DESIGN — 123

- 5.1 Safeguarding Strategies — 123
  - 5.1.1 Inherent — 124

CONTENTS

	5.1.2	Passive	125
	5.1.3	Active	125
	5.1.4	Procedural	126
	5.1.5	Characteristics of Design Solution Categories	126
	5.1.6	Safety Factor	127
	5.1.7	Safeguard Stewardship	127
5.2	Inherently Safer Design		128
	5.2.1	Minimize	129
	5.2.2	Substitute	130
	5.2.3	Moderate	131
	5.2.4	Dilution	131
	5.2.5	Simplify	131
5.3	Basic Process Control Systems		132
	5.3.1	Alarm Management	133
	5.3.2	Testing Instrumentation	134
5.4	Instrumented Safety Systems		135
5.5	Process Design / Process Chemistry		135
	5.5.1	Process Equipment Safe Operating Limits	135
	5.5.2	Consequences of Deviation	137
5.6	Plant Siting and Layout		137
	5.6.1	Site Layout	138
	5.6.2	Unit Layout	139
	5.6.3	Storage Layout	140
	5.6.4	Occupied Building Location	140
5.7	Materials of Construction		140
	5.7.1	Properties of Materials	141
	5.7.2	Corrosive Environments	141
	5.7.3	Pitfalls in Material Selection	142
5.8	Corrosion		143
	5.8.1	General Corrosion and Metallurgical Changes	143
	5.8.2	Stress-Related Corrosion	143
	5.8.3	Design Considerations	145
	5.8.4	Erosion	146
5.9	Civil / Structural / Support Design		146
	5.9.1	Site Preparation and Analysis	147
5.10	Thermal Insulation		150
	5.10.1	Properties of Thermal Insulation	150
	5.10.2	Selection of Insulation Materials	152
	5.10.3	Corrosion Under Insulation	153
5.11	Human Factors in Design		155
	5.11.1	Human Factors Tools for Project Management	157
5.12	Site Security Issues		158
	5.12.1	Physical Security	159

		5.12.2	Cyber / Electronic Security	160
	5.13		References	161

6 EQUIPMENT DESIGN — 165

	6.1	Vessels		167
		6.1.1	Past Incidents	167
		6.1.2	Failure Scenarios and Design Solutions	168
		6.1.3	Design Considerations	177
		6.1.4	References	182
	6.2	Reactors		183
		6.2.1	Past Incidents	183
		6.2.2	Failure Scenarios and Design Solutions	185
		6.2.3	Design Considerations	191
		6.2.4	References	193
	6.3	Mass Transfer Equipment		194
		6.3.1	Past Incidents	195
		6.3.2	Failure Scenarios and Design Solutions	196
		6.3.3	Design Considerations	202
		6.3.4	References	204
	6.4	Heat Transfer Equipment		204
		6.4.1	Past Incidents	204
		6.4.2	Failure Scenarios and Design Solutions	206
		6.4.3	Design Considerations	212
		6.4.4	References	213
	6.5	Dryers		214
		6.5.1	Past Incidents	214
		6.5.2	Failure Scenarios and Design Solutions	215
		6.5.3	Design Considerations	222
		6.5.4	References	222
	6.6	Fluid Transfer Equipment		223
		6.6.1	Past Incidents	223
		6.6.2	Failure Scenarios and Design Solutions	224
		6.6.3	Design Considerations	230
		6.6.4	References	235
	6.7	Solid-Fluid Separators		236
		6.7.1	Past Incidents	236
		6.7.2	Failure Scenarios and Design Solutions	238
		6.7.3	Design Considerations	242
		6.7.4	References	244
	6.8	Solids Handling and Processing Equipment		244
		6.8.1	Past Incidents	245
		6.8.2	Failure Scenarios and Design Solutions	247

CONTENTS

	6.8.3	Design Considerations	253
	6.8.4	References	255
6.9	Fired Equipment		256
	6.9.1	Past Incidents	256
	6.9.2	Failure Scenarios and Design Solutions	257
	6.9.3	Design Considerations	264
	6.9.4	References	266
6.10	Piping and Piping Components		266
	6.10.1	Past Incidents	267
	6.10.2	Failure Scenarios and Design Solutions	268
	6.10.3	Design Considerations	278
	6.10.4	References	286
6.11	Material Handling and Warehousing		290
	6.11.1	Past Incidents	291
	6.11.2	Failure Scenarios and Design Solutions	293
	6.11.3	Design Considerations	298
	6.11.4	References	304
6.12	Utility Systems		304
	6.12.1	Past Incidents	305
	6.12.2	Design Considerations	306
	6.12.3	References	314

7 PROTECTION LAYERS   315

7.1	Ignition Control	316
	7.1.1 Electrical Area Classification	316
	7.1.2 Purging and Pressurized Enclosures	319
	7.1.3 Low Energy Electrical Equipment for Hazardous Locations	320
	7.1.4 Ventilation / Exhaust	321
	7.1.5 Static Electricity	322
	7.1.6 Lightning	324
7.2	Instrumented Safety Systems	325
	7.2.1 Safety Instrumented Systems	325
	7.2.2 Engineering Aspects of Instrumented Safety Systems	328
7.3	Pressure / Vacuum Relief Systems	332
	7.3.1 Relief Design Scenarios	333
	7.3.2 Pressure Relief Devices	334
	7.3.3 Sizing of Pressure Relief Systems	337
	7.3.4 Sizing of Rupture Disks	338
	7.3.5 Other Considerations	338
	7.3.6 Methods of Overpressure Protection for Two-Phase Flows	339
7.4	Equipment Isolation / Blowdown	340

	7.4.1	Equipment Isolation	340
	7.4.2	Depressurization	340
7.5		Effluent Disposal Systems	342
	7.5.1	Flares	342
	7.5.2	Design Considerations for Flares	344
	7.5.3	Blowdown Systems	346
	7.5.4	Incineration Systems	348
	7.5.5	Vapor Control Systems	349
7.6		Emergency Response Alarm Systems	350
	7.6.1	Plant Emergency Alarm and Surveillance	351
	7.6.2	Gas / Fire Detection	353
	7.6.3	Leak Detection	357
7.7		Fire Protection	357
	7.7.1	Structural Fireproofing	358
	7.7.2	Firefighting Agents	359
	7.7.3	Fire Water Systems	359
	7.7.4	Mitigation Systems	360
	7.7.5	Portable Fire Suppression Equipment	363
	7.7.6	Fire Extinguishers	363
7.8		Deflagration / Detonation Arresters	363
	7.8.1	Selection and Design Criteria	364
7.9		Explosion Suppression	365
	7.9.1	Oxidant Concentration Reduction	366
	7.9.2	Deflagration Pressure Containment	367
	7.9.3	Explosion Venting	367
	7.9.4	Equipment and Piping Isolation	368
7.10		Specialty Mitigation Systems	369
	7.10.1	Water / Steam Curtain	369
	7.10.2	Steam Snuffing	370
	7.10.3	Mechanical Interlocks	370
	7.10.4	Inhibitor Injection	370
	7.10.5	Quench System	371
	7.10.6	Dump System	372
7.11		Effluent Handling / Post-Release Mitigation / Waste Treatment Issues	372
7.12		References	374

8 DOCUMENTATION TO SUPPORT PROCESS SAFETY 379

8.1		Process Knowledge Management	379
	8.1.1	Importance of Process Knowledge Management	381
	8.1.2	Types of Process Knowledge and Information Documentation	381
	8.1.3	Design Basis	381

		8.1.4	Managing Change	383
		8.1.5	Other Considerations	384
	8.2	Engineering Design Package		384
	8.3	Operating / Maintenance Procedures		385
		8.3.1	Need for Procedures	386
		8.3.2	Developing Procedures	386
		8.3.3	Maintaining Procedures	388
	8.4	Asset Integrity / Reliability / Predictive Maintenance Data		389
	8.5	References		390

INDEX   **393**

ACRONYMS AND ABBREVIATIONS

ACGIH	American Conference of Government Industrial Hygienists
ACI	American Concrete Institute
ACS	American Chemical Society
AEGL	Acute Emergency Guideline Levels
AGA	American Gas Association
AIChE	American Institute of Chemical Engineers
AIHA	American Industrial Hygiene Association
AISC	American Institute of Steel Construction, Inc.
AISI	American Iron and Steel Institute
AIT	Autoignition Temperature
ALARP	As Low as Reasonably Practical
ANSI	American National Standards Institute
APC	Air Pollution Control
APFA	American Pipe Fittings Association
API	American Petroleum Institute
ARC	Accelerating Rate Calorimeter
ASM	American Society for Metals
ASME	American Society of Mechanical Engineers
ASSE	American Society of Safety Engineers
ASNT	American Society of Nondestructive Testing
ASTM	American Society for Testing and Materials
AWS	American Welding Society
BLEVE	Boiling Liquid Expanding Vapor Explosion
BPCS	Basic Process Control System
Btu	British Thermal Units
BTX	Benzene, Toluene and Xylene
CAA	Clean Air Act
CAAA	Clean Air Act Amendments
CCPS	Center for Chemical Process Safety
CEM	Continuous Emissions Monitor
CERCLA	Comprehensive Environmental Response, Compensation, and Liability Act
CFR	Code of Federal Regulations
CGA	Compressed Gas Association
CIA	Chemical Industries Association

CMA	Chemical Manufacturers Association
COT	Coil Outlet Temperature
CRT	Cathode Ray Tube
CSTR	Continuous-Flow Stirred-Tank Reactor
CWA	Clean Water Act
DAF	Dissolved Air Flotation
dBA	A-Weighted Decibel Level
DCS	Distributed Control System
DDT	Deflagration to Detonation Transition
DIERS	Design Institute for Emergency Relief Systems
DIPPR	Design Institute for Physical Property Data
DOT	Department of Transportation
DOE	Department of Energy
DPC	Deflagration Pressure Containment
DSC	Differential Scanning Calorimeter
DTA	Differential Thermal Analysis
EEGL	Emergency Exposure Guidance Level
EJMA	Expansion Joint Manufacturers Association, Inc.
EPA	Environmental Protection Agency
EPRI	Electric Power Research Institute
ERPG	Emergency Response Planning Guidelines
ERS	Emergency Relief System
ERD	Emergency Relief Design
ESCIS	Expert Commission for Safety in the Swiss Chemical Industry
ESD	Emergency Shutdown Device
ECT	Eddy Current Testing
FBIC	Flexible Intermediate Bulk Containers
FEED	Front-End Engineering and Design
F&EI	Fire and Explosion Index
FMEA	Failure Modes and Effects Analysis
FMECA	Failure Modes, Effects and Criticality Analysis
FMEDA	Failure Modes, Effects and Diagnostic Analysis
FMEC	Factory Mutual Engineering Corporation
FRP	Fiber Reinforced Plastic
GFCI	Ground Fault Current Interrupter
GPM	Gallons per Minute
GSPA	Gas Processors Suppliers Association
HAZOP	Hazard and Operability Study
HEI	Heat Exchanger Institute
hp	Horsepower
HSE	Health and Safety Executive
HVAC	Heating, Ventilation, and Air Conditioning
IChemE	Institute of Chemical Engineers

ICI	Imperial Chemical Industries
IEEE	Institute of Electrical and Electronics Engineers
IDLH	Immediately Dangerous to Life or Health
IGC	Intergranular Corrosion
IPL	Independent Protection Layer
IRI	Industrial Risk Insurers
ISA	Instrument Society of America
ISGOTT	International Safety Guide for Oil Tankers and Terminals
ISO	International Standards Organization
ISS	Independent Safety System
kA	kiloampere
kV	kilovolt
LEL	Lower Explosive Limit
LFL	Lower Flammable Limit
LNG	Liquefied Natural Gas
LOC	Limiting Oxygen Concentration
LOPA	Layer of Protection Analysis
LPG	Liquefied Petroleum Gas
mA	milliampere
MAWP	Maximum Allowable Working Pressure
MCC	Motor Control Center
MEC	Minimum Explosible Concentration
MIE	Minimum Ignition Energy
mJ	millijoule
MOC	Management of Change
MSDS	Material Safety Data Sheet
MSS	Manufacturers Standardization Society
MT	Magnetic Particle Testing
NACE	National Association of Corrosion Engineers
NAS	National Academy of Science
NBIC	National Board Inspection Code
NEC	National Electrical Code
NEMA	National Electrical Manufacturers Association
NESC	National Electrical Safety Code
NDE	Nondestructive Examination
NFPA	National Fire Protection Association
NIOSH	National Institute of Occupational Safety and Health
NOAA	National Oceanic and Atmospheric Administration
NPCA	National Paint and Coatings Association
NPDES	National Pollutant Discharge and Elimination System
NPSH	Net Positive Suction Head
NRC	National Research Council
NSPS	New Source Performance Standards
NTIAC	Nondestructive Testing Information Analysis Center

OSHA	Occupational Safety and Health Administration
PAC	Protective Action Criteria
PCB	Polychlorinated Biphenyl
PEL	Permissible Exposure Limit
PES	Programmable Electronic System
PFD	Process Flow Diagram
PFR	Plug Flow Reactor
PLC	Programmable Logic Controller
P&ID	Piping and Instrumentation Diagram
PHA	Process Hazard Analysis
PID	Proportional Integral Derivative
POT	Pass Outlet Temperature
ppm	parts per million
pS	picoSiemen
PS	Process Safety
PSA	Pressure Swing Adsorption
PSD	Process Safety Device
PSV	Pressure Safety Valve
PSS	Process Safety System
PT	Liquid Penetrant Testing
PVRV	Pressure-Vacuum Relief Valve
QRA	Quantitative Risk Analysis
REST	Reactivity Evaluation Screening Tool
RC	Reactor Calorimeter
RCRA	Resource Conservation and Recovery Act
RP	Recommended Practice
RSST	Reactive System Screening Tool
RT	Radiographic Testing
RTD	Resistance Temperature Detector
SCAPA	Subcommittee on Consequence Assessment and Protective Actions
SCBA	Self-Contained Breathing Apparatus
SCC	Stress Corrosion Cracking
scf	Standard Cubic Foot
SCR	Silicon Conductor Rectifier
SAE	Society of Automotive Engineers
SFPE	Society of Fire Protection Engineers
SIF	Safety Instrumented Function
SIS	Safety Instrumented System
SLOD	Significant Likelihood of Death
SLOT	Specified Level of Toxicity
SOL	Safe Operating Limit
SPCC	Spill Prevention Control and Countermeasures
SPEGL	Short-Term Public Emergency Guidance Level
SRS	Safety Requirement Specification
SSPC	Steel Structures Painting Council

TEEL	Temporary Emergency Exposure Limits	
TEMA	Tubular Exchanger Manufacturer Association	
TLV	Threshold Limit Value	
TOC	Total Organic Compounds	
TSCA	Toxic Substance Control Act	
UBC	Uniform Building Code	
UEL	Upper Explosive Limit	
UFL	Upper Flammable Limit	
UL	Underwriters Laboratory Inc.	
UPS	Uninterruptible Power Supply	
UT	Ultrasonic Testing	
UVCE	Unconfined Vapor Cloud Explosion	
VOC	Volatile Organic Compound	
VP	Vapor Pressure	
VSP	Vent Size Package	
WEEL	Workplace Environmental Exposure Limit	

GLOSSARY

Administrative Controls — Procedural mechanisms, such as lockout / tagout procedures, for directing and / or checking human performance on plant tasks.

Auto-ignition Temperature — The autoignition temperature of a substance, whether solid, liquid, or gaseous, is the minimum temperature required to initiate or cause self-sustained combustion, in air, with no other source of ignition.

Basic Event — An event in a fault tree that represents the lowest level of resolution in the model such that no further development is necessary (e.g., equipment item failure, human failure, or external event).

Basic Process Control System (BPCS) — The control equipment which is installed to support normal production functions.

Batch Reactor — Reactor in which all reactants and solvents are introduced prior to setting the reaction conditions (temperature, pressure). Products are only taken from the reactor upon conclusion of the reaction process. Both heat generation and concentrations in the batch reactor vary during the reaction process.

Boiling Liquid-Expanding Vapor Explosion (BLEVE) — A type of rapid phase transition in which a liquid contained above its atmospheric boiling point is rapidly depressurized, causing a nearly instantaneous transition from liquid to vapor with a corresponding energy release. A BLEVE is often accompanied by a large fireball if a flammable liquid is involved, since an external fire impinging on the vapor space of a pressure vessel is a common BLEVE scenario. However, it is not necessary for the liquid to be flammable to have a BLEVE occur.

Bonding — The process of connecting two or more conductive objects together by means of a conductor.

Car Seal	Metal or plastic cable used to fix a valve in the open position (car seal open) or closed position (car seal closed). Proper authorization, controlled via administrative procedures, must be obtained before operating the valve. The physical seal should have suitable mechanical strength to prevent unauthorized valve operation.
Catastrophic Incident	An incident involving a major uncontrolled emission, fire or explosion that causes significant damage, injuries and / or fatalities onsite and have an outcome effect zone that extends into the surrounding community.
Combustible	Capable of burning.

Combustible Liquid	A term used to classify certain liquids that will burn on the basis of flash points. The National Fire Protection Association (NFPA) defines a combustible liquid as any liquid that has a closed-cup flash point above 100°F (37.8°C) (NFPA 30). There are three subclasses, as follows:
	
	• Class II liquids have flash points at or above 100°F (37.8°C) but below 140°F (60°C).
	
	• Class III liquids are subdivided into two additional subclasses:
	   - Class IIIA are those having flash points at or above 140°F (60°C) but below 200°F (93.4°C).
	   - Class IIIB are those having flash points at or above 200°F (93.4°C).
	
	The Department of Transportation (DOT) defines "combustible liquids" as those having flash points of not more than 141°F (60.5°C) and below 200°F (93.4°C).

Common Mode Failure	An event having a single external cause with multiple failure effects which are not consequences of each other.
Continuous Reactors	Reactors that are characterized by a continuous flow of reactants into and a continuous flow of products from the reaction system (e.g., Plug Flow Reactor (PFR) and the Continuous Stirred Tank Reactor (CSTR)).
Continuous Stirred Tank Reactor (CSTR)	A reaction vessel in which the feed is continuously added and the products continuously removed. The vessel (tank) is continuously stirred to maintain a uniform concentration within the vessel.
Critical Event	A critical event is an event with a specified, high consequence such as an event involving an offsite community impact, critical system damage, a severe injury or a fatality.
Critical Event Frequency	The frequency of occurrence of a critical event.

GLOSSARY

Deadheading — A blockage on the discharge side of an operating pump which results in the flow reducing to zero and an increase in the discharge pressure. The energy input from the deadheaded pump increases the temperature and pressure of the fluid in the pump.

Deflagration — The chemical reaction of a substance in which the reaction front advances into the unreacted substance at less than sonic velocity. Where a blast wave is produced that has the potential to cause damage, the term explosive deflagration may be used.

Deflagration to Detonation Transition (DDT) — The transition phenomenon resulting from the acceleration of a deflagration flame to detonation via flame-generated turbulent flow and compressive heating effects. At the instant of transition a volume of precompressed, turbulent gas ahead of the flame front detonates at unusually high velocity and overpressure.

Design Institute for Emergency Relief Systems (DIERS) — Institute under the auspices of the American Institute of Chemical Engineers founded to study relief requirements for reactive chemical systems and two-phase flow systems.

Detonation — A release of energy caused by the propagation of a chemical reaction in which the reaction front advances into the unreacted substance at greater than sonic velocity in the unreacted material.

Distributed Control System (DCS) — A system which divides process control functions into specific areas interconnected by communications (normally data highways), to form a single entity. It is characterized by digital controllers and typically by central operation interfaces.

Dow Fire and Explosion Index (F&EI) — A method (developed by Dow Chemical Company) for ranking the relative fire and explosion risk associated with a process. Analysts calculate various hazard and explosion indexes using material characteristics and process data.

Emergency Relief Device — A device that is designed to open during emergency or abnormal conditions to prevent rise of internal fluid pressure in excess of a specified value. The device also may be designed to prevent excessive internal vacuum. The device may be a pressure relief valve, a nonreclosing pressure relief device, or a vacuum relief valve.

Emergency Shutdown Device — A device that is designed to shutdown the system to a safe condition on command from the emergency shutdown system.

Emergency Shutdown System — The safety control system that overrides the action of the basic control system and shuts down the process when predetermined conditions are violated.

Equipment Reliability	The probability that, when operating under stated environment conditions, process equipment will perform its intended function adequately for a specified exposure period.
Explosion	A release of energy that causes a pressure discontinuity or blast wave.
Fail-Safe	Design features which provide for the maintenance of safe operating conditions in the event of a malfunction of control devices or an interruption of an energy source (e.g., failure direction of a motor operated value on loss of motive power).
	A feature incorporated for automatically counteracting the effect of an anticipated possible source of failure. A system is fail-safe if failure of a component, signal, or utility, initiates action that return the system to a safe condition.
Failure	An unacceptable difference between expected and observed performance.
Failure Mode and Effects Analysis (FMEA)	A systematic, tabular method for evaluating and documenting the effects of known types of component failures.
Fire Point	The minimum temperature at which a flammable or combustible liquid, as herein defined, and some volatile combustible solids will evolve sufficient vapor to produce a mixture with air that will support sustained combustion when exposed to a source of ignition, such as a spark or flame.
Fireball	The atmospheric burning of a fuel-air cloud in which the energy is mostly emitted in the form of radiant heat. The inner core of the fuel release consists of almost pure fuel whereas the outer layer in which ignition first occurs is a flammable fuel-air mixture. As buoyancy forces of the hot gases begin to dominate, the burning cloud rises and becomes more spherical in shape.
Flammability Limits	The range of gas or vapor amounts in air that will burn or explode if a flame or other ignition source is present. Importance: The range represents an unsafe gas or vapor mixture with air that may ignite or explode. Generally, the wider the range the greater the fire potential. See also Lower Explosive Limit / Lower Flammable Limit and Upper Explosive Limit / Upper Flammable Limit.

GLOSSARY

Flammable Liquid — Any liquid that has a closed-cup flash point below 100°F (37.8°C), as determined by the test procedures described in NFPA 30 and a Reid vapor pressure not exceeding 40 psia (2068.6 mm Hg) at 100°F (37.8°C), as determined by ASTM D 323, Standard Method of Test for Vapor Pressure of Petroleum Products (Reid Method). Flammable liquids are classified as Class I as follows:

- Class IA liquids include those liquids that have flash points below 73°F = (22.8°C) and boiling points below 100°F (37.8°C).

- Class IB liquids include those liquids that have flash points below 73°F (22.8°C) and boiling points at or above 100°F (37.8°C).

- Class IC liquids include those liquids that have flash points at or above 73°F (22.8°C), but below 100°F (37.8°C). (NFPA 30).

Flash Fire — The combustion of a flammable vapor and air mixture in which flame passes through that mixture at less than sonic velocity, such that negligible damaging overpressure is generated.

Flash Point — The temperature at which the vapor-air mixture above a liquid is capable of sustaining combustion after ignition from an external energy source.

Fugitive Emissions — Those emissions which could not reasonably pass through a stack, chimney, vent or other functionally-equivalent opening.

Grounding — The process of connecting one or more conductive objects to ground so that each is at the same potential as the earth. By convention, the earth has zero potential. In practice, grounding is the process of providing a sufficiently small resistance to ground so that a static hazard cannot be created at the maximum credible charging current to a system. Grounding may be referred to as "earthing" in Europe.

Hazard — An inherent chemical or physical characteristic that has the potential for causing damage to people, property, or the environment. In this document it is the combination of a hazardous material, an operating environment, and certain unplanned events that could result in an accident.

Hazard Analysis — The identification of undesired events that lead to the materialization of a hazard, the analysis of the mechanisms by which these undesired events could occur and usually the estimation of the consequences.

Hazard and Operability Study (HAZOP)	A systematic qualitative technique to identify process hazards and potential operating problems using a series of guide words to study process deviations. A HAZOP is used to question every part of a process to discover what deviations from the intention of the design can occur and what their causes and consequences may be. This is done systematically by applying suitable guidewords. This is a systematic detailed review technique, for both batch and continuous plants, which can be applied to new or existing processes to identify hazards.
Hazard Identification	The identification of causes that lead to hazardous events and an estimation of the event consequence.
Hazardous Material	In a broad sense, any substance or mixture of substances having properties capable of producing adverse effects to the health or safety of human beings or the environment. Material presenting dangers beyond the fire problems relating to flash point and boiling point. These dangers may arise from, but are not limited to, toxicity, reactivity, instability, or corrosivity
Human Factors	A discipline concerned with designing machines, operations, and work environments so that they match human capabilities, limitations, and needs. Includes any technical work (engineering, procedure writing, worker training, worker selection, etc.) related to the human factor in operator-machine systems.
Inert Gas	A nonflammable, nonreactive gas that can be used to render the combustible material in a system incapable of supporting combustion.
Inherently Safer	A condition in which the hazards associated with the materials and operations used in the process have been reduced or eliminated, and this reduction or elimination is permanent and inseparable.
Interlock System	A system that detects out-of-limits or abnormal conditions or improper sequences and either halts further action or starts corrective action.

GLOSSARY

Intrinsically Safe Equipment and wiring which is incapable of releasing sufficient electrical or thermal energy under normal or abnormal conditions to cause ignition of a specific hazardous atmospheric mixture or hazardous layer.

A protection technique based upon the restriction of electrical energy within apparatus and of interconnecting wiring, exposed to a potentially explosive atmosphere, to a level below that which can cause ignition by either sparking or heating effects. Because of the method by which intrinsic safety is achieved, it is necessary to ensure that not only the electrical apparatus exposed to the potentially explosive atmosphere but also other electrical apparatus with which it is interconnected is suitably constructed.

Likelihood A measure of the expected frequency with which an event occurs. This may be expressed as a frequency (e.g., events per year), a probability of occurrence during a time interval (e.g., annual probability), or a conditional probability (e.g., probability of occurrence, given that a precursor event has occurred).

Limiting Oxygen Concentration (LOC) The limiting oxygen concentration (LOC) is that concentration of oxygen, below which a deflagration (flame propagation in the gas, mist, suspended dust, or hybrid mixture) cannot occur. For most hydrocarbons (where oxygen is the oxidant and nitrogen is the diluent) the LOC is approximately 9 to 11 vol% oxygen. The LOC for dusts is dependent on the composition and particle size distribution of the solid. Values of LOC for most organic chemical dusts are in the range of 10 to 16 vol% oxygen, again where nitrogen is the diluent.

Lower Flammable Limit (LFL) That concentration of a combustible material in air below which ignition will not occur. It is often, interchangeably called Lower Explosive Limit (LEL) and for dusts, the Minimum Explosible Concentration (MEC).

Minimum Explosible Concentration (MEC) The lowest concentration of combustible dust necessary to produce an explosion.

Minimum Ignition Energy (MIE) Initiation of flame propagation in a combustible mixture requires an ignition source of adequate energy and duration to overcome heat losses to the cooler surrounding material. Dust and vapor clouds may be readily ignited if exposed to electric discharges that exceed the minimum ignition energy (MIE) for the combustible mixture.

Mitigation	Reducing the risk of an accident event sequence by taking protective measures to reduce the likelihood of occurrence of the event, and / or reduce the magnitude of the event and / or minimize the exposure of people or property to the event.
Net Positive Suction Head (NPSH)	The net static liquid head that must be provided on the suction side of the pump to prevent cavitation.
Oxidant	Any gaseous material that can react with a fuel (gas, dust, or mist) to produce combustion. Oxygen in air is the most common oxidant.
Piping and Instrumentation Diagram (P&ID)	A diagram that shows the details about the piping, vessels, and instrumentation.
Plug Flow Reactor (PFR)	A plug flow reactor is a tubular reactor where the feed is continuously introduced at one end and the products continuously removed form the other end. The concentration / temperature in the reactor is not uniform.
Pool Fire	The combustion of material evaporating from a layer of liquid at the base of the fire.
Pressure Relief Valve (PRV)	A relief valve is a spring-loaded valve actuated by static pressure upstream of the valve. The valve opens normally in proportion to the pressure increase over opening pressure. A relief valve is normally used with incompressible fluids.
Pressure Safety Valve (PSV)	A safety valve is a spring loaded pressure relief valve actuated by static pressure upstream if the valve and characterized by rapid opening or pop action. A safety valve is normally used with compressible fluids.
Process Flow Diagram (PFD)	A diagram that shows the material flow from one piece of equipment to the other in a process. It usually provides information about the pressure, temperature, composition, and flow rate of the various streams, heat duties of exchangers, and other such information pertaining to understanding and conceptualizing the process.
Process Hazard Analysis (PHA)	An organized effort to identify and evaluate hazards associated with chemical processes and operations to enable their control. This review normally involves the use of qualitative techniques to identify and assess the significance of hazards. Conclusions and appropriate recommendations are developed. Occasionally, quantitative methods are used to help prioritized risk reduction.
Process Safety	A discipline that focuses on the prevention of fires, explosions, and accidental chemical releases at chemical process facilities.

GLOSSARY

Process Safety Management (PSM) — A management system that is focused on prevention of, preparedness for, mitigation of, response to, and restoration from catastrophic releases of chemicals or energy from a process associated with a facility.

Process Safety System (PSS) — A process safety system comprises the design, procedures, and hardware intended to operate and maintain the process safely.

Programmable Electronic System (PES) — A system based on a computer connected to sensors and / or actuators in a plant for the purpose of control, protection or monitoring (includes various types of computers, programmable logic controllers, peripherals, interconnect systems, instrument distributed control system controllers, and other associated equipment).

Programmable Logic Controller (PLC) — A microcomputer-based solid-state control system which receives inputs from user-supplied control devices such as switches and sensors, implements them in a precise pattern determined by instructions stored in the PLC memory, and provides outputs for control or user-supplied devices such as relays and motor starters.

Purge Gas — A gas that is continuously or intermittently added to a system to render the atmosphere noncombustible. The purge gas can be inert or combustible.

Quenching — Rapid cooling from an elevated temperature, e.g., severe cooling of the reaction system in a short time (almost instantaneously), "freezes" the status of a reaction and prevents further decomposition or reaction.

Risk Based Process Safety — The CCPS's process safety management system approach that uses risk-based strategies and implementation tactics that are commensurate with the risk-based need for process safety activities, availability of resources, and existing process safety culture to design, correct, and improve process safety management activities.

Runaway Reactions — A thermally unstable reaction system which exhibits an uncontrolled accelerating rate of reaction leading to rapid increases in temperature and pressure.

Safety Instrumented System (SIS) — The instrumentation, controls, and interlocks provided for safe operation of the process.

Safety Layer	A system or subsystem that is considered adequate to protect against a specific hazard. The safety layer:

- Is totally independent of any other protective layers.
- Cannot be compromised by the failure of another safety layer.
- Must have acceptable reliability.
- Must be approved according to company policy and procedures.
- Must meet proper equipment classification.
- May be a noncontrol alternative (e.g., chemical, mechanical).
- May require diverse hardware and software packages.
- May be an administrative procedure.

Semi-Batch Reactor	In a semi-batch reactor, some reactants are added to the reactor at the start of the batch, while others are fed continuously during the course of the reaction.
Source Term	For a hazardous material and / or energy release to the surroundings associated with a loss event, the release parameters (e.g., magnitude, rate, duration, orientation, temperature, etc.) that are the initial conditions for determining the consequences of the loss event. For vapor dispersion modeling, it is the estimation, based on the release specification, of the actual cloud conditions of temperature, aerosol content, density, size, velocity and mass to be input into the dispersion model.
Task Analysis	A human error analysis method that breaks down a procedure or overall job description into individual work tasks.
Unconfined Vapor Cloud Explosion (UCVE)	When a flammable vapor is released, its mixture with air will form a flammable vapor cloud. If ignited, the flame speed may accelerate to high velocities and produce significant blast overpressure.
Upper Flammable Limit (UFL)	The highest concentration of a vapor or gas (the highest percentage of the substance in air) that will produce a flash of fire when an ignition source (heat, arc, or flame) is present. See also Lower Flammable Limit. At concentrations higher then the UFL, the mixture is too "rich" to burn.
Valve Failure Positions	In the event of instrument air or electrical power failure, valves either Fail Closed (FC), Fail Open (FO), or Fail in the last position (FL). The position of failure must be carefully selected so as to bring the system to, or leave the system in a safe operating state.

GLOSSARY xxxi

Vapor Cloud Explosion (VCE) The explosion resulting from the ignition of a cloud of flammable vapor, gas, or mist in which flame speeds accelerate to sufficiently high velocities to produce significant overpressure.

Vapor Density The weight of a vapor or gas compared to the weight of an equal volume of air; an expression of the density of the vapor or gas. Materials lighter than air have vapor densities less than 1.0 (example: acetylene, methane, hydrogen). Materials heavier than air (examples: propane, hydrogen sulfide, ethane, butane, chlorine, sulfur dioxide) have vapor densities greater then 1.0.

Importance: All vapors and gases will mix with air, but the lighter materials will tend to rise and dissipate (unless confined). Heavier vapors and gases are likely to concentrate in low places - along or under floors, in sumps, sewers and manholes, in trenches and ditches - and can travel great distances undetected where they may create fire or health hazards.

Vapor Pressure The pressure exerted by a vapor above its own liquid. The higher the vapor pressure, the easier it is for a liquid to evaporate and fill the work area with vapors which can cause health or fire hazards.

Venting Emergency flow of vessel contents out of a vessel. The pressure is controlled or reduced by venting, thus avoiding a failure of the vessel by overpressurization. The emergency flow can be one-phase or multi-phase, each of which results in different flow characteristics.

ACKNOWLEDGMENTS

The American Institute of Chemical Engineers (AIChE) and the Center for Chemical Process Safety (CCPS) express their appreciation and gratitude to all members of the Engineering Design for Process Safety, Second Edition and their CCPS member companies for their generous support and technical contributions in the preparation of these *Guidelines*. The AIChE and CCPS also express their gratitude to the team of authors from Aon Energy Risk Engineering.

SUBCOMMITTEE MEMBERS:

Committee Chairman, Pete Lodal	Eastman Chemical
Mark Davis	Eli Lilly
Americo Diniz	Braskem
Edward Dyke	Merck
Brad Fong	3M
S. Ganeshan	Toyo Engineering India Ltd
Bala Chaitanya Gottimukkala	CB&I Lummus
Chantell Lang	CB&I Lummus
Darrin Miletello	Bayer CropScience
Mikelle Moore	Buckman
Mike Moosemiller	BakerRisk
Perry Morse	DuPont
Keith Pace	Praxair
Jack Philley	Baker Hughes
Ravi Ramaswamy	Reliance Industries Limited
Ron Riselli	Nexen
Sheri Sammons	TPC Group
Narayanam Sankaran (Sank)	UOP / Honeywell
Kevin Shaughnessy	Dow Chemical
Gill Sigmon	Honeywell
James Slaugh	Lyondell Basell
Gary Solak	Bayer Material Science
Angela Summers	SIS-TECH Solutions
Scott Wallace	Olin
CCPS Staff Consultant:	Dave Belonger

CCPS wishes to acknowledge the contributions of the Aon Energy Risk Engineering staff members who wrote this book, especially John Alderman, Christy Franklyn, and Donna Pruitt.

Before publication, all CCPS books are subjected to a thorough peer review process. CCPS gratefully acknowledges the thoughtful comments and suggestions of the peer reviewers. Their work enhanced the accuracy and clarity of these guidelines.

Although the peer reviewers have provided many constructive comments and suggestions, they were not asked to endorse this book and were not shown the final draft before its release.

Peer Reviewers:

Zaheer Ahmed   Baker Hughes
Jeff Fox   Dow Corning
Stan Grossel   Process Safety and Design Consultant
Dave Krabacher   Cognis Corporation
Haluk Kopkalli   Honeywell Specialty Materials
Brook Vickery   Flint Hill Resources

FOREWORD

Engineers like to think of their discipline as a rigorous application of scientific and mathematical principles to the problem of creating a useful object. To a certain extent, this is an appropriate description of the *tools* of engineering – those techniques that we use to translate a concept in the mind of the designer into a physical object. But, where does that mental image of the object to be built come from? At its heart, engineering is intuitive, and an art form. The engineer / designer's accumulated experience, and that of others, is applied to a defined problem. By intuitive and creative problem solving processes, the engineer develops and refines a conceptual design, and uses the mathematical and scientific tools of engineering to translate a mental concept into reality.

The selection of the design basis for a process safety system is a problem like any other engineering problem. There is no equation or formula, no scientific principle, which will define the "best" design. Yes, there are scientific and mathematical tools which will help convert a design concept into something which can actually be constructed. But there is no general answer to the question "What is the best design?" Each system must be considered on its own, with a thorough evaluation of all of the details of its environment and required functions, to determine what the optimal design will be.

The number of potential solutions to any engineering problem is large, as anybody who has ever visited an automobile show quickly realizes. Sometimes, for a specific problem, there will be some solutions which clearly meet the overall objectives of nearly all stakeholders better than others. In these situations it is easy to select an optimum design. However, in other cases, different stakeholders have significantly different objectives, or will differ significantly in the relative importance of the different objectives of the design. This is one of the reasons why there are so many different kinds of cars at the automobile show, giving each potential purchaser a chance to find a design that best meets his or her objectives. But this is not possible in the design of a process plant – there is one plant which impacts many stakeholders with their different objectives and priorities. How can we best find the optimal solution? While this is not entirely a technical question, but also includes social and political aspects, I believe that the critical first step is to consider a large number of potential solutions. This increases the likelihood that the solution most acceptable to as many stakeholders as possible will be among those identified. Where do we get those potential solutions? One important source is accumulated experience our own, and that of others who have faced similar problems in the past. This book collects much of that accumulated experience from a large number of experts in the chemical process industry. Use of the tables which make up the heart of this book will allow the reader to take advantage of many years of practical experience. By considering a large number of potential solutions to the problem of specifying the design basis for safety systems, the design engineer is more likely to be able to identify the solution, or combination of solutions, which best meets most people's needs.

This book, a combination, update, and expansion of two earlier CCPS Guideline publications, emphasizes a risk-based approach to the evaluation of safety system design. Potential safety systems suggested are categorized as inherently safer / passive, active, and procedural, in decreasing order of robustness and reliability. Inherently safer approaches are often preferred, but there can be no general answer to the question of which approach or specific solution is best for a particular situation. Instead, the design engineer must take a very broad and holistic approach to the complete design, accounting for the many different, and often competing, objectives which the design must accomplish. Safety, health effects, environmental impact, loss prevention, economic and business factors, product quality, technical feasibility, and many other factors must be considered. This book challenges the engineer to adopt a risk-based approach to evaluating many competing goals when deciding among a number of potential design alternatives.

This book can be extremely useful in conducting process hazard analysis studies. The failure mode tables in Chapter 6 can be the basis for hazard identification checklists and also offer a variety of potential solutions for identified concerns. However, the book will be even more beneficial if used by the individual engineer at the earliest stages of the design process, before any formal hazard reviews.

The message of this book can be summarized very briefly:

- Consider a large number of design options
- Identify opportunities for inherent and passive safety features early
- Fully understand *all* of the hazards and resulting risks associated with design alternatives
- Use a risk-based approach to process safety systems specification

I hope that this book will find a home on the desk (not gathering dust on the bookshelf!) of every chemical process designer, particularly those involved in the earliest phases of conceptual design where the basic chemistry and unit operations are defined. It should be consulted frequently in the course of the designer's day-to-day work in specifying and designing process facilities. If you are a process safety professional, make sure that all of the process design engineers in your organization read and use this book. It will make your job a lot easier!

Dennis C. Hendershot
CCPS Staff Consultant

PREFACE

The Center for Chemical Process Safety (CCPS) was established in 1985 by the American Institute of Chemical Engineers (AIChE) for the express purpose of assisting the Chemical and Hydrocarbon Process Industries in avoiding or mitigating catastrophic chemical accidents. To achieve this goal, CCPS has focused its work on four areas:

- Establishing and publishing the latest scientific and engineering practices (not standards) for prevention and mitigation of incidents involving toxic and / or reactive materials.
- Encouraging the use of such information by dissemination through publications, seminars, symposia and continuing education programs for engineers.
- Advancing the state-of-the-art in engineering practices and technical management through research in prevention and mitigation of catastrophic events.
- Developing and encouraging the use of undergraduate education curricula which will improve the safety knowledge and consciousness of engineers.

This book, *Guidelines for Engineering Design for Process Safety, Second Edition,* is the result of multiple projects. The first project was the first edition of *Guidelines for Engineering Design for Process Safety*, which began in 1989 with volunteers from CCPS member companies working with engineers from the Stone & Webster Engineering Corporation. The intent was to produce a book that presented the process safety design issues needed to address all stages of the evolving design of a facility. The first edition discussed the impact that various engineering design choices have on the risk of a catastrophic accident, starting with the initial selection of the process and continuing through its final design.

The second project began in 1994 with volunteers from CCPS member companies working with Arthur D. Little Inc. to produce a book entitled *Guidelines for Design Solutions for Process Equipment Failures*. This book described the ways that major processing equipment could fail, causing a catastrophic accident. This second book identified available design solutions that might avoid or mitigate the failure in a series of options ranging from inherently safer / passive solutions to active and procedural solutions. By capturing industry experience in how major processing equipment can fail, this book provided a very useful tool for the selection of process safety systems. The inherently safer solutions that were suggested may, in some cases, have come as a surprise to the process and design engineer because they may have been the most cost-effective solution.

In 2009, both the Technical Steering Committee and the Planning Committee of CCPS recognized the need to consolidate these two works into one combined, expanded and updated volume. The result of this effort is the book you now hold in your hand.

Guidelines for Engineering Design for Process Safety, 2nd Edition, has been updated to provide design guidance and comprehensive references for process equipment in a number of different categories, including vessels, reactors, heat and mass transfer equipment, fluid transfer and separation equipment, fired equipment, dryers, and piping. Chapter 6 contains updated equipment failure tables from the *Design Solutions* book.

This book focuses on engineering design to reduce risk due to process hazards. It does not focus on operations, maintenance, transportation, or personnel safety issues, although improved process safety can benefit each area. Detailed engineering designs are outside the scope of this book, but the authors have provided an extensive guide to references and other literature to assist the designer who wishes to go beyond safety design philosophy to the specifics of a particular safety system design.

1
INTRODUCTION

The Center for Chemical Process Safety (CCPS) has published a number of guidelines that focus on the evaluation and mitigation of risks associated with catastrophic events in process facilities. Originally published in 1993, the purpose of *Guidelines for Engineering Design for Process Safety* was to shift the emphasis on process safety to the earliest stage of the design where process safety issues could be addressed at the lowest cost and with the greatest effect. Almost 20 years later, this 2nd edition of *Guidelines for Engineering Design for Process Safety* continues to stress the importance of emphasizing process safety during Front-End Engineering and Design (FEED) to achieve the greatest risk reduction at the lowest cost – *and also emphasizes the benefits of diligence to process safety design issues through the life of the facility.* This updated book also incorporates material from *Guidelines for Design Solutions for Process Equipment Failures,* which was originally published by CCPS in 1998 (Ref. 1-1).

This book focuses on process safety issues in the design of chemical, petrochemical, and hydrocarbon processing facilities. Enough information is provided on each topic to ensure that the reader understands:

- The concept and issues
- The design approach for process safety
- Areas of concern
- Where to go for detailed information

The scope of this book includes avoidance and mitigation of catastrophic events that could impact people and facilities in the plant or surrounding area. The scope is limited to selecting appropriate designs to prevent or mitigate the release of flammable or toxic materials that could lead to a fire, explosion, and impact to personnel and the community. Process safety issues affecting operations and maintenance are limited to cases where design choices impact system reliability. These *Guidelines* are intended to be applicable to the design of a new facility, as well as modification of an existing facility.

The scope excludes:

- Transportation safety
- Routine environmental control
- Personnel safety and industrial hygiene practices
- Emergency response
- Detailed design
- Operations and maintenance
- Security issues unrelated to process safety

These *Guidelines* highlight safety issues in design choices. For example, Section 7.1.1, Electrical Area Classification, covers the safe application of electrical apparatus in the process environment required for plant safety but does not address detailed design of the electrical supply or distribution system required to operate the plant.

It is clear that choices made early in design can reduce both the potential for large releases of hazardous materials and the severity of such releases, if they should occur.

1.1 ENGINEERING DESIGN FOR PROCESS SAFETY THROUGH THE LIFE CYCLE OF THE FACILITY

Engineering design for process safety must be an integral part of the life cycle of a facility. Process safety has been defined in previous publications as:

> *A discipline that focuses on the prevention and mitigation of fires, explosions, and accidental chemical releases at process facilities. Excludes classic worker health and safety issues involving working surfaces, ladders, protective equipment, etc.* (Ref. 1-2).

Hazard evaluations are one method used to identify, evaluate, and control hazards involved in chemical processes. Hazards can be defined as characteristics of systems, processes, or plants that must be controlled to prevent occurrence of specific undesired events. Hazard evaluation is a technique that is applied repeatedly throughout the design, construction, and operation phases of a facility (Figure 1.1). Engineering design for process safety should be considered within the framework of a comprehensive process safety management program as described in *Plant Guidelines for Technical Management of Chemical Process Safety* (Ref. 1-3).

Hazard evaluation is synonymous with process hazard analysis and process safety review. From conceptual design to decommissioning, no single method of hazard evaluation applies to all of the stages of a project. Different methods are required for different stages of a project, such as research and development, conceptual design, startup and operation. Table 1.1 presents some of the stages of facility life cycle and typical corresponding process hazard evaluation objectives. An objective shown for one stage may be applicable to another.

As illustrated in Table 1.1, different types of hazards can be identified during the stages of a facility's life cycle. Findings from the Baker Panel report (Ref. 1-4) associated with the 2005 Texas City Refinery Explosion illustrate the importance of engineering design for process safety:

> *Not all refining hazards are caused by the same factors or involve the same degree of potential damage. Personal or occupational safety hazards give rise to incidents—such as slips, falls, and vehicle accidents—that primarily affect one individual worker for each occurrence. Process safety hazards can give rise to major accidents involving the release of potentially dangerous materials, the release of energy (such as fires and explosions), or both. Process safety incidents can have catastrophic effects and can result in multiple injuries and fatalities, as well as substantial economic, property, and environmental damage. Process safety refinery incidents can affect workers inside the refinery and members of the public who reside nearby. Process safety in a refinery involves the prevention of leaks, spills,*

1. INTRODUCTION

equipment malfunctions, over-pressures, excessive temperatures, corrosion, metal fatigue, and other similar conditions. Process safety programs focus on the design and engineering of facilities, hazard assessments, management of change, inspection, testing, and maintenance of equipment, effective alarms, effective process control, procedures, training of personnel, and human factors. The Texas City tragedy in March 2005 was a process safety accident. (Ref. 1-4).

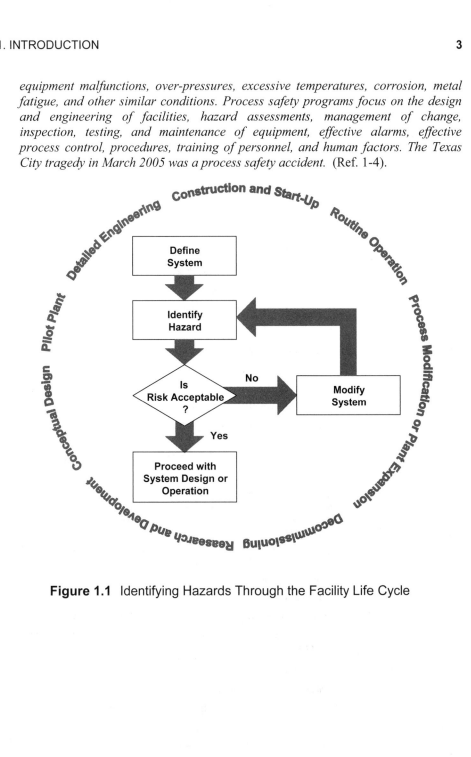

Figure 1.1 Identifying Hazards Through the Facility Life Cycle

Table 1.1 Typical Hazard Evaluation Objectives at Different Stages of a Facility Life Cycle

Stage of Facility Life Cycle	Example Hazard Evaluation Objectives
Research and Development	• Identify chemical interactions that could cause runaway reactions, fires, explosions, or toxic gas releases • Identify process safety data needs
Conceptual Design	• Identify opportunities for inherent safety • Compare the hazards of potential sites
Pilot Plant	• Identify ways for toxic gas to be released to the environment • Identify ways to deactivate the catalyst • Identify potentially hazardous operator interfaces • Identify ways to minimize hazardous wastes
Engineering	• Identify ways to prevent flammable mixtures inside process equipment • Identify how a loss of containment might occur • Identify which process control malfunctions will cause runaway reactions • Identify ways to reduce hazardous material inventories • Identify safety-critical equipment that must be regularly tested, inspected, or maintained • Identify operating conditions that effect selection of materials of construction (e.g., corrosivity) • Identify incompatibility / reactivity issues • Identify relief system and discharging location impact
Construction and Startup	• Identify error-likely situations in startup and operating procedures • Verify that all issues from previous hazard evaluations were resolved satisfactorily and that no new issues were introduced • Identify hazards that adjacent units may create for construction and maintenance workers • Identify hazards associated with the vessel-cleaning procedure • Identify any discrepancies between the as-built equipment and the design drawings
Routine Operation	• Identify employee hazards associated with the operating procedures • Identify ways an overpressure transient might occur • Identify hazards associated with out-of-service equipment
Process Modification or Plant Expansion	• Identify whether changing the feedstock composition will create any new hazards or make any existing hazards more severe • Identify hazards associated with new equipment
Decommissioning	• Identify how demolition work might affect adjacent units • Identify any fire, explosion, or toxic hazards associated with the residues left in the unit after shutdown

1. INTRODUCTION

1.2 REGULATORY REVIEW / IMPACT ON PROCESS SAFETY

The ideas presented here are not intended to replace regulations, codes, or technical and trade society standards and recommended practices. Specifically, implementation of these guidelines requires the application of sound engineering judgment because the concepts may not be applicable in all cases.

Identifying and addressing relevant process safety standards, codes, regulations, and laws over the life of a process is one of the five elements in the Risk Based Process Safety pillar of committing to process safety (Ref. 1-5). Companies should establish a process for maintaining adherence to applicable standards, codes, regulations, and laws. *Guidelines for Risk Based Process Safety* (Ref. 1-5) recommends establishing a *standards* system to achieve this objective, including:

- Establishing a system to identify, develop, acquire, evaluate, disseminate, and provide access to applicable standards, codes, regulations, and laws that affect process safety
- Promoting consistent interpretation, implementation, and efficiency in the initial identification of and ongoing monitoring of changes in standards

Safe operation and maintenance of facilities that manufacture, store, or otherwise use hazardous chemicals require robust process safety management systems. The primary objective of establishing a *standards* system is to ensure that a facility remains in conformance with applicable standards, codes, regulations, and laws, including voluntary ones adopted by the company over the life of the facility. Long-term conformance to such standards of care helps ensure that the facility is operated in a safe and legal fashion. Key principles and essential features of maintaining a dependable *standards* system include:

- Ensuring consistent implementation of the standards system
- Identifying when standards compliance is needed
- Involving competent personnel
- Ensuring that standards compliance practices remain effective

The Baker Panel also emphasizes the importance of implementation of external good engineering practices and a corporate safety management system that supports and improves process safety importance (Ref. 1-4).

For detailed information on establishing a system to comply with standards, readers are referred to Chapter 4, Compliance with Standards, of *Guidelines for Risk Based Process Safety* (Ref. 1-5).

Table 1.2 provides some examples of the types of process safety standards, codes, and regulations that many facilities comply with.

Table 1.2 Examples and Sources of Process Safety Related Standards, Codes, Regulations, and Laws

Voluntary Industry Standards
• American Chemistry Council Responsible Care ® Management System (Ref. 1-6)
• European Chemical Industry Council (Cefic) Responsible Care (Ref. 1-7)
• American Petroleum Institute Recommended Practices (Ref. 1-8)
Consensus Codes
• American National Standards Institute (Ref. 1-9)
• American Petroleum Institute (Ref. 1-8)
• American Society of Mechanical Engineers (Ref. 1-10)
• The Instrumentation, Systems and Automation Society / International Electrotechnical Commission (Ref. 1-11)
• National Fire Protection Association (Ref. 1-12)
U.S. Federal, State, and Local Laws and Regulations

• U.S. OSHA
   • Process Safety Management Standard (29 CFR 1910.119) (Ref. 1-13)
   • Flammable and Combustible Liquids Standard (29 CFR 1910.106) (Ref. 1-14)
   • PSM Covered Chemical Facilities National Emphasis Program (09-06 CPL 02) (Ref. 1-15)
   • Petroleum Refinery Process Safety Management National Emphasis Program (Ref. 1-16)

• U.S. EPA Risk Management Program Regulation (40 CFR 68) (Ref. 1-17)
• California Accidental Release Prevention Program (Ref. 1-18)
• Contra Costa County Industrial Safety Ordinance (Ref. 1-19)
• Delaware Extremely Hazardous Substances Risk Management Act (Ref. 1-20)
• Nevada Chemical Accident Prevention Program (Ref. 1-21)
• New Jersey Toxic Catastrophe Prevention Act (Ref. 1-22)
International Laws and Regulations
• Australian National Standard for Control of Major Hazard Facilities (Ref. 1-23)
• Canadian Environmental Protection Agency, Environmental Emergency Planning (Section 200) (Ref. 1-24)
• European Commission Seveso II Directive (Ref. 1-25)
• Korean OSHA PSM Standard (Ref. 1-26)
• Malaysia, Department of Occupation Safety and Health Ministry of Human Resources Malaysia, Section 16 of Act 514 (Ref. 1-27)
• United Kingdom, Health and Safety Executive COMAH Regulations (Ref. 1-28)

It is important to note that regional or local laws and regulations often mandate more stringent requirements than similar federal regulations. For example, the State of California's Accidental Release Prevention Program requires compliance by facilities with over a threshold quantity of 100 lb of chlorine, while the U.S. EPA Risk Management Program's threshold quantity for compliance is 2,500 lb of chlorine.

Different global, federal, and regional requirements pose challenges to facilities that operate in different geographic locations.

1.3 WHO WILL BENEFIT FROM THESE GUIDELINES?

Process safety is an important part of risk management and loss prevention. Although these *Guidelines* do not provide all the "answers," they do highlight the process safety issues that must be addressed in all stages of design. These *Guidelines* will benefit many different people within an organization:

- *Corporate Leadership* - Senior executives define the basis for the development of design philosophies. Their commitment and recognition of the value of integrating process safety at all levels of the design process is essential.
- *Project Managers* - Project Managers are responsible for executing projects, usually from design through startup and commissioning. A Project Manager is responsible for determining the basic protection design concepts to apply in the execution of a project. The Project Manager is responsible for implementing the decisions and abiding by the process safety systems associated with the design.
- *Engineers* - Engineers are responsible for specifying and designing process units and protection systems that meet their company's requirements. This still leaves room for making decisions when designing process units and protection systems.
- *HSE Professionals* - Health, Safety, and Environmental (HSE) Professionals provide technical guidance to engineers and typically are in an assurance role for process safety systems.

1.4 ORGANIZATION OF THIS BOOK

Figure 1.2 provides an overview of the contents of these *Guidelines* and also provides examples of how each chapter can assist in integrating process safety throughout the life of the process. Each chapter has been updated to include state-of-the-art information, industry experience, and references to other CCPS publications.

Specific references and applicable industry standards are listed at the end of each chapter. It is not the intent of this book to make specific design recommendations, but to provide a good source of references where the interested reader can obtain more detailed information.

Guideline Chapter	Questions This Chapter Will Answer
Chapter 1 Introduction	*What is process safety?* *How can this book help me?* *When is process safety incorporated into engineering design?*
Chapter 2 Foundational Concepts	*Why is incorporating process safety into a facility's lifecycle important?* *What is engineering design that incorporates process safety?* *How are unsteady state conditions included in the design?*
Chapter 3 Basic Physical Properties / Thermal Stability Data	*What basic physical properties do I need to know before beginning design?* *What flammability data is important?* *What chemical reactivity data is needed?* *What is the impact of hazards on people?*
Chapter 4 Analysis Techniques	*Why conduct hazard assessments during engineering design?* *What techniques do I use for hazard identification?* *How does risk assessment influence engineering design?*
Chapter 5 General Design	*What are safeguarding strategies?* *How does process safety influence unit or plant design?* *What materials of construction and insulation are needed to prevent corrosion?*
Chapter 6 Equipment Design	*How does process safety influence equipment design?* *What are typical failure scenarios for different types of equipment?* *What are common equipment design errors?*
Chapter 7 Protection Layers	*How do I recognize the difference between prevention and protection* *What are prevention design features?* *What are protection design features?*
Chapter 8 Documentation to Support Process Safety	*What do I need to document?* *How can this documentation help my facility?*

Figure 1.2 Overview of Guideline Contents

1.5 OTHER CCPS RESOURCES

Other CCPS Guidelines provide additional resources for topics discussed in these *Guidelines*. Some of these include:
- Continuous Monitoring for Hazardous Material Releases
- Deflagration and Detonation Flame Arresters
- Guideline for Mechanical Integrity Systems
- Guidelines for Analyzing and Managing the Security Vulnerabilities of Fixed Chemical Sites
- Guidelines for Chemical Process Quantitative Risk Analysis, Second Edition
- Guidelines for Chemical Reactivity Evaluation and Application to Process Design
- Guidelines for Developing Quantitative Safety Risk Criteria
- Guidelines for Facility Siting and Layout
- Guidelines for Fire Protection in the Chemical, Petrochemical and Hydrocarbon Processing Industries
- Guidelines for Hazard Evaluation Procedures, Third Edition
- Guidelines for Integrating Process Safety Management, Environment, Safety, Health and Quality
- Guidelines for Pressure Relief and Effluent Handling Systems
- Guidelines for Preventing Human Error in Process Safety
- Guidelines for Process Safety Documentation
- Guidelines for Process Safety in Batch Reaction Systems
- Guidelines for Risk Based Process Safety
- Guidelines for Safe and Reliable Instrumented Protective Systems
- Guidelines for Safe Handling of Powders and Bulk Solids
- Guidelines for Safe Storage and Handling of Reactive Materials
- Inherently Safer Chemical Processes a Life Cycle Approach, Second Edition
- Plant Guidelines for Technical Management of Chemical Process Safety
- Safe Operation of Process Vents and Emission Control Systems

Additional information on these publications can be found at www.aiche.org/ccps/.

1.6 REFERENCES

1-1. CCPS. *Guidelines for Design Solutions for Process Equipment Failures*. Center for Chemical Process Safety of the American Institute of Chemical Engineers. New York, NY. 1998.

1-2. CCPS. *Guidelines for Investigating Chemical Process Incidents, Second Edition*. Center for Chemical Process Safety of the American Institute of Chemical Engineers. New York, NY. 2003.

1-3. CCPS. *Plant Guidelines for Technical Management of Chemical Process Safety*. Center for Chemical Process Safety of the American Institute of Chemical Engineers. New York, NY. 1992.

1-4. Baker, et al. *The Report of the BP U.S. Refineries Independent Safety Review Panel*. January 2007.

1-5. CCPS. *Guidelines for Risk Based Process Safety*. Center for Chemical Process Safety of the American Institute of Chemical Engineers. New York, NY. 2007.

1-6. American Chemistry Council, 1300 Wilson Blvd., Arlington, VA 22209. www.americanchemistry.com

1-7. European Chemical Industry Council (Cefic), Avenue E. van Nieuwenhuyse, 4 box 1, B-1160 Brussels. www.cefic.org

1-8. American Petroleum Institute, 1220 L Street, NW, Washington, D.C. 20005. www.api.org

1-9. American National Standards Institute, 25 West 43rd Street, New York, NY, 10036. www.ansi.org

1-10. American Society of Mechanical Engineers, Three Park Avenue, New York, NY, 10016. www.asme.org

1-11. The Instrumentation, Systems, and Automation Society, 67 Alexander Drive, Research Triangle Park, NC 27709. www.isa.org

1-12. National Fire Protection Association, 1 Batterymarch Park, Quincy, MA, 023169. www.nfpa.org

1-13. Process Safety Management of Highly Hazardous Chemicals (29 CFR 1910.119), U.S. Occupational Safety and Health Administration, May 1992. www.osha.gov

1-14. Flammable and Combustible Liquids, Occupational Safety and Health Standards (29 CFR 1910.106), U.S. Occupational Safety and Health Administration. www.osha.gov

1-15. PSM Covered Chemical Facilities National Emphasis Program, OSHA Notice, 09-06 (CPL 02), U.S. Occupational Safety and Health Administration, July 2009. www.osha.gov

1-16. Petroleum Refinery Process Safety Management National Emphasis Program, OSHA Notice, CPL 03-00-010, U.S. Occupational Safety and Health Administration, August 2009. www.osha.gov

1. INTRODUCTION

1-17. Accidental Release Prevention Requirements: Risk Management Programs Under Clean Air Act Section 112(r)(7), 40 CFR Part 68, U.S. Environmental Protection Agency, June 20, 1996 Fed. Reg. Vol. 61[31667-31730]. www.epa.gov

1-18. California Accidental Release Prevention (CalARP) Program, CCR Title 19, Division 2, Office of Emergency Services, Chapter 4.5, June 28, 2004. www.oes.ca.gov

1-19. Contra Costa County Industrial Safety Ordinance. www.co.contra-costa.ca.us

1-20. Extremely Hazardous Substances Risk Management Act, Regulation 1201, Accidental Release Prevention Regulation, Delaware Department of Natural Resources and Environmental Control, March 11, 2006. www.dnrec.delaware.gov

1-21. Chemical Accident Prevention Program (CAPP), Nevada Division of Environmental Protection, NRS 459.380, February 15, 2005. http://ndep.nv.gov/bapc/capp/capp.html

1-22. Toxic Catastrophe Prevention Act (TCPA), New Jersey Department of Environmental Protection Bureau of Chemical Release Information and Prevention, N.J.A.C. 7:31 Consolidated Rule Document, April 17, 2006. www.nj.gov/dep

1-23. Australian National Standard for the Control of Major Hazard Facilities, NOHSC: 1014, 2002. www.docep.wa.gov.au/

1-24. Environmental Emergency Regulations (SOR / 2003-307), Environment Canada. www.ec.gc.ca/CEPARegistry/regulations

1-25. Control of Major-Accident Hazards Involving Dangerous Substances, European Directive Seveso II (96 / 82 / EC). http://ec.europa.eu/environment/seveso/legislation.htm

1-26. Korean Occupational Safety and Health Agency, Industrial Safety and Health Act, Article 20, Preparation of Safety and Health Management Regulations, Korean Ministry of Environment, Framework Plan on Hazards Chemicals Management, 2001-2005. http://english.kosha.or.kr/main

1-27. Malaysia, Department of Occupational Safety and Health (DOSH) Ministry of Human Resources Malaysia, Section 16 of Act 514. http://www.dosh.gov.my/doshV2/

1-28. Control of Major Accident Hazards Regulations (COMAH), United Kingdom Health & Safety Executive, 1999 and 2005. www.hse.gov.uk/comah/

2

FOUNDATIONAL CONCEPTS

Understanding basic, foundational concepts is essential in establishing a system that identifies hazards and manages risk. To be effective, this system must continuously loop-back and question "What can go wrong?" at all stages in a facility's life cycle. Identifying the hazards associated with the facility and providing engineering measures to prevent or mitigate the consequences are the basic principles of engineering design for process safety. Most effective when it is performed during conceptual and detailed design, this process also provides substantial value through construction, startup, operation, and decommissioning.

This chapter, Foundational Concepts, provides an overview of understanding hazards and risk-based design. Table 2.1 identifies the topics found in this chapter, where the reader can find more information on the topic in this book, and finally where detailed information may be found in other sources.

2.1 UNDERSTANDING THE HAZARD

2.1.1 Dangerous Properties of Process Materials

Safe handling of materials in both process and storage begins with understanding their physical and chemical properties. This concept applies to all chemical substances used by or formed in a process, including reactants, intermediates, products, catalysts, solvents, adsorbents, etc. Some important material characteristics are listed in Table 2.2 and discussed in the following pages.

2.1.1.1 General Properties

Information describing the general properties of most chemical substances is usually found on the Material Safety Data Sheets (MSDSs) which are provided by manufacturers. Information is also available in handbooks, such as the *CRC Handbook of Chemistry and Physics* (Ref. 2-12) or Perry's *Chemical Engineers' Handbook* (Ref. 2-13). The Design Institute for Physical Property Data (DIPPR®) has developed critically evaluated thermophysical property data for pure components and mixtures (Ref. 2-14) that is periodically updated.

Table 2.1   Foundational Concepts and Detailed Resources

Concept in Chapter 2	Further Information in This Book	Detailed Information in Other Resources
Section 2.1 Understanding the Hazard	Chapter 3 Basic Physical Properties / Thermal Stability Data Chapter 4 Analysis Techniques	Guidelines for Hazard Evaluation Procedures (Ref. 2-1) Practical Approach to Hazard Identification for Operations and Maintenance Workers (Ref. 2-2) Guidelines for Safe Process Operations and Maintenance (Ref. 2-3) Guidelines for Process Safety Fundamentals in General Plant Operations (Ref. 2-4) Guidelines for Chemical Reactivity Evaluation and Application to Process Design (Ref. 2-5)
Section 2.2 Risk-Based Design	Chapter 4 Analysis Techniques, Section 4.3 Risk Assessment Chapter 5 General Design Chapter 6 Equipment Design	Guidelines for Hazard Evaluation Procedures (Ref. 2-1) Guidelines for Risk Based Process Safety (Ref. 2-6) Guidelines for Chemical Process Quantitative Risk Analysis (Ref. 2-7) Inherently Safer Chemical Processes: a Life Cycle Approach (Ref. 2-8)
Section 2.3 Intentional Unsteady State Condition Evaluation	Chapter 3 Basic Physical Properties / Thermal Stability Data	Guidelines for Safe Process Operations and Maintenance (Ref. 2-3) Guidelines for Process Safety Fundamentals in General Plant Operations (Ref. 2-4)
Section 2.4 Unintentional Unsteady State Issues	Chapter 3 Basic Physical Properties / Thermal Stability Data	Safe Design and Operation of Process Vents and Emission Control Systems (Ref. 2-9) Guidelines for Safe Storage and Handling of Reactive Materials (Ref. 2-10) Guidelines for Process Safety in Batch Reaction Systems (Ref. 2-11)
Section 2.5 Non-Linearity of the Design Process	Throughout this book	Guidelines for Hazard Evaluation Procedures (Ref. 2-1) Guidelines for Risk Based Process Safety (Ref. 2-6) Guidelines for Chemical Process Quantitative Risk Analysis (Ref. 2-7) Inherently Safer Chemical Processes: a Life Cycle Approach (Ref. 2-8)

Table 2.2 Typical Material Characteristics

Property	Characteristic
General Properties	Boiling point Critical pressure and temperature Electrical conductivity Fluid density and viscosity Freezing point Molecular weight Thermal properties enthalpy, specific heat, heat of mixing Vapor pressure
Reactivity	Compatibility with materials of construction and other process materials, including heat transfer materials Heat of reaction (desired, as well as side reactions) Polymerization Potential for sudden violent reaction Reactivity with water or air Self Accelerating Decomposition Temperature (SADT) Sensitivity to mechanical or thermal shock
Flammability	Autoignition temperature Flammability limits Flash point K_{st} Minimum ignition energy Minimum / limiting oxygen concentration Self-heating
Toxicity	Emergency exposure limits, e.g., acute toxicity values Exposure effects Human threshold limit values Lethal concentration LC_{50} Lethal dose LD_{50}
Stability	Chemical stability Products of decomposition Shelf life Thermal stability, including but not limited to the following: - Differential Scanning Calorimetric (DSC) tests - Accelerating Rate Calorimetry (ARC) tests - Isothermal tests

Boiling point and freezing point data establish whether a substance is a solid, liquid, or gas at atmospheric pressure. Comparison of boiling points or volatilities relative to process conditions provides insight into a number of potentially significant issues, such as flammability or ease of separation by distillation. Vapor pressure data are more difficult to obtain but are more useful in predicting volatility-related behavior. Freezing point data reveal that some relatively common substances may require special handling for cold weather.

Molecular weight provides a quick comparison of gas densities, which indicate whether a vapor released to the atmosphere will rise and disperse or travel along the ground. Critical pressure and temperature are needed for developing thermodynamic expressions using the laws of corresponding states. Since vapors cannot be compressed into liquids at temperatures above their critical regions, substances that can exist only as vapor are indicated by critical temperatures below ambient or processing temperature.

Fluid density and viscosity determine the difficulty of transporting substances inside piping. This information is also useful in other transportation-related issues, such as overloading tank trailers with high density liquids and design of relief systems. In the event of spills, density and solubility relative to water are important issues. Electrical conductivity often indicates the degree to which static charges might build in flowing systems. Enthalpy or specific heat data predict temperature rises for heated substances, critical information when vessels containing volatile flammable liquids are subjected to fire. Heat-of-mixing data indicates pronounced thermal effects that might occur when mixing substances, such as two different concentrations of sulfuric acid.

2.1.1.2 Reactivity

The reactivity of a chemical substance not only influences process reactions, it also influences the hazard potential in accidental releases or inadvertent mixtures. Exothermic reactions can pose hazards because the heat evolved raises the temperature of the reactants leading to increased reaction rate or vaporization of materials. When high temperature is reached in an open system, the materials may ignite or explode. In a closed system, high temperature can lead to vessel rupture from overpressurization caused by gas evolution or vapor pressure.

Some materials react violently upon contact with water, generating considerable heat. For example, some strong acids may evolve large amounts of hazardous fumes when contacted with water or moisture in the air. It is important to recognize this aspect when preparing fire fighting contingencies.

Pyrophoric substances react violently with air, resulting in spontaneous ignition. Such substances are typically handled by methods that prevent contact with air, often by submerging the substance in a compatible solvent, water or oil.

Other chemicals react violently with oxidizing or reducing agents. Oxidants may generate heat, oxygen, and flammable or toxic gases. Reducing agents react with a variety of chemicals and may generate hydrogen, as well as heat, and flammable or toxic gases. Storage and usage of strong oxidizing and reducing agents require special precautions that are unique to the particular substance in question. Generally, each supplier provides complete packages of safety-related information to its customers.

2. FOUNDATIONAL CONCEPTS

Some chemicals polymerize or decompose at elevated temperature or if contaminated by polymerization initiators or catalysts. Common substances, such as water, rust, or other contaminants, can initiate polymerization reactions. When polymerization is initiated, exothermic reaction may occur leading to high temperature and pressure, possibly resulting in explosion or release of flammable or toxic substances. Such decomposition and polymerization reactions may be prevented by incorporating safety systems, inhibitors, and safe operating procedures.

Because chemical reactivity is extremely complex, hazardous materials should be examined on a specific case-by-case basis. Chemical reactivity data are available in:

- Handbook of Reactive Chemical Hazards (Ref. 2-15)
- EPA's Chemical Compatibility Chart (Ref. 2-16)
- Guidelines on Chemical Reactivity Evaluation and Applications to Process Design (Ref. 2-5)
- Sax's Dangerous Properties of Industrial Materials (Ref. 2-17)
- Chemical Reactivity Worksheet (Ref. 2-18)
- Fire Protection Handbook (Ref. 2-19)
- CCPS Reactivity Evaluation Screening Tool (Ref. 2-20)

2.1.1.3 Flammability

Another important material characteristic requiring attention in early stages of process design is flammability. The most common measures of flammability potential for materials are:

- Autoignition temperature
- Conductivity
- Fire point
- Flammable limits
- Flash point
- K_{St}
- Minimum / limiting oxygen concentration

These are discussed further in Chapter 3, Basic Physical Properties / Thermal Stability Data.

2.1.1.4 Toxicity

Toxic releases generally have a greater impact on humans than fire or explosion; therefore, recognizing the toxicity of materials is important in process design. Humans can be exposed to toxics by inhalation, ingestion, and dermal contact. Toxic exposure is influenced by the airborne concentration and the duration of exposure. Toxic exposures are described as:

- *Acute* - Acute exposures represent brief contacts with potentially lethal concentrations, typically experienced during sudden large discharges of toxic materials.
- *Chronic* - Chronic exposures occur due to prolonged exposure, usually over a period of time.

Various sources of recognized exposure limits for airborne contaminants are presented in Table 2.3. These sources can be used to determine exposure limits under a variety of circumstances. The Subcommittee on Consequence Assessment and Protective Actions (SCAPA) of the Department of Energy also maintains a hierarchal listing of chemicals' Protective Action Criteria (PAC) in the order priority of AEGL, ERPG, then TEEL, whichever has been defined (Ref. 2-21).

Table 2.3 Selected Primary Data Sources for Toxic Exposure Limits

Source	Acronym	Exposure Limit	Acronym
American Conference of Government Industrial Hygienists	ACGIH	Threshold Limit Value	TLV
American Industrial Hygiene Association	AIHA	Workplace Environmental Exposure Limit	WEEL
		Emergency Response Planning Guideline	ERPG
Department of Energy	DOE	Temporary Emergency Exposure Guidelines	TEEL
Environmental Protection Agency	EPA	Acute Exposure Guideline Levels	AEGL
National Institute of Occupational Safety and Health	NIOSH	Immediately Dangerous to Life or Health Level	IDLH
National Academy of Science / National Research Council	NAS / NRC	Short-Term Public Emergency Guidance Level	SPEGL
		Emergency Exposure Guidance Level	EEGL
Occupational Safety and Health Administration	OSHA	Permissible Exposure Limit	PEL

2.1.1.5 Effect of Impurities

Impurities in process streams may jeopardize desired reactions and possibly pose threats to plant safety. These impurities may be traces of compounds typically present in raw materials (e.g., pyrophoric iron sulfides in petroleum or catalyst poisoning agents). Sometimes impurities are the same substance in a different physical form, such as solids in a liquid stream or liquid slugs in a gas stream.

Effects of impurities should be critically analyzed before beginning process design. Engineering solutions that prevent impurities from entering the process include filters and strainers, adsorbent beds (one-time and regenerative), and guard beds.

2.1.2 Process Conditions

Process conditions, such as pressure and temperature, have their own characteristic problems and hazards. High pressures and temperatures create stresses that must be accommodated by design. Extreme temperatures or pressures individually are usually not the problem, but rather their combination. A combination of extreme conditions results in increased plant cost due to the need for material with high mechanical strength and corrosion resistance.

High pressure increases the amount of potential energy available in a process facility. For these facilities, in addition to the energy of compressed gases and of fluids kept under pressure in the liquid state, there may also be a concern of chemical reactivity under pressure or an adverse reaction from rapid depressurization. Leakage is much more pronounced in high pressure operations. Because of the large pressure difference, the amount of fluid that can discharge through a given area is greater. A high pressure difference has a considerable impact on the consequences of a release, as the hazard zone extends to a larger area.

High temperature also poses material failure problems, most frequently due to metal creep. The use of high temperature conditions usually increases plant cost, not only due to materials of construction but also due to the requirement for special supports to handle the stresses generated. Process design should take these stresses into account. The design should minimize such stresses, especially during startup and shutdown.

High temperatures are often obtained with the use of fired heaters, which have additional hazards like tube rupture and explosions. It is a good idea to consider using steam heaters, where possible, instead of fired heaters to prevent such hazards.

Low pressure operation usually does not pose much of a hazard in comparison with other operating conditions. However, in the case of vacuum applications where flammable materials are present, the potential for ingress of air does create a hazardous situation. This can result in the formation of a flammable mixture inside equipment leading to fire and / or explosion. It is essential that this aspect is reviewed and adequate measures provided in the process design to prevent air ingress. For equipment not designed for full vacuum, damage frequently occurs because of failure to vent while draining or steaming out, allowing heated equipment to cool while blocked-in, or failure of a vacuum relief device due to pluggage.

Low temperature engineering design considerations include:
- Build-up of ice on equipment and drain systems
- Low temperature caused by J-T effect (e.g., natural gas pressure reducing stations)
- Low temperature embrittlement or loss of elasticity due to inadvertent flow of low temperature fluids into systems constructed of materials not fit for low temperature services
- Low temperature in flare header application (e.g., LPG)
- Possibility of failure of refrigerant or coolant systems which are normally provided to maintain low temperature
- Thermal stresses (contraction and expansion)

Chapter 5, General Design, and Chapter 6, Equipment Design, contain details on design solutions.

2.1.3 Inventory

A common factor in major disasters in the chemical industry is a large release of a hazardous material. One of the best ways to make a plant safer is to minimize the quantity of hazardous materials. The principal approach is to minimize inventory, so that even if there is a leak or explosion, the consequences are minimized (Ref. 2-8).

Low inventories result in safer and more cost-effective process facilities. Lower inventories can be achieved by using smaller or fewer vessels. If fewer vessels are used, fewer protective devices, such as alarms, valves, trips, and smaller flare systems, may be required, further reducing facility costs.

Other methods to limit inventory include:
- Reducing reactor volumes by improving mixing conditions or better understanding reaction kinetics
- Reducing inventory by integrating plant operation, especially for storage tanks and day tanks that usually contain large inventories
- Using continuous reactors instead of batch reactors
- Reducing holdup in distillation columns by using low holdup equipment internals, e.g., packing has less holdup than conventional trays
- Reducing onsite storage by using just-in-time delivery
- Laying out equipment and pipe to reduce pipe rack toxic material holdup
- Improving the performance of the reactor (reducing by-product production) so that subsequent operations, e.g., distillation, become easier, further reducing holdup
- Making highly toxic material generation (e.g., phosgene) a subprocess just prior to using the material in the main process, shifting inventory to less toxic materials
- Producing on-demand from less hazardous materials

Substituting a less hazardous material or limiting the inventory of hazardous materials is usually the first choice in risk reduction. For example, consider using steam as heat transfer medium instead of a flammable material. If reduction of the inventory or substitution of hazardous materials is not feasible, attempts should be made to use less hazardous conditions, such as low pressure and temperature storage; use of material in its gas phase instead of its liquid phase; or use of a safer solvent.

Some secondary effects of reducing inventories may need to be considered, such as:
- A reduction in residence time could result in poor separation of materials
- Increased potential for cavitation of pumps
- Less time for operator response to a low level alarm

2.2 RISK-BASED DESIGN

Process or equipment design often involves deciding between alternative designs with differing process efficiency, safety, environmental controls, cost, and schedule

2. FOUNDATIONAL CONCEPTS

implications. To accomplish this, the formation of a multidisciplinary design team is required at the beginning of a project in order to obtain total integration of process safety with process design and environmental protection considerations. Sometimes safety considerations clearly dominate and decisions are made in the form of special design approaches (e.g., design of facilities manufacturing or using nitromethane, ethylene oxide, hydrogen fluoride, phosgene, etc.). In some instances, codes and standards exist that either mandate or suggest design approaches to known high risks.

In a majority of situations, however, no single factor dominates. In the process of arriving at a design basis decision, the risks of each option are typically dealt with judgmentally or qualitatively (Ref. 2-22). In some instances, one component of risk is quantified (i.e., either consequence or frequency) to justify the design selection. For large projects, full risk quantification is sometimes used to assess the combined impacts of multiple hazards.

Risk-based design begins at the earliest stages of a project. After the general configuration of the process has been established and the design is defined in terms of heat and material balances and basic process controls, the process design can be evaluated for quality, safety, health and environmental impact. The design team begins brainstorming how the process can deviate from normal conditions (i.e., failure scenarios) by asking questions, such as:

- What can go wrong? What failure scenarios can we realistically expect with this process?
- What impact can those failure scenarios have?
- How frequently might they occur?
- What is the risk?
- Is this risk acceptable?
- What design features can be put in place to minimize the risk?

If posed at the conceptual stage of a process design, these questions offer great opportunity for the application of inherently safer design solutions. While inherently safer solutions should emerge as recurring themes throughout the design process, the earlier the application of inherently safer solutions, the more cost-effective and easier to implement these solutions will be.

It is important to recognize that, irrespective of the specific approaches and the level of effort, engineers and technical managers are already directly or indirectly factoring risk into the selection of design options. The process used to assess risk should be systematic and comprehensive.

A systematic technique can provide a consistent risk management framework for process safety system design basis decisions. Inconsistencies in approach can develop not only between different processes and facilities, but also in the case of large, complex design projects, and different design engineers may follow different risk management philosophies.

Consistency with respect to risk acceptance decisions is necessary to assure all stakeholders (e.g., owners, employees, customers, and the general public) that risks are being properly managed. In some countries, governments are also explicit stakeholders in the effort to reduce the risk of chemical industry accidents, providing such regulations as:

- Australia, *Australian National Standard for the Control of Major Hazard Facilities*, NOHSC: 1014, 2002. (Ref. 2-23)
- Korea, Korean Occupational Safety and Health Agency, *Industrial Safety and Health Act, Article 20, Preparation of Safety and Health Management Regulations*. (Ref. 2-24)
- Malaysia, Department of Occupational Safety and Health (DOSH), *Ministry of Human Resources Malaysia, Section 16 of Act 514*. (Ref. 2-25)
- United Kingdom, *Control of Major Accident Hazards Regulations (COMAH)*, United Kingdom Health & Safety Executive. (Ref. 2-26)
- United States, Environmental Protection Agency, *Accidental Release Prevention Requirements: Risk Management Programs Under Clean Air Act*. (Ref. 2-27)
- United States, Occupational Safety and Health Administration, *Process Safety Management of Highly Hazardous Chemicals*. (Ref. 2-28)

Consequently, having a consistent, documented technique for the selection and design of process safety systems is not only prudent management; in many countries it is a regulatory requirement.

However, *systematic* does not necessarily imply *quantitative*. In many simple design situations, qualitative approaches will satisfy the requirements of the technique for selecting process safety system design bases. More complex design cases may occasionally require rigorous quantitative risk analysis approaches. But even in these complex cases, quantitative approaches should only be employed to the degree required to make an informed decision. This concept of the selective use of quantitative risk analysis has been incorporated into the technique presented later in the chapter and in Chapter 4, Analysis Techniques.

2.2.1   The Concept of Risk

The design basis selection technique for process safety systems described later in this chapter is a *risk-based* technique. Risk is defined as a measure of loss in terms of both "the incident likelihood and the magnitude of the loss" (Ref. 2-7).

This concept of risk combines an undesirable outcome, i.e., a consequence such as safety impact or financial loss, with the likelihood of that outcome. Likelihood is defined as (Ref. 2-29):

> *A measure of the expected frequency with which an event occurs. This may be expressed as a frequency (e.g., events per year), or a probability of occurrence during a time interval (e.g., annual probability).*

Inherent in the assessment of risk are the dimensions of consequences (outcomes / impacts) and likelihood (frequency / probability). Various techniques, both qualitative and quantitative, have evolved for assessment of risk. An overview of these techniques, including when to use them in the life cycle of the facility, is contained in Chapter 4. Further information can be found in *Guidelines for Chemical Process Quantitative Risk Assessment* (Ref. 2-7) and *Guidelines for Hazard Evaluation Procedures* (Ref. 2-1). Four integrated activities in risk analysis are described in Table 2.4.

2. FOUNDATIONAL CONCEPTS

Table 2.4 Four Key Integrated Activities in Risk Analysis

Activity	Description
Identify hazards	• Systematic identification of hazards and related failure scenarios that can lead to incidents • Frequently involves application of standard techniques, such as HAZOP, FMEA, What-If?, etc.
Define and document consequences	• Process used to estimate the consequence of failure scenarios • Typically involves a range of activities from simple application of qualitative damage criteria to complex computer models for characterizing impacts of hazardous materials releases that result in fires, explosions, and toxic vapor clouds • Characterization of the release conditions (i.e., source term) is a critical step in quantitative consequence analysis, having great influence on the validity of the results
Estimate likelihood	• Process used to estimate the frequency of a particular incident or outcome • Where available, historical data are used to quantify the likelihood • When historical data are unavailable, incomplete, or inappropriate, analytical approaches such as fault tree and event trees are employed to determine the likelihood of incident / outcomes based on more fundamental failure data
Estimate risk	• Process of combining consequence and likelihood estimations of all selected scenarios into a measure of overall risk, the simplest form being a risk matrix • Includes various ways of displaying risk, such as individual risk contours or overall likelihood of various levels of consequence • Prioritization of risks

2.2.2 Selection of Design Bases for Process Safety Systems

This section describes a systematic risk-based technique for selecting the design basis for process safety systems. Use of the technique imposes discipline on the thought process, yet allows for flexibility in application. This risk-based technique consists of nine steps which are discussed below and illustrated in Figure 2.1.

2.2.2.1 Step 1: Identify Failure Scenarios

Step 1 assumes the existence of a process design. Whether for a new process or a modification of an existing process, the design team has specified the major equipment, including heat and material balances. With this design established, things that can go wrong, i.e., failure scenarios, should be addressed. For example, perform hazard evaluations by employing the standard techniques described in *Guidelines for Hazard Evaluation Procedures* (Ref. 2-1).

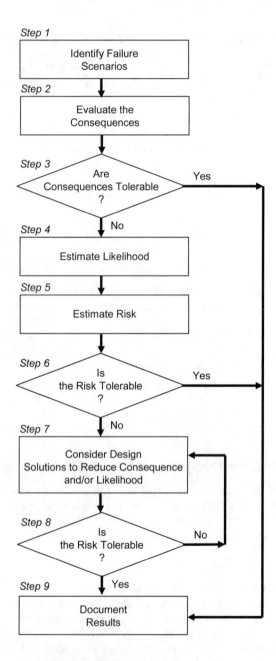

Figure 2.1 Technique for Selecting the Design Bases for Process Safety Systems

2. FOUNDATIONAL CONCEPTS 25

2.2.2.2 Step 2: Estimate the Consequences

In this step, the consequences of the failure scenarios identified in Step 1 should be estimated. In general terms, these consequences can include quality, safety, health, economic, and environmental impacts. For these *Guidelines,* consequences of interest include fires, explosions, toxic material releases, and major equipment damage. The design team may, in some cases, uncover potential consequences by direct observation, engineering judgment, or use of qualitative consequence criteria. In other cases, the use of quantitative consequence estimation techniques may be necessary.

Consequence estimation requires information on the physical, chemical, and toxic nature of the materials involved in the process, the quantity of material which could be involved in a scenario, the impact of each scenario on the surroundings (facility siting), and an economic evaluation of the impact of equipment damage and lost production.

Intrinsic chemical information can be obtained from the MSDS or other sources of process safety information. This, combined with the quantity of material in the process, can be used to assess fire, explosion, and toxic effects using appropriate source terms, dispersion calculations, and effect models for scenarios with the potential for materials release to the environment.

2.2.2.3 Step 3: Determine Tolerability of Consequences

In this step, for each failure scenario the design team should ask: *"Can we tolerate the consequences?"* Answering this question requires established tolerability criteria. Established criteria might take the form of:

- Appropriate engineering codes and standards
- Company-specific criteria (such as not exceeding a specified hazardous material concentration at the fence line)
- Government regulations
- Industry initiatives

If application of the criteria results in tolerable consequences, then no additional process safety system is needed, and no further risk assessment is required. Proceed to Step 9 and document the results. For intolerable consequences, continue the risk assessment in Step 4.

2.2.2.4 Step 4: Estimate Likelihood

The design team next estimates the likelihood of the failure scenarios identified in Step 1. Frequency is typically estimated by evaluating the layers of protection in place to prevent or mitigate the failure scenario. More credit is given to engineering controls over administrative controls.

Engineering controls in process safety design include the Basic Process Control System (BPCS), alarms and shutdowns, pressure safety valves, vacuum breakers, conservation vents, detection and mitigation systems, etc.

Reliability of administrative safeguards, on the other hand, is tied to the effectiveness of training and the strength of managerial implementation and documentation. Not only are these hard to measure, they can change significantly, in either a positive or negative manner, due to a wide variety of factors, such as personnel turnover, staffing level changes, or change in management.

Equipment failure data are available from a number of sources, and while there are uncertainties and gaps in the data, these can be objectively and consistently evaluated through the use of plant data collection and component failure testing. Also, a comprehensive risk management plan based on the results of studies such as these can provide typical component failure rates to be used for a wide range of evaluations.

At some point, quantification of likelihood may be necessary, but often it is superseded by standardization into policies, engineering standards, and standard practices. For example, failures with no or low consequences may be considered adequately controlled by normal process controls, whereas severe hazards (such as those with offsite impact) may require several independent layers of protection in order to bring the risk into an acceptable range.

2.2.2.5 Step 5: Estimate Risk

To estimate the risk, the consequence and likelihood are combined. Methods for combining likelihood and consequence estimates to obtain risk measures are presented in *Guidelines for Chemical Process Quantitative Risk Analysis* (Ref. 2-7). Facilities often use qualitative tools, such as risk matrices. Other cases may require quantified approaches, such as determining risk profiles or risk contours. Most facilities have established risk criteria that are employed during the design phase to determine if the risk is acceptable or if additional protection layers are required.

2.2.2.6 Step 6: Determine Tolerability of Risk

In this step, the design team should ask: *"Can we tolerate the estimated risk?"* Like Step 3, answering this question requires guidance in the form of established tolerability criteria. The topic of risk tolerability is discussed in more detail in Chapter 4, Analysis Techniques, and in *Guidelines for Developing Quantitative Safety Risk Criteria* (Ref. 2-30).

If application of the criteria yields tolerable risk, then no additional process safety system is needed; and the design team should proceed to Step 9 to document the results. For intolerable risk, the design team should continue with the risk reduction efforts in Step 7.

2.2.2.7 Step 7: Consider and Evaluate Design Solutions

Failure scenarios with intolerable risk require the design team to reduce risk by:
- Mitigating consequences
- Lowering the likelihood of the failure scenario
- Preventing the consequences altogether via design alternatives.

The design team should review the engineering design solutions to ensure that these proposed design changes would sufficiently reduce the risk and not introduce new hazards or risk. Each potential design solution should be evaluated for:
- *Technical Feasibility* - Will it work at all?
- *Applicability to a Specific Situation* - Will it work here?

2. FOUNDATIONAL CONCEPTS

- *Cost / Benefit* - Is it the best use of resources, or can greater risk reductions be achieved by spending the same money elsewhere?
- *Synergistic / Mutual Exclusivity Effects* - Will this solution work in conjunction with other potential enhancements, or will its implementation eliminate other potential beneficial solutions from being considered?
- *Additional New Hazards* - Will this solution create new hazards that must be evaluated?

The tables in Chapter 6, Equipment Design, are intended to suggest potential alternatives to enhance the risk tolerability of the design. Not all solutions presented in the tables will be applicable to every situation; however Chapter 6 contains detailed references.

2.2.2.8 Step 8: Determine Tolerability of Risk

After applying the design solutions selected in Step 6 above, the design reevaluates the scenario to determine if the design solutions reduce the risk to an acceptable level.

2.2.2.9 Step 9: Document Results

The results of this risk assessment should be clearly documented, including:
- The cause of the failure scenario
- The ultimate consequences
- Identified risk
- Design solutions

Some companies utilize the information from the risk assessment to build a hazard register as part of their risk management strategy. A hazard register contains all the identified hazards, their consequences, and the solutions put in place to minimize the hazards.

Documentation of the design basis captures and preserves vital information and will prove especially important during hazard evaluations, management of change situations, and other related risk management activities, including future design efforts. Without proper design documentation (Ref. 2-31), important information may not be available for consideration in future situations involving safety decisions.

Even in situations where the tolerability criteria applied in Step 3 or 6 determine that no process safety system is needed, it is important to document this decision so that the design basis is not contradicted by future operating or design changes. If for no other reason, document the rationale to avoid the need to repeat the exercise in the future.

Further discussion of documentation is contained in Chapter 8.

2.3 INTENTIONAL UNSTEADY STATE CONDITION EVALUATION

During some types of operations, a process facility expects to routinely encounter unsteady state conditions and has engineering controls (and administrative controls) in place to manage the risk during these phases of operation, such as:
- *Startup* - Startup presents unique challenges and often results in errors; however steps can be taken to minimize startup issues. During the design phase, it is

important to take startup into consideration and provide engineering solutions to problems that may be encountered during startup. A pre-startup hazard assessment can help identify issues that may be encountered and provide solutions to minimize startup issues. These solutions can include:

- Installing permissive instrumentation that prevents opening of a valve until certain conditions have been met
- Installing dedicated startup features, such as a recycle line to facilitate startup of the compressor
- Identifying potential high risk startup activities that require dedicated operator attention

Additionally, during startup, some instrumentation would be bypassed, thus the facility would not be operating with all safeguards in place.

- *Startup Following Maintenance* - Particular attention must be made to proper valve line-ups and assurance that all blinds have been removed prior to startup. Proper cleaning, draining, drying, purging, and / or evacuation of equipment is necessary to reduce the likelihood of improper mixing of materials. Communication and correct turn-over of equipment from maintenance to operations is essential. These solutions can include PSSRs and checklists.
- *Hot Startup* - Startup of a process system following an emergency shutdown presents unique hazards, including:
  - Undesired accumulation of liquids from vapors condensing
  - Relighting hot furnaces that may have a flammable mixture in the fire box
- *Temporary Operations* - Abnormal operations also provide an opportunity for errors to occur. Some examples of abnormal operations and their design solutions considerations include:
  - Summer / Winter Operations - The BPCS alarms set points may be different for different modes of operations
  - Bypassing Instrumentation For Maintenance - Equipment may be left in the bypass mode, rendering the instrument ineffective. To avoid this situation, administrative procedures should be in place to manage bypassing shutdown systems.
  - Bypassing Instrumentation For Startup - Equipment may be left in the bypass mode, rendering the instrument ineffective. PLCs can be programmed to automatically disable and enable shutdown systems. For example, if the system is designed with a low flow shutdown of a heater's fuel gas system, then logic can be programmed so that at heater light-off the shutdown is bypassed. Once the flow reaches a minimum flow, the program logic turns the interlock back on.

Design solutions can include:

- Providing permissives with appropriate controls and procedures for switching between modes of operation
- Designing bypasses such that testing can be completed online.

- *Standby Operations* - Standby operations often include the process system staying in a recycle mode. This recycle mode can introduce hazardous scenarios. For example, higher than normal temperatures may be encountered during standby operations since normal energy removal paths may be minimized.

Design solutions can include providing permissive for switching between modes of operation.

- *Shutdown* - Shutdowns are generally planned sequences of events so that the process can be brought to a safe state. Hazards might include:
  - Equipment shutdown out of sequence allowing for steps to be omitted
  - Equipment not properly cleaned before the next startup

  Design solutions can include:
  - Development of operating procedures and checklists that clearly define the steps needed.
  - Designing equipment to fail in the safe mode, hence requiring no action

- *Emergency Shutdown* - Emergency shutdowns generally occur as the result of another incident such as loss of power or activation of a shutdown system. Hazards might include:
  - Insufficient resources to address all the actions necessary in a short period of time

  Design solutions can include:
  - Indentifying the different types of emergency shutdown that the process unit may encounter, and then determining if engineering solutions can be installed to simplify or reduce that amount of actions required
  - Designing equipment to fail in the safe mode

2.3.1 Batch Reaction Systems

Understanding the behavior of all the chemicals involved in the process - raw materials, intermediates, products, and by-products - is a key aspect to identifying and understanding the relevant process safety issues. The nature of batch processes makes it more likely for the system to enter a state (pressure, temperature, and composition) where undesired reactions can take place. The opportunities for undesired chemical reactions also are far greater in batch reaction systems due to greater potential for contamination or errors in sequence of addition (Ref. 2-11).

Batch reaction systems present unique challenges for process safety. Engineering design for process safety must identify the hazard scenarios associated with batch operations and ensure that adequate layers of protection are provided. Those facilities designing, constructing, operating, and decommissioning batch operations must identify these unique hazards, such as the following examples:

- *Nature of Batch Operations* - Batch operations consist of a series of processing steps which must be carried out in the proper order and at the proper time. By their very nature, batch-type processes do not operate in a steady state. As the process is being carried out, the holdup of materials in the vessel varies with time as materials are charged, reacted, and perhaps withdrawn, thus changing mixing characteristics and effective heat transfer area. There is a continuous variation in the physical properties, chemical compositions, and physical state of the reaction mixture with time. This makes it more difficult, both for the operators and control systems, to monitor and diagnose the process. The sequence of processing steps and frequent startups and shutdowns increase the probability of human errors and equipment failures. Moreover, batch reaction systems often handle multiple processes and products in the same equipment. This can also lead to increased probability of human error.

Design Considerations - Too often, safeguards for batch operations rely on administrative safeguards, such as procedures and training. While these are important parts of a process safety program, facilities that design, own, and operate batch processes should look towards layers of engineering safeguards in combination with administrative controls. The nature of batch operations (unsteady state), frequently involving manual intervention, creates significant issues pertaining to the design of control systems, design of operating procedures, and the interaction between the control system and the operators. Design considerations should include:

- *Proper Selection of Materials* - Raw materials, intermediates, products, by-products, decomposition or unintended products which are hazardous or could be reactive with other materials handled in this equipment.

- *Avoidance of Use of Incompatible Materials, Especially Materials That React with Common Substances* - Inadvertent contact between two or more incompatible chemicals may lead to a hazardous condition. Water is of particular concern as this seemingly innocuous material can react violently with many chemicals. Some materials react rapidly and violently with water and have an NFPA reactivity rating of 2 or higher based on water reactivity alone (Ref. 2-32).

- *Human Factors* - Human factors are especially important in batch operations when much of the process is influenced by an operator's actions (or inactions). The batch operator is more involved and is often in closer proximity to the process. This close proximity puts the operator at increased risk to direct exposure to the hazards associated with larger inventory of raw materials and semi-finished products than continuous systems with comparable throughput. Special design manifolds and transfer panels can reduce the potential for human error. Automation of the batch sequence using a PLC can also reduce the potential for human error.

- *Selection of Materials of Construction* - Batch operations are often designed for general use, rather than dedicated to a specific process. The piping and layout of the equipment is often modified to meet the needs of the current process, or the process is modified to use the existing equipment. Use of the same equipment in different campaigns, complex process piping, and the use of shared auxiliary equipment, such as columns and condensers, present greater challenges in preventing cross contamination; in selecting materials of construction; and in selecting instrumentation and control systems. Additionally, the complexity of equipment and the frequency of changes complicate the process documentation task. These frequent changes often result in complex Management of Change (MOC) issues.

The issues discussed above are just a small sample of the process safety issues faced in the design process of batch operations. All of these issues make batch reaction systems unique, in terms of the challenges they pose for managing process safety. Refer to *Guidelines for Process Safety in Batch Reaction Systems* (Ref. 2-11) for more detail.

2.4 UNINTENTIONAL UNSTEADY STATE ISSUES

There are many ways a process can deviate from design intent and result in unintentional unsteady state issues. Hazard identifications conducted throughout the facility's life cycle, as well as thorough examination of incident and near-miss investigation reports, can identify these unintentional deviations and ensure that adequate layers of protection are provided. This section highlights some common process deviations and engineering design solutions to consider. It is by no means all inclusive and each facility owner and operator should ensure that their internal engineering design and hazard review process has identified hazards appropriate to their facilities and provided adequate engineering layers of protection. The hazard analysis process is discussed further in Chapter 3, Analysis Techniques, and in *Guidelines for Hazard Evaluation Procedures* (Ref. 2-1). Design considerations are discussed further in Chapter 4.

2.4.1 Runaway Reactions

By their very nature, process industries handle a wide range of materials, many of which can react energetically, either as self-reactives or with other materials. Depending on the process chemistry, off-gases may be formed that need to be collected and disposed of via an appropriate treatment device. It may also be necessary to design the emergency vent system to provide protection against a runaway reaction involving reaction rates and gas flows that may be significantly higher than normal process conditions.

Identifying potential reactivity hazards, whether due to reaction between incompatible materials or due to self-decomposition, typically involves a team that includes personnel knowledgeable in the process, the manufacturing operations, and the chemistry. Chapter 3 describes interaction matrices, which can be an effective means to identify and document the materials and conditions that could result in a reactive hazard.

Reactive chemical incidents can be categorized as either:
- *Self-reactive* - polymerizing, decomposing, isomerizing
- *Reactive with combinations of materials* - where the material may be stable by itself, but reactive with other chemicals

In the case of self-reactive materials, incidents have often been initiated by relatively small amounts of contaminants acting as catalysts; although conditions such as elevated temperatures, pH changes, or the depletion of an inhibitor may also act as initiators.

Reactivity concerns for combinations of materials tend to involve bulk mixing of incompatible materials or reaction with ubiquitous substances, such as air or water. Mechanisms can include process control failures, such as adding the wrong material to a reactor or feeding reactive materials to equipment with its agitator stopped. As in the case of self-reactive materials, these events have frequently involved a combination of materials and conditions. Therefore, the hazard identification should not be limited to single failures (See Chapter 4).

Runaway reactions occur when the heat generation rate from a reacting mass exceeds the rate at which heat can be removed, causing an uncontrolled rise in temperature. In the absence of adequate overpressure relief, if the heat of reaction exceeds the cooling capacity, the reaction rate can accelerate (runaway) and may result in a gas evolution rate that overwhelms the vent header system.

Those companies designing, constructing, operating, and decommissioning processes where runaway reactions can occur should provide engineering solutions for these unique hazards, such as the following examples:

- Use of a tempering fluid so that emergency vents can mitigate the runaway reaction by permitting the liquid to boil and for its latent heat of vaporization to "temper" the reaction
- Instrumentation and automatic responses to detect and respond to incipient runaway conditions in the event that mixing occurs
- An emergency vent header system adequate to handle the maximum vent flow rate resulting from a worst-credible event

During runaway reactions, the temperature can rise significantly, which may favor different reactions. If this occurs, the composition may shift to produce a more toxic off-gas, as occurred at Seveso, Italy (Ref. 2-33). If there is the potential for a runaway reaction, the characteristics and composition of off-gases should be understood and an appropriate treatment selected.

2.4.2 Deviating from the Design Intent

In addition to process deviations resulting from runaway reactions, there are numerous ways a process can deviate from normal operation, resulting in temperature, pressure, and level excursions. These process deviations must be considered during the design phase and throughout the life of the facility to provide the appropriate layers of protection and detection. The following are examples of deviations from design intent that should be considered during the design and operational phases of a facility:

- *Heater Overfiring* - A heater overfiring can lead to high temperature excursions that can result in heater tube failure and subsequent fire / explosion and personnel exposure.
- *Loss of Cooling* - Loss of cooling can lead to high temperature scenarios, resulting in exceeding the design temperature of equipment, and subsequent fire / explosion, toxic release, personnel exposure.
- *Loss of Agitation / Circulation* - Loss of agitation or circulation can lead to uncontrolled temperature increase and runaway reactions.
- *Loss of Reflux* - Potential increased pressure and temperature in fractionation towers, potential release, potential fire / explosion, and personnel exposure.
- *High Level Liquid Carryover* - Potential for equipment damage (carrying over to a compressor), potential environmental impact (carrying over to waste treatment), potential product quality issues (carrying over to storage), etc.
- *Low Level Gas Blow-By* - Potential to exceed the design pressure rating of downstream equipment, potential release, potential fire / explosion, toxic release, personnel exposure.
- *Process Capacity Creep* - Over time, operating rates tend to increase, potentially exceeding the relief system design.

2. FOUNDATIONAL CONCEPTS

2.5 NON-LINEARITY OF THE DESIGN PROCESS

As emphasized throughout this book, engineering design for process safety is a continuous process that is incorporated into the life cycle of a facility. As illustrated in Figure 2.2, the process of identifying hazards, examining risk, and providing engineering system modifications that reduce the risk is a continuous process for the life cycle of the facility, including:

- Research and development
- Conceptual design
- Pilot plant
- Detailed engineering
- Construction and startup
- Routine operation
- Process modification or plant expansion
- Decommissioning

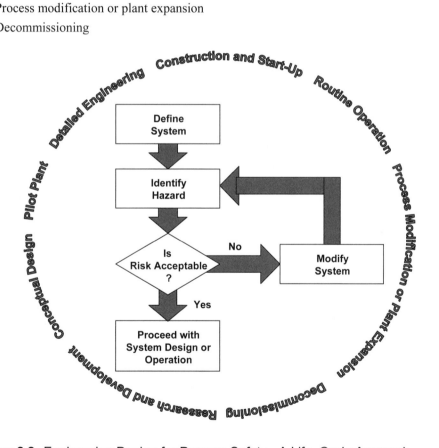

Figure 2.2 Engineering Design for Process Safety - A Life Cycle Approach

Robust management systems must be in place to successfully evaluate and manage the hazards and risk over and over for the life of the facility. These management systems influence the effectiveness and robustness of this continuous process and include:

- *Compliance with Design Standards* - Identifying and addressing relevant process safety standards, codes, regulations, and laws over the life of a process are essential parts of committing to process safety (Ref. 2-6). The primary objective is to ensure that process facilities remain in conformance with applicable standards, codes and regulations, and a company's internal standards and policies over the life of the facility.
- *Continuous Updating of Understanding of Hazard and Risk* - It is important that employees always understand the hazards and risks that are present at their facility. Too often, because something hasn't gone wrong over a period of time, people become complacent and hazards are ignored. There are many ways a facility can continuously evaluate its hazards and risks and provide engineering solutions to prevent or mitigate these hazards. Most facilities that handle, process, or store hazardous materials have management systems like these in place that continuously evaluate hazards and provide engineering solutions:
    - *Management of Change* - Managing changes to processes over the life of a facility is essential in managing risk. Management of Change (MOC) helps ensure that changes to a process do not inadvertently introduce new hazards or unknowingly increase risk of existing hazards. MOC includes a review and authorization process for evaluating proposed modifications to facility design for the life of the facility, so facilities with robust management of change programs are more likely to have a system in place that continuously evaluates hazards associated with process changes and provides a mechanism of providing engineering solutions to prevent or mitigate hazards.
    - *Process Hazards Analysis* - Process Hazards Analyses (PHAs) are required for most facilities that handle, process, or store hazardous materials in quantities greater than threshold amounts (Ref. 2-28 and Ref 2-1). The nature of this study requires a facility to systematically examine the ways that their process may deviate from normal, document what the consequences would be, identify existing safeguards, and perform a qualitative assessment of whether the risk is acceptable. Where the risk is not acceptable, recommendations, often engineered safeguards, are made to reduce the risk. When a facility takes these studies seriously, they can provide a valuable tool for hazard identification and risk management - They must be conducted and maintained for the life of the process. PHAs are discussed further in Chapter 3 and in *Guidelines for Hazard Evaluation Procedures* (Ref. 2-1).
    - *Incident and Near-Miss Investigations* - The objective of a robust incident and near-miss reporting and investigation system is preventing repeat incidents. This system should identify and eliminate root causes, often through engineering solutions and controls (Ref. 2-34).

- *Culture* - The process safety culture of an organization is a significant determinant of how it will approach process risk control issues, and process safety management system failures can often be linked to cultural deficiencies. Accordingly, enlightened organizations are increasingly seeking to identify and address such cultural root causes of process safety performance problems. A culture with a strong process safety emphasis is more likely to successfully incorporate engineering design for process safety into the life cycle of their facilities. A culture that successfully integrates process safety and risk control issues has several key attributes, including:
  - The importance of safe operations is integrated into the organization's core values.
  - Potential failures are used to provide the organization a clear understanding of risk and the means to control it.
  - Resources proportional to the perceived risks are provided.
  - An organization emphasizes learning from past experience in order to prevent future problems.
  - Employees are involved in identifying hazards and deciding how they should be addressed.
- *Management Review and Continuous Improvement* - Management review is the routine evaluation of whether management systems are performing as intended and producing the desired results as efficiently as possible. A system must also be in place for implementing any resulting plans for improvement or corrective action and verifying their effectiveness. Management review and continuous improvement are essential in integrating engineering design for process safety into all stages of a facility's life cycle and help establish and maintain a system that:
  - Defines roles and responsibilities
  - Establishes standards for performance
  - Validates program effectiveness
- *Workforce Involvement* - Promoting the active involvement of personnel at all levels of the organization is essential. Workers who are directly involved in operating and maintaining the process are most exposed to the hazards of the process. They also are potentially the most knowledgeable people with respect to the day-to-day details of operating the process and maintaining the equipment and facilities. When these employees are actively involved, they are part of a continuous improvement process for identifying hazards and reducing risk.

For more detailed information on establishing programs that effectively manage process safety, refer to *Guidelines for Risk-Based Process Safety* (Ref. 2-6).

2.6 REFERENCES

2-1. CCPS. *Guidelines for Hazard Evaluation Procedures, Third Edition.* Center for Chemical Process Safety of the American Institute of Chemical Engineers. New York, NY. 2008.

2-2. CCPS. *A Practical Approach to Hazard Identification for Operations and Maintenance Workers.* Center for Chemical Process Safety of the American Institute of Chemical Engineers. New York, NY. 2010.

2-3. CCPS. *Guidelines for Safe Process Operations and Maintenance.* Center for Chemical Process Safety of the American Institute of Chemical Engineers. New York, NY. 1995.

2-4. CCPS. *Guidelines for Process Safety Fundamentals in General Plant Operations.* Center for Chemical Process Safety of the American Institute of Chemical Engineers. New York, NY. 1995.

2-5. CCPS. *Guidelines for Chemical Reactivity Evaluation and Application to Process Design.* Center for Chemical Process Safety of the American Institute of Chemical Engineers. New York, NY. 2007.

2-6. CCPS. *Guidelines for Risk Based Process Safety.* Center for Chemical Process Safety of the American Institute of Chemical Engineers. New York, NY. 2007.

2-7. CCPS. *Guidelines for Chemical Process Quantitative Risk Analysis, Second Edition.* Center for Chemical Process Safety of the American Institute of Chemical Engineers. New York, NY. 2000.

2-8. CCPS. *Inherently Safer Chemical Processes, A Life Cycle Approach.* Center for Chemical Process Safety of the American Institute of Chemical Engineers. New York, NY. 2009.

2-9. CCPS. *Safe Design and Operation of Process Vents and Emission Control Systems.* Center for Chemical Process Safety of the American Institute of Chemical Engineers. New York, NY. 2006.

2-10. CCPS. *Guidelines for Safe Storage and Handling of Reactive Materials.* Center for Chemical Process Safety of the American Institute of Chemical Engineers. New York, NY. 1995.

2-11. CCPS. *Guidelines for Process Safety in Batch Reaction Systems.* Center for Chemical Process Safety of the American Institute of Chemical Engineers. New York, NY. 1999.

2-12. Haynes, W. *CRC Handbook of Chemistry and Physics, 91st Edition.* National Institute of Standards and Technology. Boulder, CO. 2010.

2-13. Green, D. W. and Perry, R. H. *Perry's Chemical Engineers' Handbook, Eighth Edition.* McGraw-Hill. 2008.

2. FOUNDATIONAL CONCEPTS 37

2-14. *DIPPER® Data Compilation of Pure Chemical Properties*, Design Institute for Physical Properties, American Institute of Chemical Engineers. New York, NY. 2010.

2-15. Urben, P. *Bretherick's Handbook of Reactive Chemical Hazards, Seventh Edition*. Academic Press. Oxford, UK. 2007.

2-16. EPA. EPA's Chemical Compatibility Chart, *A Method for Determining the Compatibility of Chemical Mixtures*. 1980. www.epa.gov

2-17. Lewis, R. S. *Sax's Dangerous Properties of Industrial Materials, 10th Edition*. John Wiley & Sons. Hoboken, NJ. 1999.

2-18. NOAA. *Chemical Reactivity Worksheet, Version 2.1*. National Oceanic and Atmospheric Administration. http://response.restoration.noaa.gov/CRW

2-19. NFPA. *Fire Protection Handbook, 12th Edition*. National Fire Protection Association. Quincy, MA. 2008.

2-20. CCPS. Reactivity Evaluation Screening Tool. Center for Chemical Process Safety of the American Institute of Chemical Engineers (AlCHE). New York, NY. 2010. www.aiche.org/ccps

2-21. DOE. Protective Action Criteria (PAC) Values. Subcommitee on Consequence Assessment and Protective Actions (SCAPA) of the Department of Energy (DOE). www.atlintl.com/DOE/teels/teel.html

2-22. CCPS. *Tools for Making Acute Risk Decisions*. Center for Chemical Process Safety of the American Institute for Chemical Engineers. New York, NY. 1995.

2-23. Australian National Standard for the Control of Major Hazard Facilities, NOHSC: 1014, 2002. www.docep.wa.gov.au/

2-24. Korean Occupational Safety and Health Agency, *Industrial Safety and Health Act, Article 20*, Preparation of Safety and Health Management Regulations. Korean Ministry of Environment, Framework Plan on Hazards Chemicals Management, 2001-2005. http://english.kosha.or.kr/main

2-25. Malaysia, Department of Occupational Safety and Health (DOSH) Ministry of Human Resources Malaysia, Section 16 of Act 514. http://www.dosh.gov.my/doshV2/

2-26. Control of Major Accident Hazards Regulations (COMAH), United Kingdom Health & Safety Executive, 1999 and 2005. www.hse.gov/uk/comah/

2-27. Accidental Release Prevention Requirements: Risk Management Programs Under Clean Air Act Section 112(r)(7), 40 CFR Part 68, U.S. Environmental Protection Agency, June 20, 1996 Fed. Reg. Vol. 61[31667-31730]. www.epa.gov

2-28. Process Safety Management of Highly Hazardous Chemicals (29 CFR 1910.119), U.S. Occupational Safety and Health Administration, May 1992. www.osha.gov

2-29. CCPS. *Guidelines for Chemical Transportation Safety, Security, and Risk Management*. Center for Chemical Process Safety of the American Institute of Chemical Engineers. New York, NY. 2008.

2-30. CCPS. *Guidelines for Developing Quantitative Safety Risk Criteria*. Center for Chemical Process Safety of the American Institute of Chemical Engineers. New York, NY. 2009.

2-31. CCPS. *Guidelines for Process Safety Documentation*. Center for Chemical Process Safety of the American Institute of Chemical Engineers. New York, NY. 1995.

2-32. CCPS. *Guidelines for Safe Storage and Handling of Reactive Materials*. Center for Chemical Process Safety of the American Institute of Chemical Engineers. New York, NY. 1995.

2-33. HSE. Case Study, Icmesa Chemical Company, Seveso, Italy. July 10, 1976. Health and Safety Executive. http://www.hse.gov.uk/comah/sragtech/caseseveso76.htm

2-34. CCPS. *Guidelines for Investigating Chemical Process Incidents, Second Edition*. Center for Chemical Process Safety of the American Institute for Chemical Engineers. New York, NY. 2003.

3

BASIC PHYSICAL PROPERTIES/ THERMAL STABILITY DATA

Understanding the behavior of all the chemicals involved in the process - raw materials, intermediates, products, and by-products - is a key aspect of understanding the process safety issues relevant to a given process. A knowledge of how these chemicals behave individually and how they interact with other chemicals, utilities, materials of construction, potential contaminants, or other materials that they can come in contact with during shipment, storage, and processing is essential for understanding and managing process safety.

Understanding the chemistry of the process also provides the greatest opportunity in applying the principles of inherent safety at the chemical synthesis stage. Process chemistry greatly determines the potential impact of the processing facility on people and the environment. It also determines such important safety variables as inventory, ancillary unit operations, by-product disposal, etc. Creative design and selection of process chemistry can result in the use of inherently safer chemicals, a reduction in the inventories of hazardous chemicals, and / or a minimization of waste treatment requirements.

3.1 BASIC PHYSICAL PROPERTIES

The Design Institute for Physical Properties (DIPPR®) (Ref. 3-1) is the world's best source of critically evaluated thermophysical and environmental property data. Data and estimation methods developed in DIPPR® projects are used by leading chemical, petroleum, and pharmaceutical companies throughout the world.

The mission of DIPPR® is to create and make available a database of evaluated process design data for industrially important chemicals by:
- Building upon and enhancing the value of the DIPPR® Data Compilation
- Satisfying industry needs for accurate and complete thermodynamic and physical property data for process engineering in a rapidly changing business and technical environment

DIPPR®:
- Collects data from a wide range or sources; evaluates them critically; compares them with other values; and stores them in an easily accessible form.
- Correlates the evaluated data emphasizing thermodynamic consistency, accurate reproduction of the values, and reasonable extrapolation.

- Measures property values needed by DIPPR® members that are not found in the literature. These data are added to the DIPPR® databases to replace, improve, and extend existing estimations.
- Disseminates data to the public after a period of exclusive use by members. Dissemination is via hard copy publications, computer programs and databases on diskettes and online, and multimedia.

3.2 FLAMMABILITY DATA

For something so familiar, fire is a surprisingly complex phenomenon. There are many excellent detailed references on the physics of fires, properties of burnable material, and the fundamentals of fire science.

Fire is a self-sustaining, exothermic oxidation-reduction reaction. The fire reaction usually involves oxygen, which forms the oxides of the fuel. The most common examples in petrochemical and hydrocarbon processing facilities are combustion reactions of hydrocarbons with oxygen.

The products of complete combustion of hydrocarbons in air are water and carbon dioxide. However, combustion is rarely complete and by-products are produced.

Flammability data are available in various handbooks, hazardous material data bases, and Material Safety Data Sheets (MSDS). The higher the flash point temperature is above ambient temperature, the more difficult it is to ignite the substance. Liquids with flash points below ambient temperatures are considered particularly hazardous because they generate vapor concentrations that might be rich enough to be ignited at room temperature. Extensive flash point data are available in the *Fire Protection Handbook* (Ref. 3-2).

All substances in the form of liquids (and even many solids) possess a type of molecular motion that results in the escape of molecules from their surface in the form of vapor when they are not confined. When a liquid is left in an open container at room temperature, its molecules evaporate. When the liquid is confined in a partially full container that is *closed*, the molecules will continue to escape from the surface; however, because they cannot escape from the closed container, some of the molecules will return to liquid. After a period of time, equilibrium will be achieved between the number of molecules escaping from the surface and those returning to the surface of the liquid.

When this equilibrium occurs, a certain pressure will be exerted in the empty space above the liquid in the closed container. This is called the vapor pressure of the liquid at that temperature.

Consider a container of water that is heated. If the *upward* pressure of the vapor above the bubbling surface of the water was measured, it would equal the *downward* normal atmospheric pressure applied to the liquid in the open container. The temperature at which this occurs is called the boiling point of the liquid water. At this point, the vapor pressure of the water equals the atmospheric pressure pressing upon it; as long as heat is supplied to it, the liquid boils in the attempt to release its molecules to the vapor state. The boiling points of different types of liquids vary widely. They are an important physical characteristic both of liquids and of the many solids that melt to become liquids and then boil at a certain characteristic temperature.

3. BASIC PHYSICAL PROPERTIES / THERMAL STABILITY DATA

Vapor density is a physical property of major importance to fire protection. Because the vapor density varies with the total weight of all the atoms in a molecule of the vapor of a substance, if the chemical composition of the substance comprising the vapor is known, then the weight or density of its vapor when compared to air can be determined.

The flammability hazard of a liquid is also increased by:
- Wide flammability limits
- Flash point
- Low autoignition temperature
- Low minimum ignition energy
- High maximum burning velocity
- Increasing the temperature of the fuel
- Oxygen-enriched atmosphere

The relationship of flammability terms is shown in Figure 3.1 (Ref. 3-3).

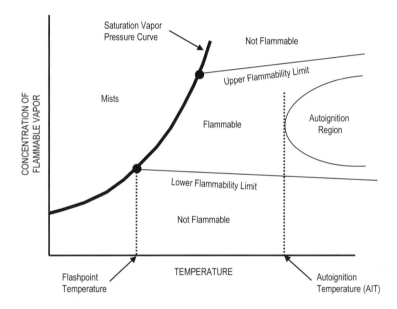

Figure 3.1   The Relationship Between Various Flammability Properties

3.2.1   Flash Point

The flash point of a substance is often treated as the principal index of flammability, especially for liquids. The lower the flash point, the more flammable the liquid.

The flash point is the minimum temperature at which a liquid gives off sufficient vapor to form an ignitable mixture with air within the test vessel used (Methods: ASTM 502) (Ref. 3-4). The flash point is less than the "fire point" at which the liquid evolves vapor at a sufficient rate for indefinite burning.

As a rule of thumb, the flash point can be thought of as the temperature above which a pool of liquid will ignite if a match or other small ignition source is dropped into it. If the temperature of the pool is below the flash point, the pool will not ignite. From a safety perspective, a release of liquid below its flash point should not ignite even if it finds an ignition source.

Because it is an indicator of the hazard of a material, the flash point of a liquid is one of its most important fire characteristics. At its flash point, a liquid continuously produces flammable vapors at the right rate and amount (volume) to give a flammable and even explosive atmosphere if a source of ignition should be brought into the mixture. Flammable liquids (like gasoline) with a flash point of -45°F (-42.8°C) continually give off vapors that can burn at ordinary temperatures. However, fuel oil (such as that used in home-heating furnaces) with a flash point of 130°F (54.4°C) does *not* give off vapor that can burn until heated above its flash point (Ref. 3-2). However, when either material is ignited, an intense fire ensues.

The flash point is when the vapor pressure of a substance is such that the concentration of vapor in air above the substance corresponds to the lower flammable limit. For flammable liquids, the term flammable is any liquid that has a closed-cup flash point below 100°F (37.8°C) and a Reid vapor pressure not exceeding 40 psia (2068.6 mm Hg) at 100°F (37.8°C). The term combustible is used for liquids that have a closed-cup flash point at or above 100°F (37.8°C) (Ref. 3-5).

The flash point and other important properties of some common materials are listed in Table 3.1 (Ref. 3-6).

3.2.2 Fire Point

A self-sustaining fire does not necessarily develop at the flash point. A closely related and less common term is fire point. The fire point is the minimum temperature at which a flammable or combustible liquid and some volatile combustible solids will evolve sufficient vapor to produce a mixture with air that will support sustained combustion when exposed to a source of ignition, such as a spark or flame.

3.2.3 Autoignition Temperature

The Autoignition Temperature (AIT) of a substance is the lowest temperature at which a solid, liquid, or gas will spontaneously ignite without the need for an external ignition source, resulting in self-sustained combustion. A material released from a process above its AIT will ignite. Autoignition temperatures of some common materials are shown in Table 3.1. It is important to determine the AIT under the test conditions that replicate the actual process conditions as close as possible.

Ignition of a mixture above its autoignition temperature is not instantaneous. The ignition time delay may be a fraction of a second for temperatures well in excess of the autoignition temperature and a few minutes when the material is just above its autoignition temperature.

The autoignition temperature is useful in determining the minimum temperature of a hot surface which will ignite a mixture. In most cases of interest, the hot surface ignition temperature is often significantly lower than the autoignition temperature.

3. BASIC PHYSICAL PROPERTIES / THERMAL STABILITY DATA

Table 3.1 Properties of Commonly Used Flammable Liquids in U.S. Customary Units

Name	Molecular Weight	Flash Point °F	Autoignition °F	LFL% by Volume	UFL% by Volume	Boiling Point °F
Acetone	58	-4	869	2.5	12.8	133
Ammonia	17	gas	1204	16	25	-28
Benzene	78	12	928	1.2	7.8	176
n-Butyl Alcohol	74	98	650	1.4	11.2	243
Carbon Disulfide	76	-22	194	1.3	50.0	115
Cyclohexane	84	-4	473	1.3	8.0	179
Ethane	30	-275	959	3.0	12.4	-128
Ethylene	287	-250	914	2.7	36	-155
Gasoline	Mix	-45	536	1.4	7.6	Range
n-Heptane	100	25	399	1.0	6.7	209
n-Hexane	86	-7	437	1.1	7.5	156
Kerosene (Fuel Oil #1)	Mix	100-162	410	0.7	5.0	Range
Methane	16	gas	1004	5	15	-259
Naphtha (VM&P Regular)	Mix	28	450	0.9	6.0	203-320
Propane	44	-220	842	2.1	9.5	-44
n-Propyl Alcohol	60	74	775	2.2	13.7	207
Toluene	92	40	896	1.1	7.1	231
Turpentine	136	95	488	0.8		300
Vinyl Acetate	86	18	756	2.6	13.4	161
o-Xylene	106	88	867	0.9	6.7	292

In general, the AIT (Ref. 3-7):
- Decreases with increasing pressure
- Increases as mixtures become rich or lean
- Decreases with increased oxygen concentration
- Decreases as the test volume increases

3.2.4 Flammable Limits

The stoichiometric ratio is the proportion of fuel and oxidizer that results in optimal combustion and maximum heat release. The optimal ratio is determined by finding the amount of air that will result in the products of the combustion reaction containing only water and carbon dioxide. Burning 100 standard cubic feet of methane requires 1,000 standard cubic feet of air for a stoichiometric mixture.

A mixture below its stoichiometric ratio of fuel to air is described as "lean". A mixture above its stoichiometric ratio of fuel to air is described as "rich". A "lean" mixture has unreacted oxygen along with the combustion products and a "rich" mixture has unreacted fuel with the combustion products.

Fuel-air mixtures at or around stoichiometric concentration have the lowest autoignition temperature, lowest minimum ignition energies, and highest burning velocities.

Flammable vapor burns in air only over a limited range of fuel-to-air concentrations. The flammable range is defined by two parameters; the Lower Flammable Limit (LFL) and the Upper Flammable Limit (UFL). These two terms are also called the Lower Explosive Limit (LEL) and the Upper Explosive Limit (UEL).

The *Lower Flammable Limit* is the minimum proportion of fuel in air that will support combustion. The *Upper Flammable Limit* is the maximum concentration of fuel in air that can support combustion. In popular terms, a mixture below the LFL / LEL is too "lean" to burn or explode and a mixture above the UFL / UEL is too "rich" to burn or explode.

For example, the lower flammable limit of methane in air at sea level is a concentration (by volume or partial pressure) of about 5%. The upper flammable limit is about 15% by volume or partial pressure. Heavier hydrocarbons tend to have lower LFLs. The LFL and UFL of some common hydrocarbons are given in Table 3.1.

Flammability limits may be altered by pressure and temperature. The general result of increasing temperature or pressure is to increase the range of flammability. A decrease in pressure or temperature may tend to narrow the flammable range by raising the lower limit and reducing the upper limit. These aspects should be remembered since published flammable and explosive limits are based on measurements taken at room temperature and atmospheric pressure, unless indicated otherwise.

An increase in temperature tends to widen the flammable range, reducing the LFL. For example, the LFL for methane in air is commonly quoted as 5%. As the temperature of methane increases to autoignition temperature, the LFL falls to around 3%. Stronger ignition sources can ignite leaner mixtures. Flammability limits also depend on the type of atmosphere. Flammability limits are much wider in oxygen, chlorine, and other oxidizers than in air (Ref. 3-2).

In general, pressure has little effect on the LFL; however, as pressure increases, the UFL generally increases.

Flammability limits can be narrowed by the addition of inert gases such as nitrogen or carbon dioxide.

3. BASIC PHYSICAL PROPERTIES / THERMAL STABILITY DATA

3.2.5 Minimum / Limiting Oxygen Concentration

Oxygen is a key ingredient in establishing the LFL in air. There is a minimum / limiting oxygen concentration required to propagate a flame. Explosions and fires can be prevented by reducing the oxygen concentration, generally by inerting. Below the minimum / limiting oxygen concentration, the reaction cannot generate enough energy to heat the entire mixture to the extent necessary for self-propagation of the flame. The minimum / limiting oxygen concentration can be obtained through experimental data or through calculation using the LFL and stoichiometry of the combustion reaction (Ref. 3-7).

3.2.6 Dust Deflagration Index - K_{St}

A deflagration index is defined for dusts in an identical fashion to vapors. For dusts, the deflagration index is denoted by K_{St} where the "St" stands for "Staub," the German word for dust. As the K_{St} value increases, the dust explosion becomes more violent. Table 3.2 shows how the K_{St} values are organized into four St-classes. The St-class number increases as the deflagration index increases, that is, as the dust explosion becomes more violent.

Table 3.2   St-Classes for Dusts

Deflagration Index K_{St} (bar-m / sec)	St-class	Example
0	St-0	Rock dust
1-200	St-1	Wheat grain dust
200-300	St-2	Organic dyes
>300	St-3	Aspirin, aluminum powder

3.2.7 Gas Deflagration Index - K_g

K_g is the deflagration index of a gas cloud. The maximum rate of pressure rise can be normalized to determine the K_g value. It should be noted, however, that the K_g value is not constant and varies, depending on test conditions. In particular, increasing the volume of the test enclosure and increasing the ignition energy can result in increased K_g values. Although the K_g value provides a means of comparing the maximum rates of pressure rise of known and unknown gases, it should be used as a basis for deflagration vent sizing only if the tests for both materials are performed in enclosures of approximately the same shape and size and if tests are performed using igniters of the same type that provide consistent ignition energy (Ref. 3-7).

In general, as the particle size or moisture content decreases, the deflagration index, K_{St}, and the maximum pressure increase while the minimum explosion dust concentration and minimum ignition energy decrease. Over a limited range of particle size, reducing the particle size has more effect on K_{St} and minimum ignition energy than on maximum pressure. As the initial pressure increases, the maximum pressure and, under certain conditions, the maximum rate of pressure rise generally increase proportionally.

Appendix E of *Understanding Explosions* (Ref. 3-3) contains combustion data for a number of dust materials. Appendix E includes the median particle size of the dust tested and the dust concentration under test conditions.

GESTIS-DUST-EX is an online database containing important combustion and explosion characteristics of more than 4,600 dust samples from virtually all sectors of industry that were determined as a basis for the safe handling of combustible dusts and for the planning of preventive and protective measures against dust explosions in dust-generating and processing plants (Ref. 3-8).

3.3   REACTIVITY / THERMAL STABILITY DATA

The general approach to safer process design and operation requires an understanding of basic principles of thermodynamics, chemical kinetics, and reaction engineering. Emphasis is placed on the need to evaluate process safety at an early stage by the process development team. A safe process is as important a goal as a more economic or productive process. The definition of a chemical reactivity hazard is:

> *A situation with the potential for an uncontrolled chemical reaction which can result directly or indirectly in a serious harm to people, property, or the environment. The uncontrolled chemical reaction might be accompanied by a temperature increase, pressure increase, gas evolution, or other form of energy release.*

There are three main parameters that determine the design of safe chemical processes:

1. The potential energy of the chemicals involved
2. The rates of their potential reactions and / or decompositions
3. The process equipment (discussed in Chapter 6)

The first key factor, energy, is involved in the production of any chemical. Design of a safe process requires an understanding of the potential energy (exothermic release / endothermic absorption) available during chemical reactions (both the desired process reaction as well as the potential undesired and side reactions). This information can come from the literature, from thermochemical calculations, or from proper use of testing equipment and procedures. The potential pressure that may be developed in the process is also a very important design consideration.

The second key process design parameter is the reaction rate, which depends on temperature, pressure, and concentrations. Rates of reaction during normal and abnormal operation (including the worst credible case) should be determined in order to design inherently safer processes. Plant process and equipment design are elements of the third key parameter. Any heat that is generated by the reaction should be removed adequately, and any gas production should be managed. The effects and requirements of scale-up (that is, the relation between bench-scale and plant equipment) should be considered. These three parameters interact. For example, a large amount of potential energy can be removed during normal operation if the rate of energy release is relatively small and is controlled by sufficient cooling capacity of the plant unit. However, if the cooling capacity of the plant unit appears insufficient because of the rate of energy release, a hazard assessment can be used to determine the necessary cooling design requirements for the operation. In most cases, data that are obtained through theoretical

3. BASIC PHYSICAL PROPERTIES / THERMAL STABILITY DATA

approaches (literature, databases, and software programs) may not be sufficient for final plant design. Experimental work is usually required on various scales depending on the extent of reactivity. Therefore, the application of well-designed experimental test methods is of prime importance to define hazardous conditions. Numerous test methods are available using a variety of sample sizes and conditions.

3.3.1 Chemical Reactivity

In the process industries, chemicals are converted into other chemicals in a defined and controlled manner. Uncontrolled chemical reactions occur as a result under abnormal conditions, for example, malfunctioning of the cooling system or incorrect charging, or as a result of insufficient or inadequately designed and maintained control instrumentation and procedures. Temperature, pressure, radiation, catalysts, and contaminants such as water, oxygen from air, and equipment lubricants can influence the conditions under which the reactions (controlled and uncontrolled) take place.

The rate at which a chemical reaction proceeds is an exponential function of temperature. In comparing reaction rates among chemicals at a certain temperature, some chemicals show a high stability and others a relatively low stability.

Almost all reactions show a heat effect. When heat is generated, liberated, or released during a reaction (exothermic), a hazardous situation might occur depending on the reaction rate, the quantity of heat that is generated, the capacity of the equipment to remove the heat, and the amount of gas produced during the reaction. Although thermal decomposition and runaway reactions are often identified with the inherent reactivities of the chemicals involved, it should be emphasized that hazards can arise from induced reactions. These induced reactions may be initiated by heat, contamination, or mechanical means (e.g., shock, friction, electrostatic spark).

3.3.1.1 *Exothermic Reactions*

A reaction is exothermic if heat (energy) is generated. Reactions in which large quantities of heat or gas are released are potentially hazardous, particularly during fast decomposition and / or oxidations. Exothermic reactions lead to a temperature rise in the material if the rate of heat generation exceeds the rate of heat removal from the material to its surroundings. The reaction accelerates due to the increasing temperature and may result in a thermal runaway reaction. The increase in temperature can be considerable if large quantities of heat are generated in a short time. Many organic compounds that decompose exothermically will liberate pressure-generating condensable and noncondensable gases at high temperatures.

In addition to thermal runaway reactions, which result from more-or-less uniform self-heating throughout the material, highly exothermic decompositions can be induced by the point source input of external energy, for example, fire, hot spots, impact, electrical sparks, and friction. In such a case, the decomposition travels through the material by either a heat or a shock wave. Therefore, the maximum quantities of both energy and gas that are generated by the exothermic reaction are prime parameters in estimating the potential reactivity hazards of a substance. Furthermore, the rates of energy generation and gas production are of utmost importance.

Even relatively small amounts of exothermic reaction or decomposition may lead to the loss of quality and product, the emission of gas, vessel pressurization, and / or environmental contamination. In the worst case, an uncontrolled decomposition may accelerate into an explosion.

3.3.1.2 Unstable Chemicals

Unstable chemicals are subject to spontaneous reactions. Situations where unstable chemicals may be present include the catalytic effect of containers, materials stored in the same area with the chemical that could initiate a dangerous reaction, presence of inhibitors, and effects of sunlight or temperature change. Examples include acetaldehyde, ethylene oxide, hydrogen cyanide, nitromethane, organic peroxides, styrene, and vinyl chloride.

As an example, styrene polymerizes at moderate temperatures and the rate of polymerization increases as temperature increases. The reaction is exothermic and becomes violent as it is accelerated by its own heat. Inhibitors are added to prevent the initiation of dangerous polymerization. When styrene is used to fabricate materials, e.g., fiberglass resin, a catalyst may be added in the manufacturing process to initiate polymerization at a controlled rate. Any unbalance of these reactions in terms of quantities or temperatures could cause hazardous fire conditions.

3.3.1.3 Chemicals That React with Common Substances

Chemicals that are water or air reactive pose a significant fire hazard because they may generate large amounts of heat and because the materials that they react with are found nearly everywhere. These materials may be pyrophoric, i.e., they ignite spontaneously on exposure to air, such as iron sulfides. They may also react violently with water and certain other chemicals. Water-reactive chemicals include anhydrides, carbides, hydrides, and alkali metals (e.g., lithium, sodium, potassium).

Air-reactive chemicals include aluminum hydride, metal alkyls, and yellow phosphorus. Other reactive chemicals include alkalis, aluminum trialkyls, anhydrides, hydrides, certain oxides, phosphorus, and sodium hydrosulfate.

Phosphorus and other metals, for example, will oxidize in air under certain conditions at a sufficient rate to heat spontaneously and ignite. The smaller the particle size, the greater the fire hazard.

3.3.1.4 Combustible Chemicals

All organic chemicals are essentially combustible. Combustion of some chemicals, such as sulfur and sulfides of sodium, potassium, and phosphorus, results in the production of hazardous gases, in this case sulfur dioxide. Carbon black, lamp black, lead sulfocyanate, nitroaniline, nitrochlorobenze, and naphthalene are examples of combustible chemicals.

3.3.1.5 Oxidizers

Oxidizers may not themselves be combustible, but they may provide reaction pathways to accelerate the oxidation of other combustible materials. Combustible solids and liquids should be segregated from oxidizers. Certain oxidizers undergo dangerous reactions with specific noncombustible materials. Some oxidizers, such as calcium hypochlorite, decompose upon heating or contamination and self-react with violent heat

3. BASIC PHYSICAL PROPERTIES / THERMAL STABILITY DATA

output. Oxidizers include nitrates, nitric acid, nitrites, peroxides, chlorates, chlorites, dichromates, hypochlorites, perchlorates, permanganates, persulfates, and the halogens.

3.3.2 Detonations and Deflagrations

An explosion is a rapid expansion of gases resulting in a rapidly moving pressure wave. The expansion can be the result of a rapid chemical reaction. If the front velocity of the shock wave exceeds the speed of sound in the material, the energy is transferred by shock compression resulting in what is termed a detonation (Ref. 3-9). At front velocities lower than the speed of sound, the energy is transferred by heat resulting in what is termed a deflagration. The effect of a detonation depends on the shock wave, that is, an immediate peak overpressure followed by a longer period with an underpressure.

The strength of the shock wave depends on the mass of the detonating materials. Detonations are mostly induced by initiation sources. In some cases, a deflagration may make a transition into a detonation. Working with chemicals and systems under plant conditions where a detonation can be induced is NOT recommended. Whether or not a chemical or chemical system can detonate can be determined only by specific tests.

A detonation rate is far higher than would be expected on the basis of kinetic data. Both preventive and defensive measures should be considered in dealing with a deflagration.

Note that the severity of a vapor cloud deflagration / detonation depends not only on the chemical properties of the materials being ignited, but also on the environment in which the ignition is propagated. This is a case where process safety design can be enhanced not only by the manner in which the process is operated, but also by the physical layout of the unit.

3.3.3 Runaway Reactions

A runaway reaction proceeds by a general temperature increase because of heat gains exceeding heat losses (e.g., caused by insufficient heat removal). This type of runaway reaction is generally encountered in large units, including storage vessels, and in nonstirred systems. A runaway reaction may be caused by a rapid decomposition or oxidation reactions in units other than reactors. In a reactor, various phenomena may cause a runaway reaction, including accumulation and / or mischarging of reactants, incorrect handling of catalysts, cooling problems, or loss of agitation.

In most cases, a thermal runaway reaction depends on the balance between heat generation and heat removal. When heat removal is insufficient, the temperature will increase according to the reaction kinetics and thermodynamics. Gases may be formed either as products of the reaction or, in later stages, as decomposition products at the elevated temperatures encountered. In general, there are two alternatives available to handle the gas production. Either the vessel should be designed to withstand the total pressure involved, or a vent system should be designed so that the vessel pressure never exceeds the design pressure during the runaway reaction. In case of a thermal runaway reaction, the use of preventive measures is recommended.

3.3.4 Calorimetric Data

The most important aspect of data is an understanding of how the values were derived. *Was the value calculated or obtained through experimental tests?* There are many sources of calorimetric data, some of which are listed in this section.

There are many published sources of chemical data. *Sax's Dangerous Properties of Industrial Materials* (Ref. 3-10) is one frequently used reference, as are the databases maintained by the Chemical Abstracts Service and the American Institute of Chemical Engineers Design Institute for Physical Property Data (DIPPR®) (Ref. 3-1). Government agencies and funded organizations like the U.S. Coast Guard, the Environmental Protection Agency, the Federal Emergency Management Agency, and the World Bank have also published chemical data. Specific threshold limits applicable to certain chemicals are included in federal, state, and local legislation and regulations.

3.3.4.1 Material Safety Data Sheets

Material Safety Data Sheets (MSDSs) are a widely used system for cataloging information on chemicals, chemical compounds, and chemical mixtures. MSDS information may include instructions for the safe use and potential hazards associated with a particular material or product. These data sheets can be found anywhere where chemicals are being used.

An MSDS should list incompatible materials that pose a reactivity hazard with the subject material. Potential incompatibles include chemicals that can trigger a violent decomposition or polymerization reaction. If a material is water reactive, it should be so indicated in the MSDS. It should also be denoted in the MSDS Section 3 (Hazards Identification) and on the NFPA 704 (Ref. 3-11) placard system for identifying hazards of materials by the symbol W with a line through it on the bottom of the placard.

The MSDS is an important component of product stewardship and workplace safety; it is intended to provide workers and emergency personnel with procedures for handling or working with that substance in a safe manner and includes information such as physical data (melting point, boiling point, flash point, etc.), toxicity, health effects, first aid, reactivity, storage, disposal, protective equipment, and spill-handling procedures.

3.3.4.2 Incompatibility Charts

Chemical incompatibility charts can provide a preliminary indication of potential reactivity hazards associated with binary combinations of chemicals or chemical families. An example is the NOAA / EPA Chemical Reactivity Worksheet (Ref. 3-12) software tool for the preparation of material-specific incompatibility charts.

3.3.4.3 Reactivity Listings in NFPA Standards and in Other References

Many National Fire Protection Association (NFPA) standards provide classification schemes for a wide range of materials. Some include:

- NFPA 30, *Flammable and Combustible Liquids Code*, 2008 Edition (Ref. 3-5)
- NFPA 55, *Compressed Gases and Cryogenic Fluids Code*, 2010 Edition (Ref. 3-13)
- NFPA 400, *Hazardous Materials Code*, 2010 Edition (Ref. 3-14)

3. BASIC PHYSICAL PROPERTIES / THERMAL STABILITY DATA

- NFPA 491, *Fire Protection Guide to Hazardous Materials*, 13th Edition (Ref. 3-15)

Perhaps the most widely utilized and comprehensive handbook for preliminary evaluations of chemical reactivity hazards is *Bretherick's Handbook* (Ref. 3-16). Other very useful references for this purpose include Sax's Handbook (Ref. 3-10), Grewer (Ref. 3-17), Pohanish and Green (Ref. 3-18), and the CCPS guidelines on reactivity hazard evaluations (Ref. 3-19).

Papers by Frurip et al. (Ref. 3-20) and Leggett (Ref. 3-21) provide excellent guidance on good current practices being followed by organizations experienced in this type of hazard evaluation. In the specific case of water-reactive and pyrophoric materials, the Gibson and Weber (Ref. 3-22) handbook contains property data for about 425 such materials.

3.3.4.4 Theoretical Considerations

Combinations of chemical compounds with known thermochemical properties are amenable to calculations of heat of reaction and of adiabatic reaction temperature for potentially self-reacting chemicals. Conceptually, it is possible to use these calculated values to provide a preliminary indication of the hazard of these compounds reacting adiabatically. However, the thermochemical equilibrium calculations do not provide any indication of the ease of reaction initiation and the rate of reaction. Therefore, these theoretical calculations are of far less value than preliminary reactivity indications based on reported experience and testing.

3.3.4.5 Government and Other Toxicity Databases and Listings

The Environmental Protection Agency (EPA) maintains perhaps the most comprehensive and extensive database for health effects of chemicals (Integrated Risk Information System (IRIS) (Ref. 3-23)). According to the EPA, "the information in IRIS is intended for those without extensive training in toxicology, but with some knowledge of health sciences."

Table 3.3 contains a list of some government and other toxicity databases and listings, along with their websites.

Table 3.3   Government and Other Toxicity Databases

Source	Website	Description
Integrated Risk Information System (IRIS)	www.epa.gov/iris/subst/index.html	The type of data covered for individual chemical includes both descriptive and quantitative information on: • Oral reference doses and inhalation reference concentrations (RfDs and RfCs, respectively) for chronic noncarcinogenic health effects • Hazard identification, oral slope factors, and oral and inhalation unit risks for carcinogenic effects
Occupational Safety and Health Administration (OSHA)	www.osha.gov/SLTC/pel/	OSHA regulations and publications include Permissible Exposure Limit (PEL) values for both short-term exposures and 8-hour exposures to numerous materials. OSHA Website searches for specific materials can be conducted at this website.
National Institute for Occupational Safety and Health (NIOSH) Registry for Toxic Effects of Chemical Substances (RTECS) (Ref. 3-24)	www.cdc.gov/niosh/rtecs/default.html	The RTECS database includes toxicity data and summaries of pertinent journal articles, government reports, and EPA test submissions. Since December 2001, responsibility for maintaining RTECS has been transferred from NIOSH to various private and foreign organizations listed at this website. These individual organizations update RTECS and make it available for purchase or lease along with software for searching and retrieving specific records.
National Institute for Occupational Safety and Health (NIOSH) Documentation for Immediately Dangerous to Life or Health Concentrations (IDLHs)	www.cdc.gov/niosh/idlh/intridl4.html	Contains a chemical listing and documentation of revised IDLH values (as of 3/1/95).
American Conference of Governmental Industrial Hygienists (ACGIH)	www.acgih.org/	Threshold Limit Values (TLV) for more than 700 chemical substances and physical agents are contained in the latest ACGIH (2003) listing. The TLV values are determined by an ACGIH committee review of pertinent scientific literature. Proposed changes and new listings can be found on the ACGIH website.
Workplace Hazardous Materials Information System (WHMIS)	www.hc-sc.gc.ca	The Canadian government provides a useful online resource for toxic material occupational exposure information called the WHMIS. The WHMIS database for carcinogenic materials includes listings and classifications from ACGIH, the California EPA, the European Union, and IARC.

3.3.5 Interaction Matrix

Chemical hazards can be identified by examining the characteristics of each chemical in a process, one at a time. Some basic information on the hazardous or reactive properties of chemicals can usually be found on Material Safety Data Sheets (MSDSs) or other common hazardous chemical data references. However, examination of the individual chemicals handled in a process may not identify all important process hazards, since many hazards are related to interactions of process chemicals, either inadvertent or intentional, with each other and with their surroundings. Thus, a complete Process Hazards Analysis (PHA) will often need to supplement a review of the chemical safety data and the process parameters with a means of systematically examining possible chemical interactions.

Developing a chemical interaction matrix is one effective means of finding the potential interactions in an operation that may lead to a fire, explosion, or hazardous material release. Matrices are usually generated for small areas or single processes in order the keep the grid size manageable. The rule of thumb is that only chemicals that can reasonably be expected to be mingled in the area should be included. Most users include ubiquitous chemicals such as water (or other liquids piped in or through the area) and jacket media as these usually meet the reasonableness criteria.

Dangerous interactions can then be assessed as part of process hazard analyses, included in emergency response plans, and incorporated into employee awareness programs. This section describes an approach to generating chemical interaction matrices. This approach is similar to the method used by one large U.S. chemical company, as summarized by Gay and Leggett (Ref. 3-25).

ASTM E2012 Standard *Guide for the Preparation of a Binary Chemical Compatibility Chart* provides guidance on chemical interaction matrix considerations (Ref. 3-26).

Chemical interactions and their consequences can be systematically studied for a storage and handling operation by considering the following in a matrix format:

- All stored / handled chemicals, which may include raw materials, intermediates, products, by-products, and catalysts
- Any other chemicals that may be introduced into the operation inadvertently, such as other raw materials that are unloaded at the same truck unloading station serving a storage operation, or piping containing other materials that is tied into the same transfer system
- All utilities (steam, compressed air, nitrogen, natural gas, heat transfer media, refrigerants, service water, etc.) that could potentially interact with the operation
- Common environmental substances: air, water / humidity, and any environmental contaminants present in significant concentrations in the actual storage / handling location
- Likely process contaminants such as dirt, rust, scale, and lubricating oil
- All materials of construction and gasket materials used in the process, including those having a reasonable likelihood of being substituted (intentionally or otherwise) sometime during the life of the facility

- Other materials that may contact process chemicals, such as absorbents and insulation
- All operating conditions that pertain to the given facility, such as elevated temperature
- In some situations, conditions such as "confinement" and "adiabatic compression" may be pertinent.

A matrix that would include all of the above items for a given process can be quite large. If it is necessary to restrict the effort involved in developing the interaction matrix, judgment can be exercised in limiting the scope of the study or including only those substances and conditions that have a reasonable likelihood of being present and causing reactivity concerns.

3.3.5.1 Developing a Chemical Interaction Matrix

To conduct a chemical interaction study, a matrix is developed that has each of the above items listed along both the horizontal and vertical axes. The cells are then filled in, either above or below the diagonal running from top left to bottom right, with the consequences expected if each interaction occurs.

3.3.5.2 Using the Chemical Interaction Matrix to Identify Scenarios

Once the interaction matrix is complete, it should be examined for severe consequences such as violent reactions, generation of toxic gases, or significant fire hazards, and particularly for interactions that were previously not recognized as having hazardous consequences. These interactions should then be studied, by a team of knowledgeable persons, to develop accident scenarios by determining what could cause each hazardous chemical interaction, where and when each interaction might occur, and what safeguards exist to prevent the occurrence of the interaction and / or deal with the consequences of the interaction. This can be accomplished as part of a Process Hazards Analysis (PHA). Any matrix cells with missing data or unknown consequences will indicate where research or testing may be required. Until missing data is resolved, such interactions should be assumed to be incompatible.

For a more complete assessment of reactivity issues within a facility, a new tool has been developed by the Center for Chemical Process Safety (CCPS) called the Reactivity Evaluation Screening Tool (REST) (Ref. 3-27). REST incorporates the NOAA chemical reactivity database and an abbreviated PHA "What-If" approach to identify warehouse segregation issues and chemical reactivity hazards. This tool will be available free of charge from the CCPS website (www.aiche.org/ccps). REST has been designed to be used by both experienced and minimally experienced users. Users can enter data directly or be walked through the process using a question and answer format. The program first asks users to enter all chemicals within the assessment area. A reactivity matrix and a warehouse segregation chart are automatically generated along with a text list of binary reactivity hazards. The user is then asked a series of questions about the type of chemical handling and processing that occur within the area to determine if a more detailed assessment should be performed. If a more detailed assessment is warranted, the user is led through a series of fault scenario building exercises. The program records and organizes these scenarios and then leads the user through the process of assessing the reactivity hazards of each scenario. Scenario assessment consists of automated responses based on user-supplied information. Guidance is given on how to obtain each

3. BASIC PHYSICAL PROPERTIES / THERMAL STABILITY DATA   55

of the requested information inputs. The tool then outputs all of the assessed scenarios along with a generalized determination of consequence severity. References are provided to guide the user to information on how to remediate each type of hazard identified.

3.3.5.3 Managing the Chemical Interaction Data

A chemical interaction matrix for relatively simple storage / handling operations may fit in a one-page table or on several pages that fit together. For larger or more complex operations with many chemicals and other matrix items, three formats that have been commonly used to capture and present the chemical interactions are database programs, spreadsheets, and word processing programs having table-generating features.

Each has its advantages:
- Database programs are most useful for very large matrices, where the power of the database program to search and retrieve a given combination is needed.
- Spreadsheets have the advantage of being able to put the entire matrix in a true matrix form, thus making it easier to see which intersections still need to be filled.
- The table format in a word processing program allows the matrix to be directly incorporated into a report and enlarges cells automatically to accommodate multiple lines of text.

Gay and Leggett (Ref. 3-25) describe a computerized approach to storing interaction information and printing compatibility charts; this approach includes a mixing hazard rating from 0 to 4 that parallels the NFPA 704 ratings for health, flammability, and reactivity (Ref. 3-11). This computer shell program, known as CHEMPAT, is available from the American Institute of Chemical Engineers and serves as an aid to organizations for establishing compatibility charts.

3.3.5.4 Chemical Interaction Data Sources

Many chemical interactions are obvious, such as acid-base reactions resulting in heat and gas generation. Many other interactions will be known to have no significant consequences. However, there are usually many potential interactions for which the results are not immediately known. In addition to materials testing in a properly equipped laboratory, data can often be obtained from chemical suppliers or from many literature sources such as the following:

- National Oceanic and Atmospheric Administration (NOAA) (http://cameochemicals.noaa.gov/) (Ref. 3-28)
- *Bretherick's Handbook of Reactive Chemical Hazards* (Ref. 3-16), a compilation of reactivity and incompatibility hazards of 4600 different elements and compounds; an electronic version is also available from the publisher
- Chemical Hazard Response Information System (CHRIS) Hazardous Chemical' Data (Ref. 3-29) COMDTINST M16565.12C is available online at www.chrismanual.com. Reactivity group, water reactivity, and reactivity with common materials are given, as well as a bulk cargo compatibility matrix
- *Dangerous Properties of Industrial Materials* (Ref. 3-10), 3-volume publication with incompatibility information on numerous hazardous materials

- *Hazardous Materials Car Placement in a Train Consist* (Ref. 3-30), 2-volume report examining all binary combinations of the top 101 hazardous commodities by rail volume movement in the U.S., plus fuming nitric acid
- Organic and inorganic chemistry textbooks
- Experience and resources of company organic and / or inorganic chemists
- Literature search on the particular chemicals. When the consequences of a given interaction are unknown, that fact should immediately raise a red flag, since unintentional chemical interactions should be identified and controlled for continued safe operation of chemical processes.
- Coast Guard compatibility charts
- EPA compatibility charts

3.3.6 Testing Methods

3.3.6.1 Assessment and Testing Strategies

Information on understanding the hazards depends on the stage of development of the process as indicated in Table 3.4. During early developmental chemistry work, only small amounts of materials will be available. In many cases, only theoretical information from the literature or from calculations is readily available.

Table 3.4 Suggested Stages in Assessment of Reactivity by Scale

Stages	Aspect
1. Development Chemistry - Characterization of materials	Characterization of process alternatives Choice of process Suitability of process Screening for chemical reaction hazards
2. Pilot Plant - Chemical reaction hazards	Influence of plant technology regarding potential hazards Definition of safe procedures Effects of expected variations in process conditions Definition of critical limits
3. Full-Scale Production - Reevaluation of chemical reaction hazards	Newly revealed reactivity hazards from plant operations Management of changes Update of safety procedures as required Ongoing hazard assessment in examining potential deviations from process conditions through interaction of process safety with engineering and production, personnel

Screening tests can be run to identify reaction hazards. Also, data for pilot plant considerations should be evaluated and obtained as necessary. In the pilot plant stage, additional material becomes available so that the reaction hazards can be investigated more extensively. Process control features and deviations from normal operating conditions should be checked. Operating procedures can be drafted and checked. Emergency procedures can be defined.

3. BASIC PHYSICAL PROPERTIES / THERMAL STABILITY DATA

During full-scale production, particularly initially, chemical reaction hazards may be reevaluated. More tests may be necessary as a consequence of increased knowledge of the process, changed production requirements, or other process changes such as the use of different feedstocks.

A typical chronology for testing is shown in Table 3.5. The tests provide either qualitative or quantitative data on onset temperature, reaction enthalpy, instantaneous heat production as a function of temperature, maximum temperature, and / or pressure excursions as a consequence of a runaway reaction. The choice of test equipment to be used depends on the conditions, such as scale, temperature, mixing, and materials of construction, at which the substance or mixture is to be handled. The interpretation of the data from each of these tests is strongly dependent on the manner in which the test is run and on the inherent characteristics of the testing device. Guidance is provided along with each test description, particularly in the detailed sections later in this chapter.

Table 3.5 Typical Testing Procedures by Chronology

Subject	Property to Be Investigated	Typical Instrument Information
Identification of exothermic activity	Thermal Stability	DSC / DTA
Explosibility of individual substances	Detonation Deflagration	Chemical structure Tube test Card gap Drop weight Oxygen balance High rate test Explosibility tests
Compatibility	Reaction with common contaminants (e.g., water)	Specialized tests
Normal reaction	Reaction profile Effect of change Gas evolution	Bench-scale reactors (e.g., RC1)
Minimum exothermic runaway temperature	Establish minimum temperature	Adiabatic Dewar Adiabatic calorimetry ARC
Consequence of runaway reaction	Temperature rise rates Gas evolution rates	Adiabatic Dewar Adiabatic calorimetry ARC VSP / RSST RC1 pressure vessel
ARC = Accelerating Rate Calorimeter DSC = Differential Scanning Calorimeter DTA = Differential Thermal Analysis RC1 = Reactor Calorimeter RSST = Reactive System Screening Tool (Ref. 3-31) VSP = Vent Size Package		

Experimental hazard evaluation includes thermal stability testing, solid flammability screening tests, explosibility testing, detailed thermal stability and runaway testing, and reactivity testing.

The recommended experimental evaluation is condensed in a number of flowcharts which, in general, follow the most reliable and internationally recognized standard test methods. Details of the strategic testing scheme are covered in the following section.

3.3.6.2 Test Strategies

The potential thermal hazards associated with thermally unstable substances, mixtures, or reaction masses are identified and evaluated in Figure 3.2 and Figure 3.3.

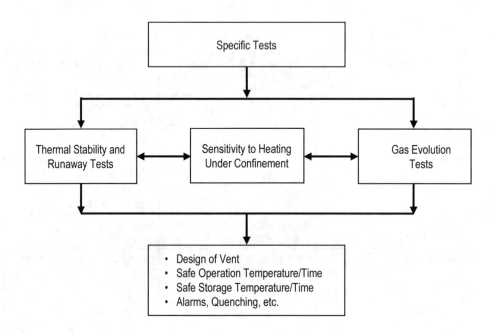

Figure 3.2   Strategy for Stability Testing Flowchart

3. BASIC PHYSICAL PROPERTIES / THERMAL STABILITY DATA

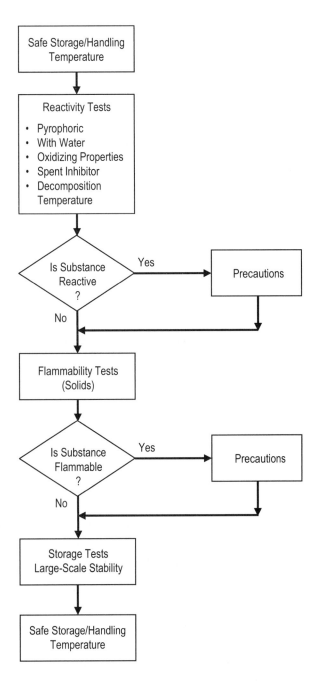

Figure 3.3 Specific Experimental Hazard Evaluation for Reactive Substances Flowchart

These tests can also be used to evaluate induction time for the start of an exothermic decomposition and compatibility with metals, additives, and contaminants. The initial part of runaway behavior can also be investigated by Dewar tests and adiabatic storage tests. To record the complete runaway behavior and often the adiabatic temperature rise, that is, the consequences of a runaway reaction, the Accelerating Rate Calorimeter (ARC) can be used, although it is a smaller scale test.

To investigate gas evolution during decomposition and / or a runaway reaction, both the ARC and RSST simultaneously record rise in temperature and pressure, which is usually proportional to the gas evolution during decomposition. Other types of equipment available to investigate the gas evolution are various autoclave tests, isoperibolic autoclave tests, and closed Dewar tests. Mass flux data are also required in designing any vent facilities. Extrapolation of data from any and all of these tests to large scale should be made with care.

3.4 REFERENCES

3-1. *DIPPR® Data Compilation of Pure Chemical Properties*. Design Institute for Physical Properties, American Institute of Chemical Engineers. New York, NY. 2010.

3-2. NFPA. *Fire Protection Handbook, 20th Edition*. National Fire Protection Association. Quincy, MA. 2008.

3-3. Crowl, D.A. *Understanding Explosions*. Center for Chemical Process Safety for the American Institute of Chemical Engineers. New York, NY. 2003.

3-4. ASTM A502-03. *Standard Specification for Rivets, Steel, Structural*. ASTM International. West Conshohocken, PA. 2009.

3-5. NFPA 30. *Flammable and Combustible Liquids Code, 2008 Edition*. National Fire Protection Association. Quincy, MA. 2008.

3-6. CCPS. *Guidelines for Fire Protection in Chemical, Petrochemical, and Hydrocarbon Processing Facilities*. Center for Chemical Process Safety of the American Institute of Chemical Engineers. New York, NY. 2003.

3-7. Crowl, D.A. and Louvar, J.F. *Chemical Process Safety: Fundamentals with Applications, 2nd Edition*. Prentice-Hall, Inc., Englewood Cliffs, NJ. 2009.

3-8. GESTIS-DUST-EX, Institute for Occupational Safety and Health of the German Social Accident Insurance (IFA). www.dguv.de/ifa/en/gestis/expl/index.jsp

3-9. NFPA 68. *Standard on Explosion Protection by Deflagration Venting, 2007 Edition*. National Fire Protection Association. Quincy, MA. 2007.

3-10. Lewis, R.S. *Sax's Dangerous Properties of Industrial Materials, 11th Edition*. John Wiley & Sons. Hoboken, NJ. 2007.

3-11. NFPA 704, *Standard System for the Identification of the Hazards of Materials for Emergency Response, 2007 Edition*. National Fire Protection Association. Quincy, MA. 2007.

3-12. National Oceanic and Atmospheric Administration, 1401 Constitution Avenue, NW, Room 5128, Washington, D.C. 20230. www.noaa.gov

3. BASIC PHYSICAL PROPERTIES / THERMAL STABILITY DATA 61

3-13. NFPA 55. *Compressed Gases and Cryogenic Fluids Code, 2010 Edition.* National Fire Protection Association. Quincy, MA. 2010.

3-14. NFPA 400. *Hazardous Materials Code, 2010 Edition.* National Fire Protection Association. Quincy, MA. 2010.

3-15. NFPA 491. *Fire Protection Guide to Hazardous Materials*, 13th Edition. National Fire Protection Association. Quincy, MA. 2001.

3-16. Urben, P. *Bretherick's Handbook of Reactive Chemical Hazards, Seventh Edition.* Academic Press. Oxford, UK. 2007.

3-17. Grewer, Th. *Thermal Hazards of Chemical Reactions. Industrial Safety Series, 4.* Elsevier. Amsterdam. 1994.

3-18. Pohanish, R.P. and Green, S.A. *Wiley Guide to Chemical Improbabilities*, Second Edition. John Wiley & Sons, Inc. Hoboken, NJ. 2003.

3-19. CCPS. *Guidelines for Safe Storage and Handling of Reactive Materials.* Center for Process Safety for the American Institute of Chemical Engineers. New York, NY. 1995.

3-20. Frurip, D.J., Hofelich, T.C., Leggett, D.J., Kurland, J.K., and Niemeier, J.K. *A Review of Chemical Compatibility Issues,* Proceedings of the 1997 AIChE Loss Prevention Symposium. American Institute of Chemical Engineers. New York, NY. 1997.

3-21. Leggett, D.J. *Chemical Reaction Hazard Identification and Evaluation: Taking the First Steps*, Proceedings, AIChE Spring National Meeting; 36th Annual Loss Prevention Symposium. New Orleans, Louisiana. 2002.

3-22. Gibson, J. and Weber, J. *Handbook of Selected Properties of Air and Water-Reactive Materials (RDTR).* U.S. Naval Ammunition Depot. Crane, IN. 1969.

3-23. EPA. *Integrated Risk Information System (IRIS).* Environmental Protection Agency. http://www.epa.gov/IRIS/

3-24. NIOSH. *Registry for Toxic Effects of Chemical Substances (RTECS).* National Institute for Occupational Safety and Health.

3-25. Gay, D.M. and Leggett, D.J. *Enhancing Thermal Hazard Awareness with Compatibility Charts.* J. Testing and Evaluation, 21, 477-480. 1993.

3-26. ASTM E2012-06. *Standard Guide for Preparation of a Binary Chemical Compatibility Chart.* ASTM International. West Conshohocken, PA. 2006.

3-27. CCPS. *Reactivity Evaluation Screening Tool (REST).* Center for Process Safety for the American Institute of Chemical Engineers. New York, NY. 2011. www.aiche.org/ccps

3-28. National Oceanic and Atmospheric Administration, 1401 Constitution Avenue, NW, Room 5128, Washington, D.C. 20230. www.noaa.gov

3-29. CHRIS. *Hazardous Chemical Data Manual.* Chemical Hazards Response Information System. 1999.
http://ocean.floridamarine.org/acp/mobacp/PDF/TACTICAL/chris.pdf.

3-30. Thompson, R.E., Zamejc, E.R., and Alhlbeck, D.R. *Hazardous Materials Car Placement in a Train Consist*, Volumes 1 and 2. Federal Railroad Administration. Washington, D.C. 1992.

3-31. Fauske, H.K. *The Reactive System Screening Tool (RSST): An Inexpensive and Practical Approach to Sizing Emergency Relief Systems.* Process Safety Symposium. Houston, TX. 1998.

4

ANALYSIS TECHNIQUES

Engineering design for process safety should consistently and systematically identify and evaluate hazards posed by a process and reduce the risk to an acceptable level. Process hazards come from many sources, including:
- Material and chemistry used (e.g., flammability, toxicity, reactivity)
- Process variables - the way the chemistry works in the process (e.g., pressure temperature, concentration)
- Equipment failures

This chapter provides an overview of:
- *Hazard Identification* - A hazard is a physical or chemical condition with the potential for harming people, property, or the environment. Hazard identification involves understanding:
  - Undesirable consequences
  - Material, system, process, and plant characteristics that could produce those consequences
- *Hazard Analysis Techniques* - A hazard analysis is an organized effort to identify and analyze the severity of hazardous scenarios associated with a process or activity. Specifically, hazard analyses are used to identify weaknesses in design and operation of facilities that could lead to hazardous material releases. Hazard analyses can also be used to identify and evaluate the effectiveness of safeguards. This chapter introduces a variety of hazard analysis techniques that can provide information to help companies improve safety and manage risk.
- *Risk Assessment* - Risk assessments used during engineering design provide a valuable tool for evaluating design concept alternatives and making risk-based decisions.

For more detailed information, refer to CCPS publications:
- *Guidelines for Hazard Evaluation Procedures* (Ref. 4-1)
- *Guidelines for Chemical Process Quantitative Risk Assessment* (Ref. 4-2)
- *Layer of Protection Analysis, Simplified Process Risk Assessment* (Ref. 4-3)

4.1 HAZARD IDENTIFICATION

In order to begin the process of hazard identification, it is important to understand the types of hazards associated with process facilities. Hazards range in complexity from simple, easily identifiable to complex hazards requiring not only the senses, but intentional focus and analysis as well. Process hazards, although unique to industry types and individual facilities, generally result from loss of containment of a hazardous material and / or energy. This loss of containment may have multiple potential ways of occurring and the consequences depend on the type of material, operating conditions, and external factors present at the time of release. The hazard may be process conditions or chemical in nature.

In some cases, the hazard is learned from the experience of others, procedures, training sessions, or plant surveys. Some causes of process hazards may be more difficult to detect:

- Design deficiencies
- Deviating from an operating procedure
- Inadequate training
- Inadequate operating procedures
- Operating equipment outside design parameters
- Incorrect MSDS (lack of information can lead to chemical hazards related to chemical instability/reactivity, inadvertent mixing, etc.)
- Equipment not fit for service
- Fatigue
- Too many tasks for the current staffing level to perform safely
- Poor communication
- Non-routine operating activities, such as startup or shutdown
- Feed composition changes
- Contamination

4.1.1 Process Hazards

The characteristics of materials that make them hazardous are often what make them valuable; consequently, it may be undesirable to eliminate all hazards. In these situations, identifying the hazard and applying safeguards to properly manage risk are key to safe operation.

Table 4.1 provides typical examples of process hazards by industry type.

4. ANALYSIS TECHNIQUES

Table 4.1 Examples of Process Hazards in Various Industries

Industry Type	Loss of Containment Hazards			Reactive Hazards	Other Process Hazards
	Flammable	**Explosive**	**Toxic**		
Breweries / Distilleries	Alcohol presents a flammability hazard.	Grain storage silos present dust explosion hazards.	Anhydrous ammonia is often used for refrigeration. Releases of anhydrous ammonia have the potential to impact both onsite workers and the community.	Inadvertent mixing of incompatible materials.	Asphyxiation hazards when employees enter confined spaces. Carbon dioxide asphyxiation.
Chemical Plant	Leak of flammable liquid, vapors, or dust may lead to fires.	If sufficient quantities of material are released and there is sufficient confinement and congestion, then a vapor cloud explosion hazard may be present. Similarly, if a flammable dust is released, then a dust explosion is possible.	A leak of toxic material could result in either chronic or acute worker exposure leading to injury and impact to the community and environment.	Inadvertent mixing of materials could lead to runaway reactions that result in a rupture of a vessel. Not following the steps of recipe could result in a runaway reaction that could rupture a vessel.	Bulk shipping presents unique hazards. Nitrogen asphyxiation. Vibration can lead to equipment failure and subsequent release.
Food / Beverages Processing	Fork-lifts may use propane or diesel as fuel. This presents a hazard to workers when refueling fork-lifts.	A release of anhydrous ammonia in an enclosed area has the potential for explosion. Processing plants such as those for grains (flour) and sugar have a risk of dust explosion.	Anhydrous ammonia is often used for refrigeration. Releases of anhydrous ammonia have the potential to impact both onsite workers and the community.	Materials of construction and repair must be compatible with anhydrous ammonia.	Contamination of product can lead to wide-scale public exposure and health effects, recall of product, and loss of reputation. Carbon dioxide or nitrogen asphyxiation.

Table 4.1 Examples of Process Hazards in Various Industries (Continued)

Industry Type	Loss of Containment Hazards			Reactive Hazards	Other Process Hazards
	Flammable	Explosive	Toxic		
Natural Gas Processing Plant	Leak of flammable liquid or vapors may lead to fires.	If sufficient quantities of material are released and there is sufficient confinement and congestion, then a vapor cloud explosion hazard may be present.	The toxic hazards at gas plants are typically limited to hydrogen sulfide.	Materials of construction and repair must be compatible with hydrogen sulfide.	Carbon dioxide may be present in some gas plants which can increase corrosion resulting in leaks. Carbon dioxide asphyxiation.
Offshore Facility	Leak of flammable liquid or vapors may lead to fires.	If sufficient quantities of material are released and there is sufficient confinement and congestion, then a vapor cloud explosion hazard may be present.	The toxic hazards at an offshore facility are typically limited to hydrogen sulfide (which may be entrained and subsequently removed from crude oil).	Materials of construction and repair must be compatible with salt water.	Most offshore facilities have some processes which are high pressure (>5,000 psi (345 bar)) that pose unique hazards. Marine operations present loading / unloading and transportation hazards. Evacuation and emergency response actions are limited because there is no place to go.
Oil Refinery	Leak of flammable liquid or vapors may lead to fires. Light hydrocarbon and hydrogen.	If sufficient quantities of material are released and there is sufficient confinement and congestion, then a vapor cloud explosion hazard may be present. Light hydrocarbon and hydrogen.	The toxic hazards at a refinery include hydrogen sulfide (which may be entrained and subsequently removed from crude oil and natural gas), sulfuric acid or hydrofluoric acid (which are used in alkylation units), and chlorine (which may be used to treat cooling water).	Common reactions between bases and acids may exist.	Large tank inventories susceptible to lightning strike and subsequent tank fires. Cogeneration units present high voltage and high pressure steam hazards. Potential for domino effects and incident escalation due to congestion of process units.

4. ANALYSIS TECHNIQUES

Table 4.1 Examples of Process Hazards in Various Industries (Continued)

Industry Type	Loss of Containment Hazards			Reactive Hazards	Other Process Hazards
	Flammable	Explosive	Toxic		
Pharmaceutical	Leak of flammable liquid, vapors, or dust may lead to fires.	Accumulation of dust presents explosion hazards.	A leak of toxic material could result in either chronic or acute worker exposure leading to injury and impact to the community and environment.	Inadvertent mixing of materials could lead to runaway reactions that result in a rupture of a vessel. Not following the steps of recipe could result in a runaway reaction that could rupture a vessel.	Off-spec or contaminated products may cause harm to the general public.
Pipelines	Leak of flammable liquid or vapors may lead to fires.	Overpressure from blocked-in pipelines or compressor stations may lead to rupture and explosion.	A leak of toxic material could result in either chronic or acute worker exposure leading to injury and impact to the community and environment.	Corrosion inhibitor chemicals not compatible with materials of construction. Materials of construction and repair not compatible with hydrogen sulfide.	In-line scraping operations can expose workers to high pressure releases. An undetected leak is a danger to the environment.
Pulp and Paper	Combustible dust may lead to fires. Turpentine. Non-condensable gases.	If a combustible dust is released, then a dust explosion is possible. Paper dust. Sawdust.	Chlorine, chlorine dioxide, and sulfur dioxide are used in the processes and have the potential to impact both onsite workers and the community.	Chlorine is a strong oxidizer and will react with most metals and organic materials. Also, chlorine reacts with water, creating the potential for worker exposure to hydrochloric acid.	Physical impact of slurry viscosity. Low pH of water causes topical exposure issues. Conveyers, moving parts, and heavy loads. Wood chippers provide exposure hazards to personnel.

Table 4.1 Examples of Process Hazards in Various Industries (Continued)

Industry Type	Loss of Containment Hazards			Reactive Hazards	Other Process Hazards
	Flammable	Explosive	Toxic		
Upstream Oil and Gas Facility	Leak of flammable liquid or vapors may lead to fires.	If sufficient quantities of material are released and there is sufficient confinement and congestion, then a vapor cloud explosion hazard may be present.	The toxic hazards at an upstream facility are typically limited to hydrogen sulfide (which may be entrained and subsequently removed from crude oil).	Materials of construction and repair must be compatible with hydrogen sulfide.	Facilities may have some processes which are high pressure (> 5,000 psi (345 bar)) that pose unique hazards. Diesel trucks and generators in proximity to light end hydrocarbon vapor. Well-blowouts present unique hazards. Evacuation and emergency response actions are limited because there is no place to go.
Water and Wastewater Treatment	Diesel driven water pumps present a potential fire hazard. Storage of dry sludge may contain pyrophoric iron sulfides.	Methane gas accumulation and its subsequent ignition present explosion hazards.	Chlorine, sulfur dioxide, and anhydrous ammonia can be used at facilities that treat water and wastewater. Releases of these materials have the potential to impact both onsite workers and the community.	Chlorine and sulfur dioxide are strong oxidizers and will react with most metals and organic materials. Sulfuric acid will react with concrete and produce a reactive by-product.	Exposure to contaminated water may cause adverse health effects. There has been a history of worker injury and fatality due to unsafe confined space entry activities in open pit treatment areas.

Process hazards can be difficult to recognize and generally have significant consequences if they are not mitigated. The potential consequences of process hazards are a function of the type of material involved, its quantity, and the process conditions.

In the simplest form, loss of containment refers to the release of any material from closed process equipment or piping. Loss-of-containment incidents may also transfer energy in the form of pressure. A worker in close proximity to a system being depressurized to atmosphere may be seriously injured.

4. ANALYSIS TECHNIQUES

Process hazards can lead to the release of a toxic or flammable material and subsequent fire, explosion, or exposure to toxics. Small events can escalate to cause significant injury, environmental impact, or asset damage. Process hazards can lead to:

- Fires
- Explosions / implosions
- Uncontrolled chemical reactions
- Exposure to:
  - Corrosive materials
  - Toxic materials
  - Ionizing and non-ionizing radiation
  - Pathogens
  - Temperature extremes

Hazardous materials can be solids, liquids, or gases. Hazards may be associated with the material size. Fine powders can form explosive atmospheres; liquids can be in the form of droplets or vapors, both of which are generally more hazardous than bulk material.

Some causes of process hazards may be easy to identify, such as:

- Equipment defects or degradation
  - External corrosion
  - Impact to piping and equipment
- Inadequate isolation of equipment or piping
- Inadequate energy isolation (lockout / tagout)

4.1.1.1 *Intrinsic*

Intrinsic hazards are characteristics that are permanently associated with the material or operation in question. They cannot be separated, and they are not dependent upon use or location, e.g., flammability, toxicity, etc.

Process conditions also create hazards or exacerbate the hazards associated with the materials in a process. For example, water is not classified as an explosion hazard based on its material properties alone. However, if a process is operated at a temperature and pressure that exceed water's boiling point, then a rapid introduction of water presents the potential for a steam explosion. Similarly, a heavy hydrocarbon may be difficult to ignite at ambient conditions, but if the process is operated above the hydrocarbon's flash point, a spill of the material may ignite. Therefore, it is not sufficient to consider only the material properties when identifying hazards; the process conditions must also be considered.

Considering the process conditions may also enable an analyst to eliminate some materials from further evaluation as significant hazards. For example, a material may have a flash point greater than 750°F (400°C). If the material is only present at ambient temperature and atmospheric pressure, then it may not be considered a significant fire hazard that warrants further evaluation. However, when identifying hazards, it is important to consider both normal and abnormal process conditions. Consider the following three cases:

- A pyrophoric material is normally processed with an inert gas blanket. The material warrants further evaluation as a fire hazard because there are many potential abnormal events that could expose the material to air.
- A combustible liquid is processed at high pressure. The material warrants further evaluation as a fire hazard because it could create a flammable mist if unintentionally sprayed into the air.
- A monomer is normally processed at relatively low temperatures and pressures. The material warrants further evaluation as an explosion hazard because it could undergo uncontrolled polymerization if a high temperature upset occurred.

These examples show how consideration of material properties and process conditions must be combined to identify process hazards. This approach is relatively quick and easy, and it can be applied to both new and existing processes.

4.1.1.2 Extrinsic

Extrinsic hazards are dependent upon where or how something is found or used, e.g., operating conditions, quantity, physical or geographical location. Extrinsic hazards can be directly related to the decisions made by the engineering team.

4.1.1.2.1 Temperature and Pressure

People (and equipment) may be exposed to high temperatures, not only as a result of fires, but also from released chemicals or process equipment that is hot.

Process conditions, such as pressure and temperature, have their own characteristic problems and hazards. Higher pressures and temperatures create stresses that must be accommodated by design. A combination of extreme conditions results in increased plant cost due to the need for material with high mechanical strength and corrosion resistance.

Higher pressures increase the amount of potential energy available in the process plant. For these plants, in addition to the energy of compressed gases and of fluids kept under pressure in the liquid state, there may also be a concern of chemical reactivity under pressure or an adverse reaction from rapid depressurization. Leakage is much more pronounced in high pressure operations. Because of the large pressure difference, the amount of fluid which can discharge through a given area is greater. This has a considerable impact on the consequences of a release, as the hazard zone extends to a larger area.

Low pressure operation usually does not pose much of a hazard in comparison with other operating conditions. However, in the case of vacuum applications where flammable materials are present, the potential for ingress of air does create a hazardous situation. This can result in the formation of a flammable mixture leading to fire and / or explosion. It is essential that this aspect is reviewed and adequate measures provided in the process design to prevent air ingress. For equipment not designed for vacuum, damage frequently occurs because of failure to vent while draining, allowing heated equipment to cool while blocked-in, or failure of a vacuum relief device due to plugging.

Higher temperatures also pose material failure problems, most frequently due to metal creep and hydrogen embrittlement. The use of high temperature conditions usually increases plant cost, not only due to materials of construction but also due to the

4. ANALYSIS TECHNIQUES

requirement for special supports to handle the stresses generated. Process design should take these stresses into account. The design should minimize stresses, especially during startup and shutdown.

High temperatures are often obtained with the use of fired heaters, which have additional hazards, such as tube rupture and explosions. Use of steam heaters, where possible, instead of fired heaters should be considered to prevent such hazards.

Process design should consider and address subfreezing temperatures and also recognize that some materials freeze well above the freezing point of water. Exposed drain valves and deadlegs have caused several major process safety incidents. The initial break in containment (such as split pipe) may not become immediately evident and can cause loss of containment release when the process unit is restarted. The process design engineer must also consider the potential impact to process fluids caused by extended low ambient temperature. Design engineering standards for insulation are based on both minimum temperature and duration of sub-freezing temperature. One common example of process fluid freezing is liquid 50% sodium hydroxide, which freezes at approximately 50°F (10°C). Viscosity and plugging process problems can occur. Fortunately, water is the only common liquid that expands when it freezes. Most process materials can freeze without damage to equipment.

4.1.1.2.2 Materials of Construction

Material failures, while relatively infrequent, can be extremely severe, resulting in catastrophic accidents. The best way to reduce the risk of material failure is to:
- Fully understand the internal process, the exterior environment, and failure modes.
- Select proper materials for the intended application.
- Apply proper fabrication techniques and controls.
- Follow good maintenance, inspection, and repair techniques.

Materials of construction are discussed further in Chapter 5, Section 5.7, Materials of Construction.

4.1.1.2.3 Physical Location

Evaluating siting options early in the design phase can help a company site a new process in the most desirable location. Conducting risk assessments ensures the layout and siting of new process will provide:
- The least impact to offsite receptor
- The least impact to people and buildings onsite
- The least impact to adjacent units
- Adequate drainage and retention

Siting and layout are discussed further in Chapter 5, Section 5.6, Plant Siting and Layout.

4.1.1.2.4 Utility Systems

Utility systems are often overlooked as being a process hazard. However, without utilities to support the process, the process may quickly become a hazard. The loss of ancillary systems, such as utility air, water, fuel gas, inerting gas, etc., can lead to an uncontrollable situation, e.g., loss of cooling to a reactor could result in a runaway reaction and explosion.

Another issue with utilities is cross-contamination, e.g., air connected to nitrogen systems could result in a flammable mixture in a conveying system.

Utility systems are discussed further in Chapter 6, Equipment Design.

4.1.2 Chemical / Material Hazards

There are hazards that are inherent to the chemicals and other materials used in processing facilities. A clear understanding of the intrinsic and extrinsic hazards is essential in managing the associated risks. Initial hazard identification can be performed by simply comparing the material properties available from these diverse resources to the consequences of concern. For example, if the process design engineer is concerned about the consequences of a fire, they can identify which process materials are flammable or combustible. The process design engineer could then classify all of those materials as fire hazards and perform more detailed hazard evaluations.

> Every proposed or existing process is based on a certain body of knowledge. An important part of this process knowledge is data on all of the chemicals and waste products used or produced in the process, including chemical intermediates that can be isolated. This information is the foundation of all hazard identification efforts.

4.1.2.1 Intrinsic

4.1.2.1.1 Reactive Hazards

Intrinsic reactive explosion hazards are caused by reaction of two or more materials. Large explosions can be caused by mixtures of reactive chemicals. Contamination leading to chemical explosion can occur in a number of ways. Manifolds and other multiple connection points are sometimes the cause of reactive explosion hazards.

Incompatibilities between chemicals and also with materials of construction should be evaluated and documented, most often using a binary chemical / material interaction matrix. Incompatibilities should be evaluated through the range of expected process conditions (temperature, pressure, and composition) and modes of operation (startup, shutdown, temporary operations, standby operations, etc.). See Section 3.3.5 for more information on chemical incompatibility.

Reactors may require heating or cooling for proper control. Leakage between the process and utility services can result in cross-contamination and subsequent reaction.

Batch processes (Ref. 4-4) offer a means of introducing errors in mixing reactant chemicals. Some examples include:
- When adding various process chemicals and other ingredients at various stages of the batch, the potential exists to omit an ingredient, add an ingredient in an incorrect sequence, or add an incorrect amount.

- Liquid ingredients are piped to manifolds at reactors with manually operated valves for charging to the reactor. Manifold valves inadvertently left open can result in cross-contamination. Depending on the nature of the process chemicals, there can be severe reactions.
- Batch reactors often require cleaning between batches, particularly when they are used for making different products. Cleaning fluids that are incompatible with the process chemicals (like water or solvent) may remain trapped in the system and cause inadvertent mixing and severe reactions.

4.1.2.1.2 Fire Hazards

Fires in process facilities produce four major outputs: gases, flame, heat, and smoke. The materials involved in the fire will determine the combination of these four outputs. For example, crude oil will produce a very dark thick smoke cloud, and ethylene does not produce much smoke but does have a very large flame. A hydrogen fire can have an invisible flame, making it particularly hazardous. Fire hazards in process facilities can impact personnel, the environment, structures, and equipment.

Fire hazards to personnel include:

- *Thermal Radiation* - When there is a line-of-sight between a person and the flame, the main impact is thermal radiation. The primary potential effects of thermal radiation are burns to exposed skin and ignition or melting / burning of clothing.
- *Exposure to Smoke and Gas* - Smoke is comprised of combustion gases, soot (solid carbon particles), and unburned fuel. For outdoor fires, the impact of smoke is usually a secondary consideration after the heat transfer. In many circumstances, the immediate thermal threat from the fire plume (jet, pool, or flash fire) overwhelms the smoke threat, particularly for personnel close to the event. There may be circumstances where personnel are in a downwind smoke plume where there is no immediate thermal threat. As a rule-of-thumb, all people within a smoke plume may be immediately or nearly immediately affected and at risk from a life safety standpoint (be it from lack of visibility or by toxic products).

Impact on the environment may result from both unwanted fires, improper control of fire effluent, or improper use of suppression system agents. Environmental considerations impact decisions on whether to provide protection for a hazard and whether this protection should be provided automatically or manually. Scenarios to be considered include uncontrolled fires, potential hazardous situations, firefighting training, and fixed or mobile vehicle suppression system discharge testing.

Steel, aluminum, concrete, and other materials that form part of a process or building frame are subject to structural failure when exposed to fire. Bare metal elements are particularly susceptible to damage. A structural member undergoes any combination of three basic types of stress: compression, tension, and shear. The time to failure of the structural member will depend on the amount and type of heat flux (i.e., radiation, convection, or conduction) and the nature of the exposure (one-sided flame impingement, flame immersion, etc.). Cooling effects from suppression systems and effects of passive fire protection will reduce the impact.

A heat flux of 8,000 Btu/hr/ft² (25 kW/m²) has been published as a general rule-of-thumb for damage to process equipment (Ref. 4-5). Clearly, this excludes electrical and electronic equipment, which may fail to operate at much lower heat fluxes and resulting temperatures.

4.1.2.1.3 Toxicity Hazards

Exposure to toxic chemicals can result in illness, disease, or death by interfering with the body's biological processes. Chemicals may be inhaled, absorbed, ingested, or injected. The toxic effects typically vary with contact time and type of exposure (e.g., skin contact versus ingestion or inhalation). Large spills of toxic chemicals can put neighboring facilities and communities at risk of injury, illness, or fatality.

4.1.2.1.4 Overpressure Hazards

There are many types of overpressure that can occur in process facilities. Some common overpressure sources include:

- *Physical Explosions* - A pressure vessel contains stored energy due to its internal pressure and represents an explosion hazard. If the vessel is pressurized beyond its mechanical strength or the integrity of the vessel is lost, the stored energy can be released suddenly and significant damage can result. The damage is caused by the pressure wave from the sudden gas release which propagates rapidly outward from the vessel. This pressure wave may be a shock wave, depending on the nature of the failure. Flying fragments from the vessel wall or structure can also cause damage. If the vessel contents are flammable, a subsequent fire or vapor cloud explosion might result.

- *BLEVE* - A BLEVE, or Boiling Liquid Expanding Vapor Explosion, occurs when a vessel containing liquid above its normal boiling point fails catastrophically. A BLEVE is a type of rapid phase transition in which a liquid contained above its atmospheric boiling point is rapidly depressurized, causing a nearly instantaneous transition from liquid to vapor with a corresponding energy release. A BLEVE is often accompanied by a large fireball if a flammable liquid is involved, since an external fire impinging on the vapor space of a pressure vessel is a common BLEVE scenario. However, it is not necessary for the liquid to be flammable to have a BLEVE occur.

- *Rapid Phase Transition Explosions* - A rapid phase transition explosion occurs when a liquid or solid undergoes a very rapid change in phase. If the phase change is from liquid to gas or from solid to gas (sublimation), the volume of the material will increase hundreds or thousands of times, frequently resulting in an explosion. This is the process that causes popcorn to pop when the moisture within the kernel changes phase and expands rapidly.

- *Vapor Cloud Explosions* - A Vapor Cloud Explosion, or VCE, results from the ignition of a cloud of flammable vapor, gas, or mist in which flame speeds accelerate to sufficiently high velocities to produce significant overpressure. The resulting explosion produces an overpressure which propagates outward from the explosion site as a blast wave. Significant damage from the resulting fire ball is also possible due to thermal radiation.

4. ANALYSIS TECHNIQUES

Further details on overpressure sources can be found in the CCPS Concept Book *Understanding Explosions* (Ref. 4-6) and *Guidelines for Vapor Cloud Explosion, Pressure Vessel Burst, BLEVE and Flash Fire Hazards, Second Edition* (Ref. 4-7).

4.1.2.1.5 Corrosivity / pH Hazards

Early in the design process raw materials, by-products, products, and other streams that could be used in the process need to be evaluated to identify noncompatible characteristics and potential interactions that may affect corrosion. Properties to consider include:

- Temperature
- Pressure
- Composition
- Wet vs. dry
- Reactive vs. non-reactive
- Corrosive vs. non-corrosive

Corrosion is discussed further in Chapter 5, Section 5.8, Corrosion.

4.1.2.2 Extrinsic

4.1.2.2.1 Intentional Mixing

Intentional mixing normally occurs when a material is added during a batch operation. For example, an additive is required to initiate a reaction. Adding too much additive could cause an exothermic reaction, whereas adding the additive too early in the process or too late could result in a useless material requiring disposal.

4.1.2.2.2 Inadvertent Mixing

The reactivity matrix is a tool to rapidly visualize the consequences of the intentional or unintentional mixing of various chemicals stored or used within a specified area. The format is usually a list of chemicals along the X-axis and the same chemicals along the Y-axis of a grid. Interaction consequences are recorded at the intersection block of two chemicals. Matrices are usually generated for small areas or single processes in order to keep the grid size manageable. The rule of thumb is only chemicals that can reasonably be expected to be mingled in the area should be included. Most users include ubiquitous chemicals such as water (or other liquids piped in or through the area) and jacket media as these usually meet the reasonableness criteria. The most widely utilized tool for generating reactivity matrices is the NOAA Chemical Reactivity Worksheet (Ref. 4-8).

Delayed starting of agitation can cause a significant deviation in intended ratios of reactants and can have a significant impact on reaction rates, reaction kinetics, and reaction chemistry (quality and by-products) up to and including runaway reactions.

See Chapter 3, Section 3.3.5, for more discussion of inadvertent mixing.

4.1.2.2.3 Ignition Sources

All potential ignition sources must be identified, although some may be difficult to analyze or control. Therefore, it is common practice to minimize the occurrence of such sources while taking all necessary steps to protect the equipment should such a source be present. These steps may involve control to protect against flammable atmospheres, design to contain any explosion within the equipment, or incorporation of devices to intercept, suppress, or vent a flame reaction zone. Even if all internal ignition sources were eliminated within the process equipment, an external pool fire or impingement flame might still damage the equipment or initiate an uncontrolled internal reaction. Therefore, external fire protection measures such as thermal insulation and sprinkler systems may be used in addition to prudent design and layout to minimize the severity of damage caused by external fires.

In addition to protecting equipment, measures should be taken to minimize the probability of a flash fire or vapor cloud explosion should a leak occur. Many ignition sources are obvious, such as flares, burn pits, furnaces, and other flame sources. Less obvious ignition sources include internal combustion engines, atmospheric static charges, and equipment that might not be recognized as "fixed" ignition sources on a site plan.

Often, ignition sources are insidious. For example, a poorly designed liquid transfer system might regularly give rise to static sparks but not cause ignition because the vapor is outside its flammable range. Any change in the vapor concentration might quickly give rise to an explosion. As another example, after years of uneventful operation, a fire might develop in a spray dryer due to accumulation of an unusually thick powder layer which spontaneously ignites (the accumulated heat reaches the autoignition temperature of the material). This fire might in turn ignite a powder suspension in the dryer causing an explosion. Measures to avoid ignition sources must often be taken at the design stage. However, to do this it is necessary to gather appropriate information on the ignition behavior of the materials concerned. Discovery of this behavior once a unit is operational means costly retrofit, redesign, or add-on safety measures.

Further details can be found in:
- API RP 2003, *Protection Against Ignitions Arising out of Static, Lightning and Stray Currents* (Ref. 4-9)
- NFPA 55, *Compressed Gas Code* (Ref. 4-10)
- NFPA 400, *Hazardous Material Code* (Ref. 4-11)
- NFPA 69, *Explosion Prevention Systems* (Ref. 4-12)
- NFPA 70, *National Electrical Code* (Ref. 4-13)
- NFPA 77, *Static Electricity* (Ref. 4-14)
- NFPA 78, *Lightning Protection Code* (Ref. 4-15)
- NFPA 497M, *Manual for Classification of Gases, Vapors and Dusts for Electrical Equipment in Hazardous (Classified) Locations* (Ref. 4-16)

4.1.2.3 Types of Ignition Source

Apart from obvious ignition sources such as flames, there are several types of ignition sources in process facilities, including:
- High temperature sources that may give rise to spontaneous ignition
- Electrical sources such as powered equipment, electrostatic accumulation, stray currents, radio frequency pick-up, and lightning

4. ANALYSIS TECHNIQUES

- Physical sources such as compression energy, heat of adsorption, friction, and impact
- Chemical sources such as catalytic materials, pyrophoric materials, and unstable species formed in the system

Ignition sources are often considered only in the context of the "Fire Triangle," whose sides comprise a fuel, an oxidant, and an ignition source (the three essential ingredients for most fires). However, it is important to recognize that some materials can be "ignited" in the absence of an oxidant. Examples include acetylene and ethylene oxide (decomposition flames) and some metal dusts (reaction with nitrogen). Also, under process conditions, some materials may be "ignited" in the absence of oxidant even though at ambient conditions they may have a significant Limiting Oxygen Concentration (LOC). An example is ethylene at elevated temperature and pressure, which may be ignited by many of the mechanisms discussed in this section.

4.1.2.4   Ignition by Flames

Ignition by flames includes both obvious ignition sources such as fired heaters and less obvious ignition sources such as internal combustion engines. An important feature of flames, as opposed to sparks and other brief ignition sources, is that they can readily ignite flammable or combustible materials of high ignition energy. Specifically, flammable mixtures can be ignited throughout their flammable ranges, since flames are at least equivalent to the ignition sources used to establish these ranges. Types of ignition sources include:

- Flares, burn pits, furnaces
- Hot work: welding, cutting
- Internal combustion engines
- Vacuum trucks

4.1.2.5   Spontaneous Ignition (Autoignition)

Spontaneous ignition is defined as the ignition and sustained combustion of a substance, whether gas, liquid, or solid, without introduction of any apparent ignition source such as a spark or flame. It is synonymous with "autoignition" and "self-ignition." Ignition is the result of self-reaction from any initial condition (temperature, pressure, volume) at which the rate of heat gain exceeds the rate of heat loss from the reacting system. Examples of autoignition include:

- Gas-phase autoignition
- Spontaneous ignition of liquids in absorbent solids
- Spontaneous ignition of powders (and other solids)
- Ignition of fibrous insulation and liquids on structured packing

4.1.2.6   Chemical Reactions

There are numerous possible routes to ignition via local chemical reactions which cannot occur in the system as a whole. Examples include:

- Catalysis
- Reaction with powerful oxidants

- Reactions of metals
- Thermite reactions
- Thermally unstable materials
- Accumulation of unstable materials
- Pyrophoric materials

4.1.2.7 Other Ignition Sources

Other ignition sources include:
- Electrical sources
  - Static electricity
  - Lightning
  - Stray currents
- Physical sources
  - Compression ignition
  - Mechanical: sparks, friction, impact, and vibration
  - Heat of adsorption

4.1.2.8 Design Alternatives

In some cases ignition is predictable and avoidable at the design stage. For example, knowing ignition characteristics of bulk powder, container temperature, size, geometry, or hold-up time may be designed to avoid spontaneous ignition. To assess such alternatives, it is essential to conduct appropriate material tests prior to design. This can avoid primary reliance on more active control measures such as inertion and flame mitigation.

A common shortcoming in solid-phase systems subject to self-heating is provision of inadequate temperature monitoring. Examples include purification beds, catalyst beds, and storage containers. Thermocouples, especially when mounted in heavy thermowells, may fail to respond to exothermic reactions occurring elsewhere in the system. Thermocouples mounted in the gas outlet will tend to average out any exothermic reaction in the solid phase. Large volumes should be monitored by many thermocouples or by commercially available temperature profiling systems. For purification beds, such as molecular sieve or activated carbon, special attention should be paid to exothermic activity during and after regeneration and pre-loading.

NFPA 69 (Ref. 4-12) provides recommendations on the following alternatives to minimize the probability of ignition or to mitigate an ignition event inside equipment:
- Reduce oxidant concentration.
- Reduce combustible concentration.
- Detect and extinguish sparks.
- Chemically suppress the incipient flame.
- Isolate the section of equipment containing the flame event.
- Construct equipment to contain the flame event.

4. ANALYSIS TECHNIQUES

Further alternatives, such as deflagration venting, are described in Chapter 7, Protection Layers. It is often important to determine the most probable site for ignition in a system. The ignition site can determine the severity of any flame event, since in pipes and other equipment of large length-to-diameter ratio, run-up to a detonation might occur in the available flame acceleration space. The ignition site can also influence the effectiveness of flame arresters under deflagrative conditions. In deflagration venting of enclosures, the ignition site influences the amount of unburned material that will be vented ahead of the flame and therefore the severity of explosions external to the equipment (this can be significant especially when the unburned material is vented into a partially confined space).

In reactive chemical systems in particular, every effort should be made to identify and evaluate the cause of unexpected observations, such as solid deposits in equipment. Simple observations, such as mild electric shocks experienced by personnel, should be seriously assessed in any area that might contain flammable gas or powder suspensions. Years of uneventful operation usually occur before a hazardous condition is recognized. A major objective is to recognize this condition before it becomes only too obvious. The ideal solution is to recognize and eliminate the potential at the design stage.

4.1.3 Human Impact Data

This section discusses the impact of fire, explosion, and toxic release on people and the environment.

4.1.3.1 Individual, Tabular, and Graphical

Data comes in many forms including individual, tabular, graphical, and some not measured / collected yet. It was once said "if you search long enough you will find the numbers you need to justify your results."

This statement is certainly true for data relating to human impact for fires, explosions, and toxic release. There is a multitude of data available; however, such data may not be presented in a useful or consistent format. In fact, the data may be contradictory and the user will need to decide what data and sources to use.

The CCPS book *Guidelines for Chemical Process Quantitative Risk Analysis* (Ref. 4-2) has a chapter on all types of data and can be used as a source for further information.

4.1.3.2 Probit Analysis

Probit analysis defines the relationship between the "dose" of a chemical received by an individual and the effects that result. There are several ways to represent dose, depending on the pathway by which the exposure occurs. One way is in terms of the quantity administered to the test organism per unit of body weight, which is usually associated with ingestion of the chemical. Another method expresses dose in terms of quantity per skin surface area, which relates to dermal contact. With respect to inhaled vapors, the dose can be represented as a specified vapor concentration administered over a period of time.

It is difficult to evaluate precisely the human response caused by an acute, hazardous exposure for a variety of reasons. Humans experience a wide range of acute adverse health effects, including irritation, narcosis, asphyxiation, sensitization, blindness, organ system damage, and death. In addition, the severity of many of these effects varies with intensity and duration of exposure. For example, exposure to a substance at an intensity that is sufficient to cause only mild throat irritation is of less concern than one that causes severe eye irritation or dizziness, since the latter effects are likely to impede escape from the area of contamination.

There is also a high degree of variation in response among individuals in a typical population. Withers and Lees (Ref. 4-17) discuss how factors such as age, health, and degree of exertion affect toxic responses (in this case, to chlorine). Generally, sensitive populations include the elderly, children, and persons with diseases that compromise the respiratory or cardiovascular system. As a result of the variability in response of living organisms, a range of responses is expected for a fixed exposure. Suppose an organism is exposed to a toxic material at a fixed dose and the responses determined. Some of the organisms will show a high level of response while some will show a low level. The results are frequently modeled as a Gaussian, or "bell-shaped," curve.

The experiment is repeated for a number of different doses and Gaussian curves are drawn for each dose. The mean response and standard deviation are determined at each dose. A complete dose-response curve is produced by plotting the cumulative mean response at each dose. This form typically provides a much straighter line in the middle of the dose range. The logarithm form arises from the fact that in most organisms there are some subjects who can tolerate rather high levels of the causative variable and, conversely, a number of subjects who are highly sensitive to the causative variable.

4.1.3.3 Probit Functions

A useful expression for performing the conversion from probits to percentage is given by the equation below, where the probit variable, Y, is based on a causative variable, V (representing the dose), and at least two constants (Ref. 4-18):

$$Y = k_1 + k_2 \bullet \log_e(V)$$

where

k_1 and k_2 are constants.

Probit equations of this type are derived as lines of best fit to experimental data (percentage fatalities versus concentration and duration) using log-probability plots or standard statistical packages.

Probit equations are available for a variety of exposures, including exposures to toxic materials, heat, pressure, radiation, impact, and sound, to name a few. For toxic exposures, the causative variable is based on the concentration; for explosions, the causative variable is based on the explosive overpressure or impulse, depending on the type of injury or damage. For fire exposure, the causative variable is based on the duration and intensity of the radiation exposure.

4.1.3.3.1 Probit for Fires

A probit is used to estimate the likely injury or damage to people from thermal radiation from incident outcomes.

4. ANALYSIS TECHNIQUES

Experiments have shown that the threshold of pain occurs when the skin temperature at a depth of 0.1 mm is raised to 840°F (450°C). When the skin surface temperature reaches about 1025°F (550°C), blistering occurs.

The inputs to most thermal effect models are the thermal flux level and duration of exposure. Thermal flux levels are provided by one of the fire consequence models and durations by either the consequence model (e.g., for BLEVEs) or an estimate of the time to extinguish the fire or escape from the fire. More detailed models use thermal energy input after a particular skin temperature is reached.

4.1.3.3.2 Probit for Toxic Materials

The analysis of toxic effects requires input at two levels:
1. Predictions of toxic gas concentrations and durations of exposure at all relevant locations
2. Toxic criteria for specific health effects for the particular toxic gas

Predictions of gas cloud concentrations and durations are available from neutral and dense gas dispersion models. IDLH and other acute toxic criteria are available for many chemicals and are described by AIChE / CCPS (Ref. 4-19).

Strength of the probit method is that it provides a probability distribution of consequences and it may be applicable to all types of incidents in risk assessment (fires, explosions, toxic releases). It is generally the preferred method of choice for risk assessment studies. A weakness of this approach is the restricted set of chemicals for which probit coefficients are published. Probit models can be developed from existing literature information and toxicity testing.

The potential for error arises both from the dispersion model and the toxicity measures. Interpretation of animal experiments is subject to substantial error due to the limited number of animals per experiment and imprecise applicability of animal data to people.

A useful equation for lethal toxicity is: (Ref. 4-2)

$$Y = a + b \bullet \log_e(C^n t_c)$$

where

- Y is the probit
- a, b, n are constants
- C is the concentration in ppm by volume
- t_c is the exposure time in minutes

Probit constants for a number of different vapor exposures are provided in Table 4.2.

Table 4.2 Probit Equation Constants for Lethal Toxicity

Substance	U.S. Coast Guard (1980)			World Bank (1988)		
	a	b	n	a	b	n
Acrolein	-9.931	2.049	1	-9.93	2.05	1.0
Acrylonitrile	-29.42	3.880	1.43			
Ammonia	-35.9	1.85	2	-9.82	0.71	2.00
Benzene	-109.78	5.3	2			
Bromine	-9.04	0.92	2			
Carbon Monoxide	-37.98	3.7	1			
Carbon Tetrachloride	-6.29	0.408	2.50	0.54	1.01	0.5
Chlorine	-8.29	0.92	2	-5.3	0.5	2.75
Formaldehyde	-12.24	1.3	2			
Hydrogen Chloride	-16.85	2.00	1.00	-21.76	2.65	1.00
Hydrogen Cyanide	-29.42	3.008	1.43			
Hydrogen Fluoride	-25.87	3.354	1.00	-26.4	3.35	1.0
Hydrogen Sulfide	-31.42	3.008	1.43			
Methyl Bromide	-56.81	5.27	1.00	-19.92	5.16	1.0
Methyl Isocyanate	-5.642	1.637	0.653			
Nitrogen Dioxide	-13.79	1.4	2			
Phosgene	-19.27	3.686	1	-19.27	3.69	1.0
Propylene Oxide	-7.415	0.509	2.00			
Sulfur Dioxide	-15.67	2.10	1.00			
Toluene	-6.794	0.408	2.50			

4.1.3.4 Fire Impact

There are innumerable situations where gases, liquids, and hazardous chemicals are produced, stored, or used in a process that, if released, could potentially result in a hazardous fire condition. It is important to analyze all materials and reactions associated with a particular process, including production, manufacturing, storage, or treatment facilities. The mode of burning depends on characteristics of the material released, temperature and pressure of the released material, ambient conditions, and time to ignition. Types of process fires include:

- Jet fire
- Flash fire
- Pool fire
- Running liquid fire
- Boiling liquid expanding vapor explosion (BLEVE) or fireball
- Solid fires, for example, cellulose fires involving material such as wood, paper, dust, etc.

4. ANALYSIS TECHNIQUES

The main mechanisms of heat transfer in a process facility are thermal radiation and direct flame contact. Heat transfer to personnel can cause burns. Heat transfer to equipment and structures can lead to failure equipment containing flammable or combustible material, which can further feed the fire.

Radiant energy that strikes a surface can be:
- Reflected
- Absorbed
- Transmitted (for transparent material)

Flames of some materials, such as natural gas, contain relatively little soot, whereas heavier hydrocarbons, such as kerosene and crude oil, generate copious amounts of soot and smoke.

Radiant heat transfer can result in burns to personnel and can heat up unprotected process equipment and structural elements. If the heat is not dissipated by the application of cooling or conduction, the process equipment or structure may fail.

4.1.3.4.1 Data Tables and Plots

A substantial body of experimental data exists and forms the basis for effect estimation. Two approaches are used:
- Simple tabulations or charts based on experimental results
- Theoretical models based on the physiology of skin burn response. Continuous bare skin exposure is generally assumed for simplification.

API Standard 521 (Ref. 4-20) provides a short review of the effects of thermal radiation on people. The data on time for pain threshold is summarized in Table 4.3 (Ref. 4-20). It is stated that burns follow the pain threshold "fairly quickly." The values in Table 4.3 may be compared to solar radiation intensity on a clear, hot summer day of about 320 Btu/hr ft^2 (1 kW/m^2). Based on these data, API suggests the thermal criteria shown in Table 4.4, excluding solar radiation, to establish exclusion zones or determine flare height for personnel exposure. Other criteria for thermal radiation damage are shown in Table 4.5.

Table 4.3 Exposure Time Necessary to Reach the Pain Threshold

Radiation Intensity		Time to Pain Threshold (s)
Btu/hr/ft^2	kW/m^2	
500	1.74	60
740	2.33	40
920	2.90	30
1500	4.73	16
2200	6.94	9
3000	9.46	6
3700	11.67	4
6300	19.87	2

Table 4.4 Recommended Design Flare Radiation Levels Excluding Solar Radiation

Permissible Design Level (K)		Condition
Btu/hr/ft²	kW/m²	
5000	15.77	Heat intensity on structures and in areas where operators are not likely to be performing duties and where shelter from radiant heat is available, for example, behind equipment.
3000	9.46	Value of K at design flare release at any location to which people have access, for example, at grade below the flare or on a service platform of a nearby tower. Exposure must be limited to a few seconds, sufficient for escape only.
2000	6.31	Heat intensity in areas where emergency actions lasting up to 1 min may be required by personnel without shielding but with appropriate clothing.
1500	4.73	Heat intensity in areas where emergency actions lasting several minutes may be required by personnel without shielding but with appropriate clothing.
500	1.58	Value of K at design flare release at any location where personnel are continuously exposed.

Table 4.5 Effects of Thermal Radiation

Radiation Intensity		
Btu/hr/ft²	kW/m²	Observed effect
11,900	37.5	Sufficient to cause damage to process equipment.
8,000	25	Minimum energy required to ignite wood at indefinitely long exposures (nonpiloted).
4,000	12.5	Minimum energy required for piloted ignition of wood, melting of plastic tubing.
3,000	9.5	Pain threshold reached after 8 sec; second degree burns after 20 sec.
1,200	4	Sufficient to cause pain to personnel if unable to reach cover within 20 sec. however blistering of the skin (second degree burns) is likely; 0% lethality.
500	1.6	Will cause no discomfort for long exposure.

The effect of thermal radiation on structures depends on whether they are combustible or not and the nature and duration of the exposure. Thus, wooden materials will fail due to combustion, whereas steel will fail due to thermal lowering of the yield stress. Many steel structures under normal load will fail rapidly when raised to a temperature of 932-1112°F (500-600°C), whereas concrete will survive for much longer. Flame impingement on a structure is more severe than thermal radiation.

4. ANALYSIS TECHNIQUES

4.1.3.5 Bio Impact (Toxic Gas Impact)

Toxic effect models are employed to assess the consequences to human health as a result of exposure to a known concentration of toxic gas for a known period of time. This section does not address the release and formation of nontoxic, flammable vapor clouds that do not ignite but pose a potential for asphyxiation.

Concentration-time information is estimated using dispersion models. Probit models are used to develop exposure estimates for situations involving continuous emissions (approximately constant concentration over time at a fixed downwind location) or puff emissions (concentration varying with time at a downwind location). It is much more difficult to apply other criteria that are based on a standard exposure duration (e.g., 30 or 60 minutes) particularly for puff releases that involve short exposure times and varying concentrations over those exposure times. The objective of the toxic effects model is to determine whether an adverse health outcome can be expected following a release and, if data permit, to estimate the extent of injury or fatalities that are likely to result. For the overwhelming majority of substances encountered in industry, there are not enough data on toxic responses of humans to directly determine a substance's hazard potential. Frequently, the only data available are from controlled experiments conducted with laboratory animals. In such cases, it is necessary to extrapolate from effects observed in animals to effects likely to occur in humans. This extrapolation introduces uncertainty and normally requires the professional judgment of a toxicologist or an industrial hygienist with experience in health risk assessment.

4.1.3.5.1 Regulatory and Industry Standard Endpoints

Many useful measures are available to use as benchmarks for predicting the probability that a release event will result in injury or death. The American Institute of Chemical Engineers (Ref. 4-2) reviews various toxic effects and discusses the use of various established toxicological criteria.

The Department of Energy (DOE) has developed a website that allows users to access DOE's current data set of Protective Action Criteria (PAC) values in a variety of ways: as a searchable database, as an Excel file, and as a series of tables in PDF format (Ref. 4-21). Emergency exposure limits are essential components of planning for the uncontrolled release of hazardous chemicals. These limits, combined with estimates of exposure, provide the information necessary to identify and evaluate accidents for the purpose of taking appropriate protective actions. During an emergency response to an uncontrolled release, these limits may be used to evaluate the severity of the event, to identify potential outcomes, and to decide what protective actions should be taken. In anticipation of an uncontrolled release, these limits may also be used to estimate the consequences of an uncontrolled release and to plan emergency responses. PACs for emergency planning of chemical release events are based on the chemical exposure limit values provided in:

- Acute Exposure Guideline Levels (AEGLs) are developed by the U.S. Environmental Protection Agency (EPA) (Ref. 4-22). AEGLs are defined for five time periods: 10 minutes, 30 minutes, 60 minutes, 4 hours, and 8 hours. The 60-minute AEGL values have been selected for use in the PAC database.

- Emergency Response Planning Guidelines (ERPGs) are produced by the American Industrial Hygiene Association (AIHA) Emergency Planning Committee (Ref. 4-23).
- Temporary Emergency Exposure Limit (TEEL) data sets are developed by the DOE Office of Emergency Management (Ref. 4-24).

Other criteria and methods include:
- Immediately Dangerous to Life or Health (IDLH) levels established by the National Institute for Occupational Safety and Health (NIOSH).
- Emergency Exposure Guidance Levels (EEGLS) and Short-Term Public Emergency Guidance Levels (SPEGLs) issued by the National Academy of Sciences National Research Council.
- Threshold Limit Values (TLVs) established by the American Conference of Governmental Industrial Hygienists (ACGIH) including Short-Term Exposure Limits (STELs) and ceiling concentrations (TLV-Cs).
- Permissible Exposure Limits (PELs) promulgated by the Occupational Safety and Health Administration (OSHA).
- Various state guidelines, for example, the Toxicity Dispersion (TXDs) method used by the New Jersey Department of Environmental Protection (NJ-DEP).
- Toxic endpoints promulgated by the U.S. Environmental Protection Agency.

4.1.3.5.2 Emergency Response Planning Guidelines (ERPGs)

Emergency Response Planning Guidelines (ERPGs) are prepared by an industry task force and are published by the American Industrial Hygiene Association (AIHA). Three concentration ranges are provided as a consequence of exposure to a specific substance:

- The ERPG-1 is the maximum airborne concentration below which it is believed that nearly all individuals could be exposed for up to 1 hour without experiencing any symptoms other than mild transient adverse health effects or without perceiving a clearly defined objectionable odor.
- The ERPG-2 is the maximum airborne concentration below which it is believed that nearly all individuals could be exposed for up to 1 hour without experiencing or developing irreversible or other serious health effects or symptoms that could impair their abilities to take protective action.
- The ERPG-3 is the maximum airborne concentration below which it is believed nearly all individuals could be exposed for up to 1 hour without experiencing or developing life-threatening health effects (similar to EEGLs).

ERPG data (Ref. 4-23) are shown in Table 4.6. ERPGs are generally an acceptable industry / government norm.

4. ANALYSIS TECHNIQUES

Table 4.6   Emergency Response Planning Guidelines

Chemical	ERPG-1	ERPG-2	ERPG-3
Acetaldehyde	10	200	1000
Acrolein	0.1	0.5	3
Acrylic Acid	2	50	750
Acrylonitrile	NA	35	75
Allyl Chloride	3	40	300
Ammonia	25	200	1000
Benzene	50	150	1000
Benzyl Chloride	1	10	25
Bromine	0.2	1	5
1,3 - Butadiene	10	50	5000
n-Butyl Acrytate	0.05	25	250
n-Butyl Isocyanate	0.01	0.05	1
Carbon Disulfide	1	50	500
Carbon Tetrachloride	20	100	750
Chlorine	1	3	20
Chlorine Trifluoride	0.1	1	10
Chloroacetyl Chloride	0.1	1	10
Chloropicrin	NA	0.2	3
Chlorosulfonic Acid	2 mg/m^3	2 mg/m^3	2 mg/m^3
Chlorotrifluoroethylene	20	100	300
Crotonaldehyde	2	10	50
Diborane	NA	1	3
Diketene	1	5	50
Dimethylamine	1	100	500
Dimethylchlorosilane	0.8	5	25
Dimethyl Disulfide	0.01	50	250
Epichlorohydrin	2	20	100
Ethylene Oxide	NA	50	500
Formaldehyde	1	10	25
Hexachlorobutadiene	3	10	30
Hexafluoracetone	NA	1	50
Hexafluoropropylene	10	50	500
Hydrogen Chloride	3	20	100
Hydrogen Cyanide	NA	10	25
Hydrogen Fluoride	54	20	50
Hydrogen Sulfide	0.1	30	100
Isobutyronitrile	10	50	200
2-Isocyanatoethyl Methacrylate	NA	0.1	1
	25 μgm/m^3	100 μgm/m^3	500 μgm/m^3
Lithium Hydride	200	1000	5000
Methyl Chloride	NA	400	1000
Methylene Chloride	200	750	4000
Methyl Iodide	25	50	125
Methyl Isocyanate	0.025	0.5	5
Methyl Mercaptan	0.005	25	100

Table 4.6   Emergency Response Planning Guidelines (Continued)

Chemical	ERPG-1	ERPG-2	ERPG-3
Methyltrichlorosilane	0.5	3	15
Monomethylamine	10	100	500
Perfluoroisobutylene	NA	0.1	0.3
Phenol	10	50	200
Phosgene	NA	0.2	1
Phosphorus Pentoxide	5 mg/m^3	25 mg/m^3	100 mg/m^3
Propylene Oxide	50	250	750
Styrene	50	250	1000
Sulfonic Acid (Oleum, Sulfur Trioxide, and Sulfuric Acid)	2 mg/m^3	10 mg/m^3	30 mg/m^3
Sulfur Dioxide	0.3	3	15
Tetrafluoroethylene	200	1000	10,000
Titanium Tetrachloride	5 mg/m^3	20 mg/m^3	100 mg/m^3
Toluene	50	300	1000
Trimethylamine	0.1	100	500
Uranium Hexafluoride	5 mg/m^3	15 mg/m^3	30 mg/m^3
Vinyl Acetate	5	75	500

Default value is in ppm unless noted.

4.1.3.5.3 Immediately Dangerous to Life or Health (IDLH)

The National Institute for Occupational Safety and Health (NIOSH) publishes Immediately Dangerous to Life and Health (IDLH) concentrations to be used as acute toxicity measures for common industrial gases. Updated IDLH levels can be found on the NIOSH website at www.cdc.gov/niosh/idlh/intridl4.html. An IDLH exposure condition is defined as a condition "that poses a threat of exposure to airborne contaminants when that exposure is likely to cause death or immediate or delayed permanent adverse health effects or prevent escape from such an environment" (Ref. 4-25). IDLH values also take into consideration acute toxic reactions, such as severe eye irritation, that could prevent escape. The IDLH is considered a maximum concentration above which only a highly reliable breathing apparatus providing maximum worker protection is permitted. If IDLH values are exceeded, all unprotected workers must leave the area immediately.

Because IDLH values were developed to protect healthy worker populations, they must be adjusted for sensitive populations, such as older, disabled, or ill populations. For flammable vapors, the IDLH is defined as 1/10 of the lower flammable limit (LFL) concentration.

4.1.3.5.4 Emergency Exposure Guidance Levels (EEGLs)

Since the 1940s, the National Research Council's Committee on Toxicology has submitted Emergency Exposure Guidance Levels (EEGLs) for 44 chemicals of special concern to the Department of Defense. An EEGL is defined as a concentration of a gas, vapor, or aerosol that is judged to be acceptable and that will allow healthy military personnel to perform specific tasks during emergency conditions lasting from 1 to 24 hours.

4. ANALYSIS TECHNIQUES

Exposure to concentrations at the EEGL may produce transient irritation or central nervous system effects but should not produce effects that are lasting or that would impair performance of a task. In addition to EEGLs, the National Research Council has developed Short-Term Public Emergency Guidance Levels (SPEGLs), defined as acceptable concentrations for exposures of members of the general public. SPEGLs are generally set at 10-50% of the EEGL and are calculated to take account of the effects of exposure on sensitive, heterogeneous populations. The advantages of using EEGLs and SPEGLs rather than IDLH values are (1) a SPEGL considers effects on sensitive populations, (2) EEGLs and SPEGLs are developed for several different exposure durations, and (3) the methods by which EEGLs and SPEGLs were developed are well documented in National Research Council publications.

4.1.3.5.5 Other

Some states have their own exposure guidelines. For example, the New Jersey Department of Environmental Protection (NJ-DEP) uses the Toxic Dispersion (TXDS) method of consequence analysis for the estimation of potentially catastrophic quantities of toxic substances as required by the New Jersey Toxic Catastrophe Prevention Act (TCPA) (Ref. 4-26). An Acute Toxic Concentration (ATC) is defined as the concentration of a gas or vapor of a toxic substance that will result in acute health effects in the affected population and one fatality out of 20 or less (5% or more) during 1 hour exposure. ATC values as proposed by the NJ-DEP are estimated for 103 "extraordinarily hazardous substances" and are based on the lowest value of one of the following:

- The Lowest Reported Lethal Concentration (LCLO) value for animal test data
- The Median Lethal Concentration (LC50) value from animal test data multiplied by 0.1
- The IDLH value

The EPA (Ref. 4-27) published a set of toxic endpoints to be used for air dispersion modeling of toxic gas releases as part of the EPA Risk Management Plan (RMP). The toxic endpoint is, in order of preference: (1) the ERPG-2, or (2) the Level of Concern (LOC) promulgated by the Emergency Planning and Community Right-to-Know Act. The LOC is considered "to be the maximum concentration of an extremely hazardous substance in air that will not cause serious irreversible health effects in the general population when exposed to the substance for relatively short duration" (Ref. 4-28). Toxic endpoints are provided for 77 chemicals under the RMP rule (Ref. 4-28).

In general, the most directly relevant toxicological criteria currently available, particularly for developing emergency response plans and conducting risk assessments, are ERPGs, SPEGLs, and EEGLs. These were developed specifically to apply to general populations and to account for sensitive populations and scientific uncertainty in toxicological data. For incidents involving substances for which no SPEGLs or EEGLs are available, IDLHs provide alternative criteria.

A much more extensive list of toxic chemical characteristics has been prepared by the Health and Safety Executive in the UK. The HSE uses two levels of impact, "SLOT" and "SLOD." These terms have several definitions, most notably:

- *SLOT (Specified Level of Toxicity)* - Highly susceptible people possibly being killed
- *SLOD (Significant Likelihood of Death)* - 50% mortality in exposed population

There is no direct comparison between the HSE data and the earlier approaches but the results seem comparable. The HSE values can be used as a basis for estimating probabilities of fatality for the broader range of chemicals that the HSE reports.

Table 4.7 contains an excerpt from the beginning of the almost 100 SLOT / SLOD Dangerous Toxic Load (DTL) values provided by the HSE (Ref. 4-29).

Table 4.7 Example of SLOT DTL and SLOD DTL Values for Various Substances

Substance	n Value	SLOT DTL ppm^n min	SLOD DTL ppm^n min
Acetic Acid	1	7.5×10^4	3×10^5
Acetonitrile	1	8.1×10^4	1.6×10^5
Acetyl Chloride	1	9900	3.96×10^4
Acrolein	1	420	1680
Acrylamide	1	1.3×10^5	5.2×10^5
Acrylonitrile	1	9600	2.52×10^4
Adiponitrile	1	8.1×10^4	1.6×10^5

The user calculates the integral concentration of toxic material (in ppm), raised to the n power with respect to the exposure duration (in minutes). The result is then compared to the SLOT and SLOD values in the table above to determine if the specific impact level has been reached.

There is no specified method for converting a SLOT/SLOD form into a probit form in order to facilitate interpolation or extrapolation from the SLOT / SLOD values to other impact magnitudes. Therefore, if SLOT / SLOD data are used for impact levels other than those defined above, the basis for doing so must be described by the analyst.

4.1.3.6 Temperature Impact

The human body's thermal regulation system tries to maintain a relatively stable internal (core) temperature of 97-99°F (36.1-37.2°C). The core temperature must stay within a narrow range to prevent serious damage to health and performance (Ref. 4-30). The body maintains heat balance by increasing or decreasing blood circulation to the skin. The body also exchanges heat with the environment through:

- Convection: absorbing from or losing heat to the surroundings through the skin
- Conduction: by contacting sources of heat or cold directly

4. ANALYSIS TECHNIQUES

- Perspiration: by losing heat through the evaporation of water vapor on the skin
- Radiation: receiving radiation from an external source or radiating heat from our body

Clearly, some methods are more effective than others.

Thermal hazards can include both heat and cold hazards. This section discusses temperature hazards associated with the process or chemical properties of process materials.

4.1.3.7 Overpressure Impact

The types of explosions that may occur depend on the confinement of the reactive material, its energy content, its kinetic parameters, and the mode of ignition (self-heating or induced by external energy input). Explosions are characterized as physical or chemical explosions and as homogeneous or heterogeneous as described in Figure 4.1.

A physical explosion, for example, a boiler explosion, a pressure vessel failure, or a BLEVE (Boiling Liquid Expanding Vapor Explosion), is not necessarily caused by a chemical reaction. Chemical explosions are characterized as detonations, deflagrations, and thermal explosions. In the case of a detonation or deflagration (e.g., explosive burning), a reaction front is present that proceeds through the material. A detonation proceeds by a shock wave with a velocity exceeding the speed of sound in the unreacted material. A deflagration proceeds by transport processes such as by heat (and mass) transfer from the reaction front to the unreacted material. The velocity of the reaction front of a deflagration is less than the velocity of sound in the unreacted material. Both types of explosions are often called heterogeneous explosions because of the existence of a reaction front which separates completely reacted and unreacted material.

A thermal explosion is the third type of chemical explosion. In this case, no reaction front is present, and it is therefore called a homogenous explosion. Material can have a uniform temperature distribution or hot spots. If the temperature in the bulk material is sufficiently high so that the rate of heat generation from the reaction exceeds the heat removal, then self-heating begins. The bulk temperature will increase at an increasing rate, and local hot spots may develop as the thermal runaway reaction proceeds. The runaway reaction can lead to overpressure and possible rupture of the vessel.

Explosion phenomena have occurred in all types of confined and unconfined units: reactors, separation and storage units, filter systems, pipe lines, and so forth. Typical reactions that may cause explosions are oxidations, decompositions, nitrations, and polymerizations. Examples of chemical and processing system characteristics that increase the potential for an explosion are the following:

- High decomposition or reaction energies
- High rates of energy generation,
- Insufficient heat removal (i.e., too large a quantity of the substances)
- The presence of an initiation source
- Substances with an oxygen balance close to zero
- Confinement
- Large amounts and / or high rates of gas production

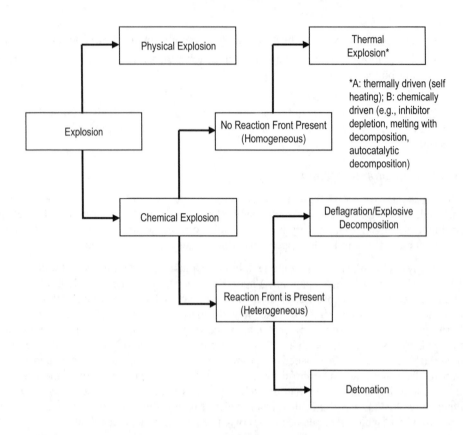

Figure 4.1 Types of Explosions

Explosion effect models predict the impact of blast overpressure and projectiles on people and objects. Most effect models for explosions are based on either the blast overpressure alone or a combination of blast overpressure, duration, or impulse. The blast overpressure, impulse, and duration are determined using a variety of models, including TNO multi-energy and Baker-Strehlow methods. See *Guidelines for Vapor Cloud Explosion, Pressure Vessel Burst, BLEVE and Flash Fire Hazards* (Ref. 4-7) for details on these models.

Since the blast overpressure decreases rapidly as the distance from the source increases, significant offsite damage from blasts is usually not expected. Most studies are directed toward onsite damage.

4.1.3.7.1 Pressure Effects on Structures

The use of Building Damage Levels (BDLs) is a common siting criterion. Building damage increases as the severity of the blast load increases and may be represented as a continuous or discrete function. When a continuous function is used, the scale is "percentage of damage" (Ref. 4-31). When the discrete approach is used, BDLs are categorized into a number of damage states ranging from minimal damage to collapse.

4. ANALYSIS TECHNIQUES

The continuous damage function is the approach used by the U.S. Department of Defense Explosive Safety Board (Ref. 4-31). The limitations of this approach are that it does not readily allow for the identification of what type of damage has occurred and which building components may be governing the percentage of damage to the structure.

Typical discrete BDLs used in the process industry are shown in Table 4.8. One advantage of this approach is that the nature of the damage is indicated by the damage description. Pressure-impulse diagrams serve to define the boundaries between the damage states when discrete BDLs are used.

Table 4.8   Typical Industry Building Damage Level Descriptions

Building Damage Level (BDL)	BDL Name	Damage Description
1	Minor	Onset of visible damage to reflected wall of building.
2A	Light	Reflected wall components sustain permanent damage requiring replacement, other walls and roof have visible damage that is generally repairable.
2B	Moderate	Reflected wall components are collapsed or very severely damaged. Other walls and roof have permanent damage requiring replacement.
3	Major	Reflected wall has collapsed. Other walls and roof have substantial plastic deformation that may be approaching incipient collapse.
4	Collapse	Complete failure of the building roof and a substantial area of walls.

Overpressure duration is important for determining effects on structures. The positive-pressure phase of the blast wave can last from 10 to 250 milliseconds or more for typical VCEs. The same overpressure level can have markedly different effects depending on the duration. Therefore, some caution should be exercised in application of simple overpressure criteria for buildings or structures. These criteria can in many cases cause overestimation of structural damage. If the blast duration is shorter than the characteristic structural response times it is possible the structure can survive higher overpressures. Baker et al. (Ref. 4-32) discuss design issues relating to the response of structures to explosion overpressures. AIChE / CCPS (Ref. 4-7) provides an extensive review of risk criteria and risk reduction methods for structures exposed to explosions and a discussion of blast-resistant building design.

4.1.3.7.2 People

People outside of buildings or structures are susceptible to:
- Direct blast injury (blast overpressure)
- Indirect blast injury (missiles or whole-body translation)

Relatively high blast overpressures (>15 psig) are necessary to produce fatality (primarily due to lung hemorrhage).

It is generally believed that fatalities arising from whole-body translation are due to head injury from impact. Baker et al. (Ref. 4-32) present tentative criteria for probability of fatality as a function of impact velocity. Lees (Ref. 4-18) provides probit equations for whole-body translation and impact. Injury to people due to fragments usually occurs either because of penetration by small fragments or blunt trauma by large fragments. Injury from blunt projectiles is a function of the fragment mass and velocity. Very limited information is available for this effect.

4.1.3.8 Effects on Environment

The effects on the environment from fires, explosions, and toxic releases are much harder to measure. For a fire there will most likely be smoke plume that will carry offsite. The plume may have unburned toxic materials and particles that could impact people or the environment. Water runoff from firefighting could contain toxic materials harmful to the environment. Water collection systems should be designed to collect and process water used during emergencies.

A toxic release can be in the form of a spill or vapor cloud. If a spill, then the water collection systems should be able to collect and neutralize the material water used during emergencies. If a vapor cloud, then depending on the material being released there could be effects on people, animals, and vegetation.

4.2 HAZARD ANALYSIS TECHNIQUES

Hazard analysis is an organized effort to identify and evaluate the severity of hazardous scenarios associated with a process or activity. Specifically, hazard analyses are used to identify weaknesses in the design and operation of facilities that could lead to harm to people and the environment. These studies provide information to aid in making decisions for improving safety and managing the risk of operations. Hazard analyses usually focus on process safety issues, like the acute effects of unplanned chemical releases on plant personnel or the public. Although primarily directed at providing safety-related information, many hazard evaluation techniques can also be used to investigate operability, economic, and environmental concerns.

This section provides an overview of hazard evaluation techniques. For detailed information, see *Guidelines for Hazard Evaluation Procedures* (Ref. 4-1).

4.2.1 A Life Cycle Approach

Hazard analyses should be performed throughout the life of a process as an integral part of an organization's PSM program. Hazard analyses can be performed to help manage process risks from the earliest stages of research and development; in detailed design and construction; periodically throughout the operating lifetime; and continuing until the process is decommissioned and dismantled. By using this "life cycle" approach in concert with other PSM activities, hazard analyses can efficiently reveal deficiencies in design and operation before a unit is sited, built, or operated, thus making the most effective use of resources devoted to ensuring the safe and productive life of a facility. Table 4.9 identifies typical hazard evaluation objectives and their appropriate process stages as well as suggested hazard analysis techniques.

4. ANALYSIS TECHNIQUES

Table 4.9 Typical Hazard Evaluation Objectives at Different Stages of a Process Lifetime

Process Phase	Example Objectives	Hazard Analysis Technique
Research and development	Identify chemical reactions or interactions that could cause runaway reactions, fires, explosions, or toxic gas releases. Identify process safety data needs.	Hazard Identification What-If
Conceptual design	Identify opportunities for inherent safety. Compare the hazards of potential sites. Provide input to facility layout and buffer zones.	Checklist Hazard Identification What-If What-If / Checklist
Pilot plant	Identify ways for hazardous materials to be released to the environment. Identify ways to deactivate the catalyst. Identify potentially hazardous operator interfaces. Identify ways to minimize hazardous wastes.	Checklist Hazard Identification What-If What-If / Checklist Hazard and Operability Study Failure Mode and Effects Analysis Fault Tree Analysis Event Tree Analysis
Detailed engineering	Identify ways for a flammable mixture to form inside process equipment. Identify how a loss of containment might occur. Identify which process control malfunctions will cause runaway reactions. Identify ways to reduce hazardous material inventories. Evaluate whether designed safeguards are adequate to control process risks to tolerable, required or as low as reasonable practical (ALARP) level. Identify safety-critical equipment that must be regularly tested, inspected, or maintained.	What-If What-If / Checklist Hazard and Operability Study Failure Mode and Effects Analysis Fault Tree Analysis Event Tree Analysis
Construction and startup	Identify error-likely situations in the startup, and operating procedures. Verify that all issues from previous hazard evaluations were resolved satisfactorily and that no new issues were introduced. Identify hazards that adjacent units may create for construction and maintenance workers. Identify hazards associated with vessel cleaning procedures. Identify any discrepancies between as-built equipment and the design drawings.	Safety Review Checklist What-If What-If / Checklist Critical Task Analysis

Table 4.9 Typical Hazard Evaluation Objectives at Different Stages of a Process Lifetime (Continued)

Process Phase	Example Objectives	Hazard Analysis Technique
Routine operation	Identify employee hazards associated with the operating procedures. Identify ways an overpressure transient might occur. Update previous hazard evaluation to account for operational experience. Identify hazards associated with out-of-service equipment.	Checklist What-If What-If / Checklist Hazard and Operability Study Critical Task Analysis
Process modification or plant expansion	Identify whether changing the feedstock composition will create any new hazards or worsen any existing ones. Identify hazards associated with new equipment.	Safety Review Checklist Hazard Identification What-If What-If / Checklist Hazard and Operability Study Failure Mode and Effects Analysis Fault Tree Analysis Event Tree Analysis
Decommissioning	Identify how demolition work might affect adjacent units. Identify any fire, explosion, or toxic hazards associated with the residues left in the unit after shutdown.	Safety Review Checklist What-If What-If / Checklist

4.2.2 Qualitative

Hazard analysis is the cornerstone of an organization's overall PSM program. Although hazard analyses typically involve the use of qualitative techniques to analyze potential equipment failures and human errors that can lead to incidents, the studies can also highlight gaps in the management systems of a process safety program. Qualitative hazard evaluation techniques, often referred to as Process Hazard Analyses (PHA), include:

- Hazard Identification
- Checklist Analysis
- What-If Analysis
- Hazard and Operability Study (HAZOP)

4.2.2.1 Hazard Identification

Hazard Identification, sometimes referred to as Preliminary Hazard Analysis, focuses in a general way on the hazardous materials and major process areas of a plant. It is most often conducted early in the development of a process when there is little information on design details or operating procedures and is often a precursor to further hazard analyses. It can be a cost-effective way to identify hazards early in a plant's life.

4. ANALYSIS TECHNIQUES

Hazard Identification is generally applied during the research and development or conceptual design phase of a process plant and can be very useful when making site selection decisions. It is also commonly used as a design review tool before a process P&ID is developed.

Hazard Identification formulates a list of hazards and generic hazardous situations by considering various process characteristics. As each hazardous situation is identified, the potential causes, effects, and possible corrective and / or preventive measures are listed.

Table 4.10 provides an overview of Hazard Identification requirements and results.

Table 4.10 Hazard Identification Overview

Typically Used During	Resource Requirements	Type of Results	Advantages and Disadvantages
Research and development. Conceptual design. Pilot plant operation.	Material, physical, and chemical data. Basic process chemistry. Process flow diagram.	Rough screening of general hazards. Ranking of hazardous areas or processes.	Provides a quick focus on big issues. Potential to miss something.

Detailed information on performing Hazard Identification can be found in *Guidelines for Hazard Evaluation Procedures* (Ref. 4-1).

4.2.2.2 Checklist Analysis

A Checklist Analysis uses a written list of items or procedural steps to verify the status of a system. Traditional checklists vary widely in level of detail and are frequently used to indicate compliance with standards and practices. In some cases, analysts use a more general checklist in combination with another hazard evaluation method to discover common hazards that the checklist alone might miss. The Checklist Analysis approach is easy to use and can be applied at any stage of the process life cycle.

A detailed checklist provides the basis for a standard evaluation of process hazards. It can be as extensive as necessary to satisfy the specific situation, but it should be applied conscientiously in order to identify problems that require further attention.

Table 4.11 provides an overview of Checklist Analysis requirements and results.

Table 4.11 Checklist Overview

Typically Used During	Resource Requirements	Type of Results	Advantages and Disadvantages
Conceptual design Pilot plant operation Detailed engineering Construction / startup Routine operation Decommissioning Expansion or modification During What-If or HAZOP studies to provide compliance with items such as facility siting, human factors, and other general issues.	Material, physical, and chemical data Basic process chemistry Process flow diagram Operating procedures Piping and Instrumentation Diagrams (P&IDs)	Response to pre-defined questions Documentation of compliance	Can be used with less experienced personnel if the experience is captured in the checklist. Quality of the analysis is only as good as the quality of the checklist. Checklists that are too long or don't relate specifically enough to the process being analyzed may have a tendency to be completed without thorough evaluation.

Detailed information on performing a Checklist Analysis can be found in *Guidelines for Hazard Evaluation Procedures* (Ref. 4-1).

4.2.2.3 What-If Analysis

The What-If Analysis technique is a brainstorming approach in which a group of experienced people familiar with the subject process ask questions or voice concerns about possible undesired events. The purpose of a What-If Analysis is to identify hazards, hazardous situations, or specific event sequences that could produce undesirable consequences. An experienced group of people identifies possible abnormal situations, their consequences, and existing safeguards and then suggests alternatives for risk reduction where obvious improvement opportunities are identified or where safeguards are judged to be inadequate. The method can involve examination of possible deviations from the design, construction, modification, or operating intent. It requires a basic understanding of the process intention, along with the ability to mentally combine possible deviations from the design intent that could result in an incident. This is a powerful technique if the staff is experienced; otherwise, the results are likely to be incomplete.

The What-If Analysis concept encourages the hazard evaluation team to think of questions that begin with "What-If." For example:

- What if the wrong material is delivered?
- What if pump A stops running during startup?
- What if the operator opens valve B instead of valve A?

Table 4.12 provides an overview of What-If Analysis requirements and results.

4. ANALYSIS TECHNIQUES

Table 4.12 What-If Analysis Overview

Typically Used During	Resource Requirements	Type of Results	Advantages and Disadvantages
Research and development Conceptual design Pilot plant operation Detailed engineering Construction / startup Routine operation Decommissioning Expansion or modification During HAZOP studies to address issues such as loss of utilities.	Material, physical, and chemical data Basic process chemistry Process flow diagram Piping and Instrumentation Diagrams	Scenario-based documentation of What-If questions, consequences, safeguards, risk ranking, and recommendations, if any.	Allows an experienced facilitator to efficiently address issues of concern, such as loss of cooling water or lube oil. Inexperienced facilitators may miss potential process deviations if they don't brainstorm all potential What-If questions.

Detailed information on performing a What-If Analysis can be found in *Guidelines for Hazard Evaluation Procedures* (Ref. 4-1).

4.2.2.4 Hazard and Operability Study

The Hazard and Operability (HAZOP) technique systematically reviews a process or operation to determine whether deviations from the design or operation intent can lead to undesirable consequences. This technique can be used for continuous or batch processes and can be adapted to evaluate written procedures. In a HAZOP study, an interdisciplinary team uses a systematic approach to identify hazard and operability problems resulting from deviations from the process's design intent that could lead to undesirable consequences. An experienced team leader systematically guides the team through the plant design using a fixed set of words (called "guidewords"). These guidewords are applied at specific points or "study nodes" in the plant design and are combined with specific process parameters to identify potential deviations from the plant's intended operation. Typical steps in a HAZOP are:

- Choose study node
- Apply a deviation (guideword + parameter) (e.g., no flow)
- Brainstorm causes of the deviation
- Develop each cause to its ultimate consequence(s)
- Identify existing safeguards
- Qualitatively assess the risk of the scenario
- If warranted, make recommendation(s) to reduce risk and / or improve the operability of the facility

Table 4.13 provides an overview of HAZOP requirements and results.

Table 4.13 HAZOP Overview

Typically Used During	Resource Requirements	Type of Results	Advantages and Disadvantages
Pilot plant operation Detailed engineering Routine operation Expansion or modification	Material, physical, and chemical data Basic process chemistry Process flow diagram Piping and Instrumentation Diagrams	Scenario-based documentation of deviations, causes, consequences, safeguards, risk ranking, and recommendations, if any.	Provides a structured methodology to systematically and consistently analyze hazard scenarios. Provides input to Layer of Protection Analysis by identifying high consequence scenarios. Potential for redundancy.

Detailed information on performing a HAZOP, along with worked examples, can be found in *Guidelines for Hazard Evaluation Procedures* (Ref. 4-1).

4.2.3 Semi-Quantitative

Semi-quantitative hazard evaluations are more focused than a qualitative risk assessment, but not as rigorous as a quantitative approach. Semi-quantitative techniques include:

- Layer of Protection Analysis
- Failure Modes and Effects Analysis

4.2.3.1 Layer of Protection Analysis

Layer of Protection Analysis (LOPA) is a semi-quantitative tool for analyzing and assessing risk. LOPA typically uses order-of-magnitude categories for initiating event frequency, consequence severity, and the probability of failure of Independent Protection Layers (IPLs) to approximate the risk of a scenario. LOPA is an analysis tool that typically builds on information developed during a qualitative hazard evaluation, such as a HAZOP.

Like many other hazard analysis methods, the primary purpose of LOPA is to determine if there are sufficient layers of protection against an accident scenario (can the risk be tolerated?). A scenario may require one or many protection layers depending on the process complexity and potential severity of a consequence.

LOPA provides a consistent basis for judging whether there are sufficient IPLs to control the risk of an accident for a given scenario. If the estimated risk of a scenario is not acceptable, additional IPLs may be added. Alternatives encompassing inherently safer design can be evaluated as well. LOPA does not suggest which IPLs to add or which design to choose, but it assists in judging between alternatives for risk mitigation. LOPA is not a fully quantitative risk assessment approach, but is rather a simplified method for assessing the value of protection layers for a well-defined accident scenario.

4. ANALYSIS TECHNIQUES

LOPA can be effectively used at any point in the life cycle of a process or a facility, but it is most frequently used during:
- The design stage when the process flow diagram and P&IDs are essentially complete
- Modifications to an existing process or its control or safety systems
- The regular cycle of Process Hazard Analyses performed on a process

Table 4.14 provides an overview of LOPA requirements and results.

Table 4.14   LOPA Overview

Typically Used During	Resource Requirements	Type of Results	Advantages and Disadvantages
Detailed engineering Routine operation Expansion or modification	Material, physical, and chemical data Basic process chemistry Process flow diagram Piping and Instrumentation Diagrams High consequence scenarios (often identified during a HAZOP) Established LOPA criteria Cause and Effect Diagram Interlock description	Scenario-based documentation initiating cause, consequence, severity ranking, IPLs, and identification of whether additional IPLs are required to mitigate the risk. Screening tool to identify the need for more thorough, detailed analysis, such as QRA.	Requires less time than quantitative risk analysis. Provides a consistent basis for determining the estimated frequency of consequence based on event frequencies and reliability of independent protection layers. Provides more defensible comparative risk judgments than qualitative methods. Helps identify operations and practices that were previously thought to have sufficient safeguards. LOPA is not intended to be a scenario identification tool. LOPA is not intended to be used as a replacement for detailed quantitative analysis.

Detailed information on LOPA can be found in *Layer of Protection Analysis - Simplified Process Risk Assessment* (Ref. 4-3).

4.2.3.2   Failure Modes and Effects Analysis

The purpose of an FMEA is to identify single equipment and system failure modes and each failure mode's potential effect(s) on the system or plant. This analysis typically generates recommendations for increasing equipment reliability, thus improving process safety.

The failure mode describes how equipment fails to provide the function the user expects. Using a pump as an example, the pump fails to start, stop, or pump at expected head, fails to contain the process, or fails to run at expected intervals without maintenance. The effect of the failure mode is the evidence a failure has occurred (e.g., visible leak, low pressure, etc.). An FMEA identifies single failure modes that either directly result in or contribute significantly to an incident. Human operator error is usually not examined directly in an FMEA; however, the consequences of inadequate design, improper installation, lack of maintenance, or improper operation are usually manifested as an equipment failure mode.

Failure Modes and Effects Analysis evaluates how equipment can fail (or be improperly operated) and the consequences these failures can have on a process. These failure descriptions provide analysts with a basis for determining where changes can be made to improve a system design. During an FMEA, hazard analysts describe potential consequences and relate them only to equipment failures; they rarely investigate damage or injury that could arise if the system operated successfully. An FMEA is not as efficient as other methods such as HAZOP studies in identifying an exhaustive list of combinations of equipment failures that lead to incidents, since it examines all failure modes that result in safe outcomes as well as those that can lead to or contribute to loss events.

Each individual failure is considered as an independent occurrence, with no relation to other failures in the system, except for the subsequent effects that it might produce. However, in special circumstances, common cause failures of more than one system component may be considered.

The results of an FMEA are usually listed in tabular format, equipment item by equipment item. Generally, hazard analysts use FMEA as a qualitative technique, although it can be extended to give a priority ranking based on failure severity. Proactive tasks, put in place as a result of an FMEA, reduce the likelihood of an initiating event and thus lower the likelihood of a process safety incident. Table 4.15 provides an overview of FMEA requirements and results.

Table 4.15   FMEA Overview

Typically Used During	Resource Requirements	Type of Results	Advantages and Disadvantages
Conceptual engineering Detailed engineering Routine operation Expansion or modification	Material, physical, and chemical data Basic process chemistry Process flow diagram Piping and Instrumentation Diagrams	Identified failures and safeguards.	Designed to analyze potential equipment failures. Not a team approach. Experience of analyst is essential.

Detailed information on performing an FMEA can be found in *Guidelines for Hazard Evaluation Procedures* (Ref. 4-1).

4. ANALYSIS TECHNIQUES

4.2.4 Quantitative

Process quantitative risk analysis is a methodology designed to provide management with a tool to help evaluate overall process safety in the chemical process industry. Management systems such as engineering codes, checklists, and Process Safety Management (PSM) provide layers of protection against accidents. However, the potential for serious incidents cannot be totally eliminated. Quantitative risk analysis provides a quantitative method to evaluate risk and to identify areas for cost-effective risk reduction. This section provides an overview of quantitative risk analysis. For further detail, see *Guidelines for Hazard Evaluation Procedures* (Ref. 4-1) and *Guidelines for Chemical Process Quantitative Risk Assessment* (Ref. 4-2).

A quantitative risk analysis examines a range of possible incident outcomes for a given loss event, such as by the use of event trees to evaluate the probability of success or failure of each applicable mitigative safeguard and the overall risk of each resulting scenario. Techniques used as inputs to Quantitative Risk Analyses (QRAs) include:
- Fault Tree
- Event Tree

4.2.4.1 Fault Tree

Fault Tree Analysis (FTA) is a deductive technique that focuses on one particular incident or main system failure and provides a method for determining causes of that event. The purpose of an FTA is to identify combinations of equipment failures and human errors that can result in an incident. FTA is well suited for analyses of highly redundant systems. For systems particularly vulnerable to single failures that can lead to incidents, it is better to use a single-failure-oriented technique such as FMEA or HAZOP Study. FTA is often employed in situations where another hazard evaluation technique (e.g., HAZOP Study) has pinpointed an important incident of interest that requires more detailed analysis.

The fault tree is a graphical model that displays the various combinations of equipment failures and human errors that can result in the main system failure of interest (called the top event). This allows the hazard analyst to focus preventive or mitigative measures on significant basic causes to reduce the likelihood of an incident.

Fault Tree Analysis is a deductive technique that uses Boolean logic symbols (i.e., AND gates, OR gates) to break down the causes of a top event into basic equipment failures and human errors (called basic events). The analyst begins with an incident or undesirable event that is to be avoided and identifies the immediate causes of that event. Each of the immediate causes (called fault events) is further examined in the same manner until the analyst has identified the basic causes of each fault event or reaches the boundary established for the analysis. The resulting fault tree model displays the logical relationships between basic events and the selected top event.

Top events are specific hazardous situations that are typically identified through the use of a more broad-brush hazard evaluation technique (e.g., What-If Analysis, HAZOP study). A fault tree model can be used to generate a list of the failure combinations (failure modes) that can cause the top event of interest. These failure modes are known as cut sets. A Minimal Cut Set (MCS) is a smallest combination of component failures which, if they all occur or exist simultaneously, will cause the top event to occur. Such combinations are the "smallest" combinations in that all of the failures in a MCS must occur if the top event is to occur as a result of that particular MCS. For example, a car will not operate if the cut set "no fuel" and "broken windshield" occurs. However, the MCS is "no fuel" because it alone can cause the top event; the broken windshield has no bearing on the car's ability to operate. Sometimes analysts may include special conditions or circumstantial events in a fault tree model (e.g., the existence of a certain plant operating condition). Thus, a list of minimal cut sets represents the known ways the undesired consequence can occur, stated in terms of equipment failures, human errors, and associated circumstances.

The fault tree is a graphical representation of the relationships between failures and a specific consequence. Fault events and basic events representing failures of equipment or humans (hereafter, both equipment and humans are referred to as components) can be divided into failures and faults. A component failure is a malfunction that requires the component to be repaired before it can successfully function again. For example, when a pump shaft breaks, it is classified as a component failure. A component fault is a malfunction that will "heal" itself once the conditions causing the malfunction are corrected. An example of a component fault is a switch whose contacts fail to operate because they are wet and when the contacts are dried they operate properly.

Whether a component malfunction is classified as a fault or a failure, a basic assumption of Fault Tree Analysis is that all components are in either a failed state or a working state. Analysis of several degraded operating states is generally not practical. Analysts must define the conditions of failure and success for each event used in a fault tree model.

Detailed information on performing a Fault Tree Analysis can be found in *Guidelines for Hazard Evaluation Procedures* (Ref. 4-1).

4.2.4.2 Event Tree

An event tree graphically shows all of the possible outcomes following the success or failure of protective systems given the occurrence of a specific initiating cause (equipment failure or human error). Event trees are also used to study other events, such as starting at a loss event and evaluating mitigation systems.

Event trees are used to identify various outcomes that can result from a specific initiating event. After these individual event sequences are identified, the specific combinations of failures that can lead to the outcomes can then be determined using Fault Tree Analysis.

Detailed information on performing an Event Tree Analysis can be found in *Guidelines for Hazard Evaluation Procedures* (Ref. 4-1).

4.2.5 Human Factors

Human factors involve designing machines, operations, and work environments so that they match human capabilities, limitations, and needs. It is based on the study of operators, managers, maintenance staff, and other people in the work environment and of factors that generally influence humans in their relationship with a technical installation. It is now recognized that such factors go well beyond basic ergonomics and operator-machine interface considerations and include aspects of a safety culture such as

4. ANALYSIS TECHNIQUES

management leadership and commitment, clear communication of expectations, and operating discipline.

Although control systems achieve a high degree of automation, the process operator still has the overall immediate responsibility for safe and economic operation of the process. Opinions differ as to the extent to which the function of safety shutdown or other response to abnormal situations should be removed from the operator and assigned to an instrumented protective system. In general, the greater the hazards are, the stronger is the argument for protective instrumentation. Whatever approach is adopted, the operator still has the vital function of running the plant so that control is maintained when possible and operator action is taken when needed to avoid a loss event. The job of the process operator is therefore a crucial one and therefore should be considered when conducting a hazard evaluation.

In today's world, there are tools to allow a thorough review of human factors as they relate to the process being studied. Most often, facilitators feel that the "human" aspect of process safety will be covered throughout the study as they go through their nodes, deviations, and guide words and therefore feel there is no reason to review this topic separately. This may be the case in very detailed hazard evaluations done on procedure-based operations and batch process operating procedures; however, in a hazard evaluation on a continuous process, it's important to address human factors specifically as a separate line item. Some companies recommend a human factors engineer be part of the evaluation team. API developed a tool to assist facilitators and teams in identifying and evaluating human factors issues associated with specific equipment types (Ref. 4-33).

There are other techniques used to conduct hazard analysis, including some that focus on human factor issues, such as:
- Critical Task Analysis
- Human Reliability Analysis

4.2.5.1 Critical Task Analysis

Critical Task Analysis is a systematic method of identifying critical tasks within a process facility, prioritizing their importance, analyzing those tasks that are considered most critical, and identifying appropriate safeguards to mitigate the risk. This human error analysis method requires breaking down a procedure or overall task into unit tasks. It involves determining the detailed performance required of people and equipment and determining the effects of environmental conditions, malfunctions, and other unexpected events. A Critical Task Analysis consists of several steps:
- Identify each sequential activity to be performed for the task
- Document possible errors in performing the tasks
- Document the consequences of those errors
- Identify human factors concerns
- Identify potential solutions and mitigations

For further information, see *Ergonomic Solutions for the Process Industries* (Ref. 4-30).

4.2.5.2 Human Reliability Analysis

A Human Reliability Analysis (HRA) is a systematic evaluation of the factors that influence the performance of operators, maintenance staff, technicians, and other plant personnel. It involves one of several types of task analyses; these types of analyses describe a task's physical and environmental characteristics, along with the skills, knowledge, and capabilities required of those who perform the tasks. An HRA will identify error-likely situations that can cause or lead to incidents. An HRA can also be used to trace the causes of human errors. An HRA is usually performed in conjunction with other hazard evaluation techniques.

The purpose of conducting an HRA is to identify potential human errors and their effects or to identify the underlying causes of human errors. An HRA systematically lists the errors likely to be encountered during normal or emergency operation, factors contributing to such errors, and proposed system modifications to reduce the likelihood of such errors. The results are qualitative, but may be quantified. The analysis includes identifying system interfaces affected by particular errors, and ranking these errors in relation to the others, based on probability of occurrence or severity of consequences. The results are easily updated for design changes or system, plant, or training modifications. A worked example showing HRA results can be found in *Guidelines for Hazard Evaluation Procedures* (Ref. 4-1).

4.2.6 Selecting the Appropriate Technique

Each hazard evaluation technique has unique strengths and weaknesses. Understanding these strengths and weaknesses is important in selecting the appropriate hazard evaluation technique. The process of selecting an appropriate hazard evaluation technique may be difficult for an inexperienced facilitator because the "best" technique may not be apparent. As hazard analysts gain experience with various hazard evaluation methods, the task of choosing an appropriate technique becomes easier and somewhat instinctive. Factors to consider when selecting an appropriate technique include:

- Motivation for the study
- Type of results needed
- Type of information available to perform the study
- Characteristics of the analysis problem
- Perceived risk associated with the subject process or activity
- Resource availability and analyst / management preference

Figure 4.2 illustrates the process for selecting the appropriate hazard analysis technique. This selection process can be used in conjunction with a decision flowchart in *Guidelines for Hazard Evaluation Procedures* (Ref. 4-1).

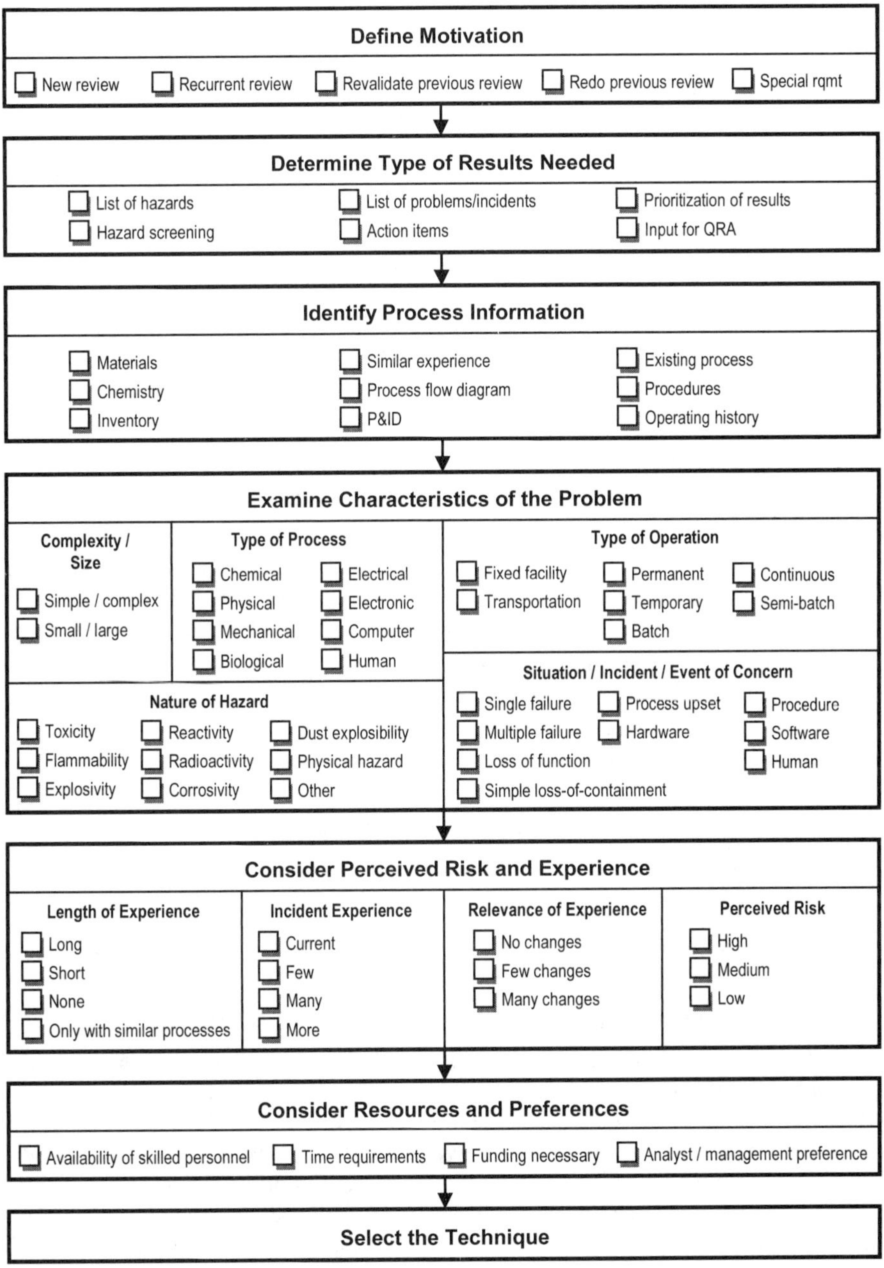

4. ANALYSIS TECHNIQUES

Define Motivation
- ☐ New review ☐ Recurrent review ☐ Revalidate previous review ☐ Redo previous review ☐ Special rqmt

Determine Type of Results Needed
- ☐ List of hazards ☐ List of problems/incidents ☐ Prioritization of results
- ☐ Hazard screening ☐ Action items ☐ Input for QRA

Identify Process Information
- ☐ Materials ☐ Similar experience ☐ Existing process
- ☐ Chemistry ☐ Process flow diagram ☐ Procedures
- ☐ Inventory ☐ P&ID ☐ Operating history

Examine Characteristics of the Problem

Complexity / Size
- ☐ Simple / complex
- ☐ Small / large

Type of Process
- ☐ Chemical ☐ Electrical
- ☐ Physical ☐ Electronic
- ☐ Mechanical ☐ Computer
- ☐ Biological ☐ Human

Type of Operation
- ☐ Fixed facility ☐ Permanent ☐ Continuous
- ☐ Transportation ☐ Temporary ☐ Semi-batch
- ☐ Batch

Nature of Hazard
- ☐ Toxicity ☐ Reactivity ☐ Dust explosibility
- ☐ Flammability ☐ Radioactivity ☐ Physical hazard
- ☐ Explosivity ☐ Corrosivity ☐ Other

Situation / Incident / Event of Concern
- ☐ Single failure ☐ Process upset ☐ Procedure
- ☐ Multiple failure ☐ Hardware ☐ Software
- ☐ Loss of function ☐ Human
- ☐ Simple loss-of-containment

Consider Perceived Risk and Experience

Length of Experience
- ☐ Long
- ☐ Short
- ☐ None
- ☐ Only with similar processes

Incident Experience
- ☐ Current
- ☐ Few
- ☐ Many
- ☐ More

Relevance of Experience
- ☐ No changes
- ☐ Few changes
- ☐ Many changes

Perceived Risk
- ☐ High
- ☐ Medium
- ☐ Low

Consider Resources and Preferences
- ☐ Availability of skilled personnel ☐ Time requirements ☐ Funding necessary ☐ Analyst / management preference

Select the Technique

Figure 4.2 Criteria for Selecting Hazard Evaluation Technique

4.3 RISK ASSESSMENT

Risk assessment plays an important role in the engineering design process by providing a tool to evaluate design concept alternatives, determine the suitability of a proposed location given the surrounding populations, evaluate locations for emergency isolation, etc.

The CCPS *Guidelines for Chemical Process Quantitative Risk Analysis* (*CPQRA Guidelines*) defines risk management as (Ref. 4-2):

- *Risk Management* - The systematic application of management policies, procedures, and practices to the tasks of analyzing, assessing, and controlling risk in order to protect employees, the general public, and the environment as well as company assets while avoiding business interruptions.

The keys to the implementation of a risk management program are the activities of risk analysis and risk assessment, which are defined in the *CPQRA Guidelines* as:

- *Risk Analysis* - The development of a quantitative estimate of risk based engineering evaluations and mathematical techniques for combining estimates of incident consequences and frequencies.
- *Risk Assessment* - The process by which the results of a risk analysis (i.e., risk estimates) are used to make decisions, either through relative ranking of risk reduction strategies or through comparison with risk targets (*risk criteria*).

These activities are discussed in greater detail in the *CPQRA Guidelines*. While the distinction between risk analysis and risk assessment is important within the context of the *CPQRA Guidelines*, these *Guidelines* use the single term 'risk assessment' to aggregate the following activities that are essential to understanding the hazards in a chemical process:

- What are the hazards / what can go wrong (scenario)?
- How severe could it be (consequence)?
- How often could it happen (frequency)?
- How do consequence and frequency combine (risk)?
- Is the current level of risk tolerable?
- If not, what needs to be done to reduce and control the risk?

In a QRA, the estimated consequences and the estimated frequency of each scenario are combined to estimate the associated risk. Having an estimate of the level of risk associated with an activity does not directly control or reduce that risk. Improvements in safety (i.e., risk reduction), when necessary, require making a decision to change something, followed by action to effect that change. The primary reason to examine risk is to assist in making such decisions. Risk criteria are useful in examining and judging the significance of risk. Producing risk results without having understandable criteria for judging them would be like a teacher giving grades without providing a way to understand what an A, B, or F means.

Industry performs risk assessments because society increasingly expects industry to be cognizant of, and to do a responsible job of controlling, the risks of its operations.

Specific benefits from risk assessment as part of a risk management system include:
- Providing a clear process and concrete criteria, increasing confidence that risk management decisions are rationally determined and not the result of arbitrary decisions
- Providing a basis for prioritizing / apportioning finite resources (providing the best mix of expenditures to minimize total risk across the company)
- Assisting in the evaluation of the relative benefits of risk reduction alternatives
- Helping define which level of the organization should take responsibility for the decisions that affect the risk (i.e., higher risk decisions made at higher levels)
- Helping protect the organization's permission to operate (actual or figurative) and enhancing the sustainability of the business
- Yielding a better understanding of the management of the risk

4.3.1 Technical Aspects of QRA

As discussed in the *Guidelines for Chemical Process Quantitative Risk Analysis* (Ref. 4-2), QRA is a methodology designed to give management a tool to help evaluate overall process risk. Other aspects of risk management, such as the implementation of a risk-based process safety management system as described in the *Guidelines for Risk Based Process Safety* (Ref. 4-34), may provide layers of protection against process incidents. However, the potential for serious incidents cannot be eliminated. QRA provides a quantitative method to evaluate risk and to identify areas for effective risk reduction.

A basic understanding of QRA methodology may be of value in helping understand the application of risk criteria. Figure 4.3 illustrates the basic steps in a QRA.

4.3.1.1 *Consequence / Impact Assessment*

The incidents of concern within the process industries are often, but not always, associated with the loss of containment of material from the process. The material has hazardous properties, which might include toxicity and energy content (e.g., thermal, pressure, or potential combustion energy). Typical incident scenarios might include the rupture or break of a pipeline, a hole in a tank, a runaway reaction in a vessel, fire external to the vessel causing a relief valve to open, an operator erroneously opening a vent or drain valve, etc.

Once the incident is defined, a source model(s) is selected to describe how materials are discharged from the process. The source model provides a description of the rate of discharge, the total quantity discharged (or total time of discharge), and the state of the discharge (solid, liquid, vapor, or a combination). Typically, a dispersion model is subsequently used to describe how the material is transported downwind and mixes with air to some concentration level.

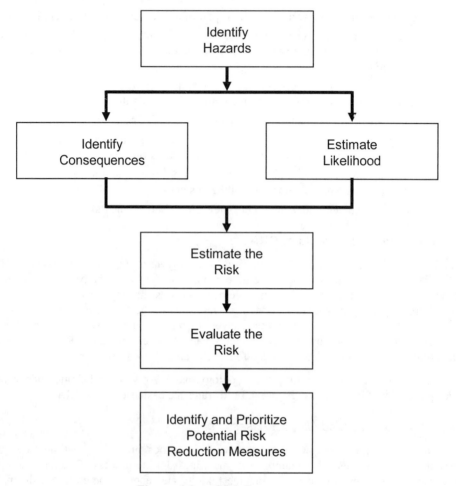

Figure 4.3 QRA Process

For toxic releases, effect models consider the concentration and duration of exposure and the mode of physiological impact to convert these incident-specific results into effects on people (injury or death). For flammable releases, fire and explosion models convert information on the concentration and mass of material present (and, perhaps, information describing the physical environment of the flammable cloud) into energy hazard potentials such as thermal radiation and explosion overpressures. Other effect models are then used to estimate effects on people and structures.

Additional refinement to consequence estimates may be provided by consideration of mitigation factors, such as isolation systems that might reduce the duration of the release or water sprays, foam systems, and sheltering or evacuation that may reduce the magnitude of potential effects.

Figure 4.4 shows an overall logic diagram for consequences models for releases of volatile, hazardous substances.

4. ANALYSIS TECHNIQUES

For additional guidance on consequence modeling, refer to *Guidelines for Chemical Process Quantitative Risk Assessment* (Ref. 4-2), *Guidelines for Evaluating the Characteristics of Vapor Cloud Explosions, Flash Fires, and BLEVEs* (Ref. 4-7), and *Guidelines for Evaluating Process Plant Buildings for External Explosions and Fires* (Ref. 4-35).

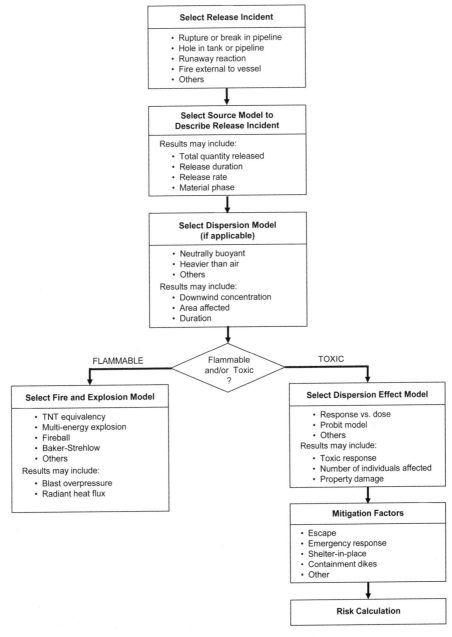

Figure 4.4 Consequence Analysis Flowchart

4.3.1.2 Frequency Assessment

Guidelines for Chemical Process Quantitative Risk Assessment (Ref. 4-2) provides detailed information on the most common techniques to answer the question: *"How often might this incident scenario occur?"* Frequency assessment techniques include:

- Review of historical records of similar events
- Fault tree analysis
- Event tree analysis
- Layer of protection analysis
- External event analysis
- Common cause failure analysis
- Human reliability analysis

In performing frequency analyses, it is often difficult to determine the appropriate level of detail needed to provide sufficient information to make the necessary risk-based decision. Typically, a phased approach, such as the following, should be considered in performing frequency analysis:

- Perform a qualitative study using, for example, HAZOP or What-if Analysis to identify potential initiating events that could lead to incident scenarios of interest.
- For initiating events of interest, prepare an event tree to further develop the scenarios; e.g., showing the various outcomes which could result based upon the success or failure of relevant protective features.
- Use techniques such as fault tree analysis or the review of historical records to estimate the initiating event frequencies and branch point probabilities for each scenario.
- Calculate the frequency estimates for each scenario outcome by multiplying the initiating event frequency by the appropriate branch point probabilities.

Often simplified frequency analyses are performed by providing estimates at the branch point levels without using detailed Fault Tree Analysis.

4.3.1.3 Developing a Comprehensive QRA

Terms like "worst case scenario" or "worst credible scenario" often creep into discussions of risk assessments. Every scenario has a frequency and consequence (and, therefore, risk) associated with it, and the significance of that scenario cannot be determined until the risk has been quantified (or at least estimated). As mentioned earlier, most high consequence accident scenarios occur at a relatively low frequency. Examples of events that are normally categorized as "worst credible scenarios" include full-bore ruptures of liquid lines and pressure vessel failures. Whether or not these accident scenarios may actually be the worst from a consequence standpoint, their risk significance would not be clear unless a full spectrum of typical accident scenarios is assessed (i.e., including more moderate consequence events that are more frequent). For example, a medium size pipe rupture that results in a vapor cloud explosion may have a lower consequence compared to a pressure vessel rupture, but the risk may be determined to be higher once all scenarios are quantified. For this reason, it is recommended that a representative range of identified scenarios should be evaluated in a QRA.

4.3.1.4 Standardization of Approach

Consequence and frequency assessments are complex and evolving topics and, often, there can be divergent opinions, even among the experts, as to the best way to model a particular scenario. In fact, past comparative studies involving multiple independent analysts modeling the same scenario have yielded results with outliers which range one or two orders of magnitude. However, once the teams were coached to use similar assumptions, the results converged to within an acceptable range (i.e., within a factor of 5) (Ref. 4-2).

Certainly, if an organization commits to the use of QRA, well-defined guidelines regarding assumptions and techniques need to be established to promote consistency. Companies are increasingly striving to standardize across all locations so that any company-wide risk-based decision making is less likely to become an exercise in comparing apples to oranges. Corporate QRA guidance could provide standardized data and analytical approaches addressing topics such as:

- Equipment failure rate data
- Human error rates
- Toxicity dose-response relationships
- Physiological response to thermal exposures (fires) and explosion overpressure
- Structural analyses to determine building damage in response to explosions and resultant occupant vulnerability
- Analytical techniques for source term and dispersion modeling
- Assumptions made regarding credit to be given for mitigation design features / activities such as sheltering-in-place, remote isolation of leaks, water spray systems, etc.
- Selection of modeling software
- Training / qualification requirements for QRA analysts

Similarly, in some jurisdictions where QRA is mandated, regulatory authorities may require and implement such standardization through the prescription of standardized protocols (perhaps including scenario definitions) to minimize variability of results for similar situations between different organizations and analysts.

4.3.2 Risk Criteria

4.3.2.1 Qualitative Risk Criteria

The previous section discussed determining the severity of the consequences and potential frequency of the identified hazard scenario. After a hazard evaluation team or process risk analyst has estimated the severity and frequency, a scenario risk estimate can be made.

Scenario Frequency x Scenario Severity = Scenario Risk

This scenario risk estimate can then be used to determine whether the existing safeguards are adequate to control the scenario risk. This must, of course, be repeated for every scenario with consequences of concern. Many companies use risk matrices to perform this qualitative approach. Figure 4.5 contains an example risk matrix. Further detail on qualitative risk criteria can be found in *Guidelines for Hazard Evaluation Procedures, Chapter 7, Risk-Based Determination of the Adequacy of Safeguards* (Ref. 4-1).

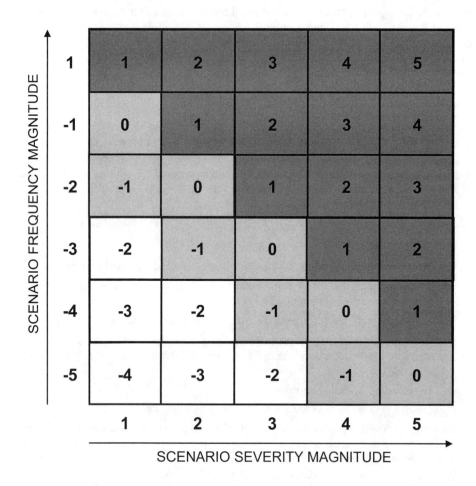

Figure 4.5 Example Risk Matrix Using Order-of-Magnitude Frequency and Severity Categories

4.3.2.2 Quantitative Risk Criteria

Before beginning a detailed discussion of risk criteria, it is necessary to define the risk measures to which the risk criteria apply. Experience has shown that, to get a balanced perspective of the risks associated with process plant operations, it is necessary to evaluate risks from two perspectives:

1. The risk to individuals
2. The risk to groups of people

These are referred to, respectively, as individual and societal risk.

4. ANALYSIS TECHNIQUES

There are many diverse measures of individual and societal risk. Those addressed here are the most commonly applied in the process industries. Readers seeking a broader perspective may wish to consult *Guidelines for Developing Quantitative Safety Risk Criteria* (Ref. 4-36) for other examples of risk measures and formats for their presentation.

4.3.2.3 Individual Risk

Individual risk expresses the risk to a single person exposed to a hazard; i.e., an individual in the potential effect zone of an incident or set of incidents. The scale of any incident, in terms of the number of people impacted by a single event, does not affect individual risk. Individual risk measures can be single numbers, tables of numbers, or various graphical summaries. Commonly used individual risk measures include (Ref. 4-2):

- Individual risk
- Maximum individual risk
- Average individual risk (exposed population)
- Average individual risk (total population)
- Average individual risk (exposed hours / worked hours)

Other bases for calculating individual risk have been used. Considering the multiplicity of individual risk measures, it is important that there is consistency between the manner in which individual risk is calculated and the basis upon which the risk criteria are defined.

The individual risk measures described above are normally expressed as the frequency of fatal injuries per year. While all injuries are of concern, effect models for predicting degrees of injury often entail additional uncertainties; thus risk analysts often estimate the risk of fatal injury (death) as a less equivocal measure. As will be noted later, there also are more comparative historical data available that can be used to calibrate the risk criteria against the risk of death associated with other hazardous activities.

Other individual risk measures that have been used include the *Fatal Accident Rate* (FAR) (Ref. 4-37). FAR is the estimated number of fatalities per 10^8 exposure hours. The FAR is a single number index that is directly proportional to the average individual risk. To calculate the FAR, multiply the average individual risk by a factor of $10^8/(24 \times 365) = 1.14 \times 10^4$.

Individual risk can be calculated for either onsite personnel (e.g., employees or contractors) or members of the offsite public. Calculations of public risks can introduce considerations that are either unique or more severe than for onsite risk calculations. For example:

- Offsite populations may include individuals who, because of their age or general health condition (e.g., young children or the elderly), may be more vulnerable to health impacts and may be less capable of responding to protect themselves.
- Offsite individuals are less likely to be aware of process hazards and the means to protect themselves from those hazards and, most likely, will not have the same sorts of protective equipment that onsite personnel would have.

- Offsite individuals, while perhaps further removed from the hazards, may be exposed for a greater percentage of the time (e.g., stay-at-home residents who may be at risk nearly 100% of the time). Conversely, depending upon the nature of offsite developments, there may be individuals whose risk exposure is transient and brief (e.g., visitors to a park adjacent to a chemical or petroleum facility).

The calculation of *individual risk* is made with the understanding that the contributions of all incident outcome cases (i.e., event sequences) are additive. For example, the total *individual risk* to an individual working at a facility is the sum of the risks from all potentially harmful incidents considered separately, i.e., the sum of all risks due to fires, explosions, toxic chemical exposures, etc., to which the individual might be exposed.

4.3.2.4 Societal Risk

Societal risk measures the potential for effects to a group of people located in the effect zone of an incident or set of incidents. Thus, societal risk estimates include a measure of incident scale in terms of the number of people impacted. Some societal risk measures are designed to reflect the observation that society tends to be more concerned about the risk of large (i.e., multi-fatality) incidents than small (fewer fatality) incidents, and may assign disproportionately greater significance to large incidents. This potential risk aversion will be discussed further when addressing risk criteria formulation.

Societal risk measures can be expressed as single number measures, tabular sets of numbers, or graphical summaries, with the most common graphical representation being the F-N (Frequency-Number) curve. An F-N curve is a plot of the frequency distribution of multiple casualty events, where F is the cumulative frequency of all events leading to N or more casualties (typically expressed as the number of fatalities). F-N curves typically use log-log plots since the frequencies and number of fatalities often range over several orders of magnitude.

The calculation of *societal risk* requires the same frequency and consequence information as *individual risk*. Whereas *individual risk* requires details of an individual's occupancy within hazard zones, *societal risk* estimation requires a definition of the number of exposed population within hazard zones. This definition can include factors such as the:

- Number and geographical distribution of the population
- Population type (e.g., residential, school, industrial)
- Probability of people being present (i.e., the number of hours a day people are present)

Traditional emphasis has been on the calculation of societal risk for offsite populations; however, companies are increasingly recognizing the importance of the consideration of group risk for onsite personnel. As with individual risk, societal risk estimates are typically the summation of risk contributions from many incident outcome cases.

4. ANALYSIS TECHNIQUES

4.3.3 Quantitative Risk Assessment

From a manager's perspective, the objective of a QRA is to provide information to decision makers to allow rational risk management decisions. Examples of the application of QRA include:

- Identifying major contributors to risk. As noted above, both individual and societal risk estimates are commonly the summation of risk contributions from many scenarios. It is not unusual for the risks associated with a relatively few scenarios to dominate the sum.
- Identifying and addressing the most significant contributors are effective means of stewarding risk reduction resources.
- Comparison of risk management alternatives. QRAs are often used to evaluate the risk reduction benefits of one alternative relative to others. Options that represent the most prudent investment of risk reduction resources can be identified.
- Comparison to risk of an existing operation. A company may seek to compare the risk of several of its operations without making absolute judgments.
- Defining approval levels under the risk elevation principle. Organizations may require that higher risk activities be sanctioned at higher levels of authority within the organization.
- Making hard decisions about the tolerability of risk. QRAs can form the basis for absolute go / no-go decisions regarding a particular course of action or provide significant input into the deliberation of *"Have we done enough to reduce the risk?"*
- Achieving regulatory compliance. Some regulatory authorities may require QRAs to justify the initiation or continuation of certain hazardous activities.

Of the seven general applications above, it should be noted that only the last two require the use of risk criteria (in the former case, promulgated by the company and, in the latter, promulgated by the regulatory authority). While this book is focused on the creation of risk criteria, it is appropriate to acknowledge the importance of those applications of QRA where only relative evaluations of risk are required. In fact, because of the uncertainties inherent in the assumptions made and the models used in QRA, some analysts may be more comfortable with comparative situations where any inaccuracies apply, presumably equally, to all risk estimates so that relative judgments remain valid. However, in the absence of risk criteria, it is often more difficult to determine whether all (or none) of the alternatives being offered represent a tolerable risk.

4.3.4 Risk Tolerance / Decision Making Criteria

Utilization of risk estimates is the process by which the results from a risk analysis are used to make decisions, either through relative ranking of risk reduction strategies or through comparison with specific risk targets.

Refer to *Guidelines for Developing Quantitative Safety Risk Criteria* for more information (Ref. 4-36).

4.4 RELIABILITY / MAINTAINABILITY ANALYSIS

Reliability, maintainability, and quality considerations are usually considered outside the scope of most hazard evaluations, except as they may also have personnel safety, environmental damage, or property loss consequences. However, the reason for their exclusion is generally in an attempt to reduce the scope and time requirements for the hazard evaluations. Most scenario-based hazard evaluation methodologies are good tools for studying reliability, maintainability, and quality issues in process facilities. This has always been the case, for example, with the hazard and operability study method, by its very name. The original development and usage of HAZOP studies was for the purpose of studying both hazard and operability considerations, and the cost-effectiveness of using the method to find and correct (or avoid) operability problems was demonstrated many times over.

A variation on HAZOP studies known as HAZROP (for hazard, reliability, and operability) has been used to combine HAZOP studies with reliability-centered maintenance. Likewise, modifications to the FMEA methodology known as FMECA (Failure Modes, Effects, and Criticality Analysis) and FMEDA (Failure Modes, Effects, and Diagnostic Analysis) are used for product quality and reliability purposes and as six-sigma tools. FMEA can be used with a quality, reliability, and / or safety focus, making it possible to meet multiple objectives with one FMEA.

The most effective means of considering reliability, operability / maintainability, and quality impacts in the context of hazard evaluations is to quantify the consequences of abnormal situations in the same terms or categories that property damage impacts from fires and explosions would be considered. For example, if consequence severity categories of order-of-magnitude total costs were used (such as $10,000, $100,000, $1 million, etc.), then the same scale should be used for assessing the loss potential of events such as compressor bearing failures or product batches being off-quality. The impact should also take into consideration mitigating factors such as the availability of a spare compressor or the ability to rework bad product.

Recommendations for reliability, quality, or operability / maintainability improvements are often recorded and tracked separately from safety and environmental hazard evaluation findings. In this way, their benefit can be assessed by the business unit on each recommendation's own cost-benefit merits, taking feasibility and resource constraints into account. Many recommendations will be made not on a risk reduction basis but as straightforward changes or improvements to a final design (e.g., installing a low point drain where one was omitted) or an operating procedure (e.g., adding a means of improving batch-to-batch consistency).

4.5 REFERENCES

4-1. CCPS. *Guidelines for Hazard Evaluation Procedures*. Center for Chemical Process Safety of the American Institute of Chemical Engineers. New York, NY. 2008.

4-2. CCPS. *Guidelines for Chemical Process Quantitative Risk Analysis, Second Edition*. Center for Chemical Process Safety of the American Institute of Chemical Engineers. New York, NY. 2000.

4-3. CCPS. *Layer of Protection Analysis - Simplified Process Risk Assessment*. Center for Chemical Process Safety of the American Institute of Chemical Engineers. New York, NY. 2001.

4-4. CCPS. *Guidelines for Process Safety in Batch Reaction Systems*. Center for Chemical Process Safety of the American Institute of Chemical Engineers. New York, NY. 1999.

4-5. CCPS. *Guidelines for Fire Protection in Chemical, Petrochemical, and Hydrocarbon Processing Facilities*. Center for Chemical Process Safety of the American Institute of Chemical Engineers. New York, NY. 2003.

4-6. Crowl, D.A. *Understanding Explosions*. Center for Chemical Process Safety for the American Institute of Chemical Engineers. New York, NY. 2003.

4-7. CCPS. *Guidelines for Vapor Cloud Explosion, Pressure Vessel Burst, BLEVE and Flash Fire Hazards, Second Edition*. Center for Chemical Process Safety. New York, NY. 2010.

4-8. NOAA. *Chemical Reactivity Worksheet, Version 2.1*. National Oceanic and Atmospheric Administration. http://response.restoration.noaa.gov/CRW.

4-9. API RP 2003. *Protection Against Ignitions Arising out of Static, Lightning and Stray Currents*. American Petroleum Institute, Washington, D.C. 1991.

4-10. NFPA 55. *Compressed Gas Code*. National Fire Protection Association, Quincy, MA. 2010.

4-11. NFPA 400. *Hazardous Material Code*. National Fire Protection Association, Quincy, MA. 2010.

4-12. NFPA 69. *Explosion Prevention Systems*. National Fire Protection Association, Quincy, MA. 1986.

4-13. NFPA 70. *National Electrical Code*. National Fire Protection Association, Quincy, MA. 2011.

4-14. NFPA 77. *Static Electricity*. National Fire Protection Association, Quincy, MA. 1988.

4-15. NFPA 78. *Lightning Protection Code*. National Fire Protection Association, Quincy, MA. 1989.

4-16. NFPA 497M. *Manual for Classification of Gases, Vapors and Dusts for Electrical Equipment in Hazardous (Classified) Locations*. National Fire Protection Association, Quincy, MA. 1991.

4-17. Withers, R.M.J. and Lees, F.P., *The Assessment of Major Hazards: The Lethal Toxicity of Chlorine, Parts 1 and 2*. Journal of Hazardous Materials, 12(3). 1985.

4-18. Lees, F.P. *Loss Prevention in the Process Industries, Third Edition*. Elsevier, Inc. Oxford, UK. 2005.

4-19. CCPS. *Guidelines for Safe Storage and Handling of High Toxic Hazard Materials*. Center for Chemical Process Safety of the American Institute of Chemical Engineers. New York, NY. 1988.

4-20. ANSI / API STD 521. *Pressure-Relieving and Depressuring Systems, Fifth Edition*. American Petroleum Institute. Washington, D.C. 2007.

4-21. DOE. *Protective Action Criteria (PAC) Values*. Subcommittee on Consequence Assessment and Protective Actions (SCAPA) of the Department of Energy (DOE). www.atlintl.com/DOE/teels/teel.html

4-22. EPA. *Acute Exposure Guideline Levels (AEGLs)*. Environmental Protection Agency. www.epa.gov/oppt/aegl/index.htm

4-23. AIHA. *Emergency Response Planning Guidelines and Workplace Environmental Exposure Level Guides*. American Industrial Hygiene Association. Fairfax, VA. www.aiha.org

4-24. DOE. *Temporary Emergency Exposure Limit (TEEL) Data Sets*. Department of Energy Office of Emergency Management. http://orise.orau.gov/emi/scapa/chem-pacs-teels/default.htm

4-25. NIOSH. *Publication No. 94-116: NIOSH Pocket Guide to Chemical Hazards*. US Department of Health and Human Services. Washington, D.C. 1994.

4-26. Baldini. R., Komosinsky, P. *Consequence Analysis of Toxic Substance Clouds*. New Jersey Department of Environmental Protection. Trenton, NJ. 1988.

4-27. EPA. *RMP Offsite Consequence Analysis Guidance*. Environmental Protection Agency. Washington, D.C. 1996.

4-28. EPA. *Accidental Release Prevention Requirements: Risk Management Programs Under Clean Air Act Section 112(r)(7)*. 40 CFR Part 68, U.S. Environmental Protection Agency, June 20, 1996 Fed. Reg. Vol. 61[31667-31730]. www.epa.gov

4-29. HSE. Health and Safety Executive, UK. http://www.hse.gov.uk/hid/haztox.htm (referenced March, 2010)

4-30. Attwood, D.A., Deeb, J.M., and Danz-Reece, M.E. *Ergonomic Solutions for the Process Industries*. Elsevier, Inc. Oxford, UK. 2004.

4-31. U.S. Department of Defense Explosive Safety Board. Technical Paper 14. 2009.

4-32. Baker, W.E., Cox, P.A., Westine, P.S., Kulesz, J.J., and Strehlow, R.A. *Explosion Hazards and Evaluation*. Elsevier. New York, NY. 1983.

4-33. API. *Tool for Incorporating Human Factors during Process Hazard Analysis (PHA) Reviews of Plant Design*. American Petroleum Institute. Washington, D.C. 2004.

4-34. CCPS. *Guidelines for Risk Based Process Safety*. Center for Chemical Process Safety of the American Institute of Chemical Engineers. New York, NY. 2007.

4. ANALYSIS TECHNIQUES

4-35. CCPS. *Guidelines for Evaluating Process Plant Buildings for External Explosions and Fires.* Center for Chemical Process Safety. New York, NY. 1996.

4-36. CCPS. *Guidelines for Developing Quantitative Safety Risk Criteria.* Center for Chemical Process Safety of the American Institute of Chemical Engineers. New York, NY. 2009.

4-37. Vinnem, J.E. *Offshore Risk Assessment: Principles, Modeling and Applications of QRA Studies.* Kluwer Academic Publishers Group. Dordrecht, The Netherlands. 1999.

5

GENERAL DESIGN

This chapter provides design considerations for general design issues. Chapter 6 provides design considerations for specific pieces of equipment. Chapter 7 provides design information on protection layers used to prevent and mitigate incidents.

5.1 SAFEGUARDING STRATEGIES

Process risk management is the term given to collective efforts to manage process risks through a wide variety of strategies, techniques, procedures, policies, and systems that can reduce the hazard of a process, the probability of an accident, or both. In general, the strategy for reducing risk, whether directed toward reducing the frequency or the consequences of potential accidents, can be classified into one of four categories:

- *Inherent* - Eliminating the hazard by using materials and process conditions that are more benign; e.g., substituting water for a flammable solvent.
- *Passive* - Minimizing the hazard through process and equipment design features that reduce either the frequency or consequence of the hazard without the active functioning of any device; e.g., providing a dike around a storage tank of flammable liquids.
- *Active* - Using controls, alarms, safety instrumented systems, and mitigation systems to detect and respond to process deviations from normal operation; e.g., a pump which is shut off by a high level switch in the downstream tank when the tank is 90% full. These systems are commonly referred to as engineering controls, although human intervention is also an active layer.
- *Procedural* - Using policies, operating procedures, training, administrative checks, emergency response, and other management approaches to *prevent* incidents or to *minimize* the effects of an incident; e.g., hot work procedures and permits. These approaches are commonly referred to as administrative controls.

All four categories can contribute to the overall safety of a process. Ideally, the steps of analyzing, reducing, and managing risk will be considered in a hierarchical manner as shown in Figure 5.1.

Inherent safety uses the properties of a material or process to eliminate or reduce the hazard. The fundamental difference between inherent safety and the other three categories is that inherent safety seeks to remove the hazard at the source, as opposed to accepting the hazard and attempting to mitigate the effects. If implementing inherently safer approaches alone to meet project risk goals is feasible, other layers of protection - and their associated costs in time, capital, and expenses - may not be required.

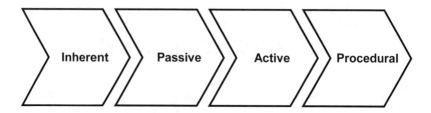

Figure 5.1 Hierarchy of Process Risk Management Strategies

5.1.1 Inherent

Inherently safer design solutions eliminate or mitigate the hazard by using materials and process conditions that are less hazardous. For additional information on the concept of inherently safer chemical processes, see Section 5.2.

Examples of inherently safer solutions include:
- Substituting water for a flammable solvent
- Reducing or eliminating inventories of hazardous intermediates

Continuous metal equipment, such as a steel pipe, is inherently bonded and once it is grounded permanently (such as via multiple steel pilings anchoring the equipment) requires minimal maintenance of ground connections. This is an inherently safer design than one incorporating rubber boots, swivel joints, or other potential breaks in electrical continuity that would require external bond connections and associated maintenance.

A vessel designed to contain the maximum pressure predicted due to any credible upset, such as an internal explosion, is inherently safer than one designed to mitigate the event via other protective means.

In both the above examples, the systems described are inherently safer than some alternative design options. However, they would be better described as passive systems rather than inherently safer. As discussed, true inherently safer designs reduce the hazard by using materials or process conditions that are less hazardous. In the examples, higher levels of inherent safety might be provided by designing the process to eliminate flammable atmospheres that require bonding or equipment reinforcement.

Frequently, both active and procedural design solutions are used to complement each other. For example, in a tank truck bonding procedure, an "active" ground indicating device could be installed to show the presence of a positive ground connection. In such a case, it would still be necessary to ensure that the system is not defeated by simple neglect of an alarm or even bypassing of the indicating device. A ground indicating device might additionally be interlocked with a pump to prevent operator error. For a flame arrester, a complementing procedural system might be monitoring the pressure drop periodically and performing maintenance when a specific differential has been reached (Ref. 5-1).

5. GENERAL DESIGN

5.1.2 Passive

Passive design solutions do not require any device to sense and / or actively respond to a process variable and have very reliable mechanical design. Examples of passive design solutions include:
- Using incompatible hose couplings, non-splash filling using permanently installed dip pipes, permanent grounding, and bonding via continuous metal equipment and pipe rather than with removable cables
- Containing hazardous inventories with a dike that has a bottom sloped to a remote impounding area, which is designed to minimize surface area

Passive designs may be complemented by procedural or active systems, especially where transient conditions are routinely experienced. As an example, a passive system might comprise a permanent dip pipe going to the bottom of a flammable liquid storage tank to avoid splash filling.

Other examples of passive safeguards include:
- Spacing
- Bollards for collision protection

While passive designs typically require less ongoing maintenance than active systems, maintenance is still critical for them to function as intended. For example, a remote impound area to capture a hazardous spill will not be effective if the impound area is allowed to fill with rainwater or breached due to poor maintenance practices.

5.1.3 Active

Active design solutions require devices to monitor a process variable and function to mitigate a hazard.

Frequently active solutions involve a considerable maintenance and procedural component and are therefore typically less reliable than inherently safer or passive solutions. To achieve necessary reliability, redundancy is often used to eliminate conflict between production and safety requirements (such as having to shut down a unit to maintain a relief valve).

Active solutions are sometimes referred to as engineering controls. Examples of active solutions include:
- Using a pressure safety valve or rupture disk to prevent vessel overpressure
- Interlocking a high level sensing device to a vessel inlet valve and pump motor to prevent liquid overfill of the vessel
- Installing a deluge system

Active solutions include pressure relief valves, deflagration vents, explosion suppression systems, fast-acting valves, check valves, and regulators. All these devices require maintenance, operate by responding to a process variable, or both.

5.1.4 Procedural

Procedural design solutions require human intervention to avoid a hazard. This would include following a standard operating procedure or responding to an indication of a problem such as an alarm, an instrument reading, a noise, a leak, or a sampling result. Since an individual is involved in performing the corrective action, consideration needs to be given to human factors issues (Ref. 5-2), e.g., over-alarming, improper allocation of tasks between machine and person, and inadequate support culture. Because of the human factors involved, procedural solutions are generally the least reliable of the four categories.

Procedural solutions are sometimes referred to as administrative controls. Examples of procedural solutions include:
- Following standard operating procedures to keep process operations within established equipment mechanical design limits.
- Completing checklists with sign-offs for certain operations
- Manually closing a feed isolation valve in response to a high level alarm to avoid tank overfilling.
- Executing preventive maintenance procedures to prevent equipment failures.
- Manually attaching bonding and grounding systems.

5.1.5 Characteristics of Design Solution Categories

An illustrative comparison of the four categories of design solutions with respect to several cost and functional attributes appears in Figure 5.2. While procedural solutions can be less complex, they are usually the least reliable. For active solutions, as compared to inherently safer / passive solutions, reliability is typically lower and complexity is greater. Inherently safer / passive solutions tend to have higher associated initial capital outlays; however, operating costs are usually lower than those for the other design solutions. Operating costs are likely to be the greatest for active solutions.

Inherently safer and passive design solutions often overlap. For this reason, the inherently safer and passive solution categories have been combined in the tables presented in the equipment sections of Chapter 6.

An important aspect in the classification of design solutions is the distinction between inherently safer / passive and active systems. It is generally accepted that a containment dike is a passive solution (Ref. 5-3). What about safety devices such as a rupture disk or end-of-line flame arresters? In the case of the rupture disk, it can be argued that it should sense pressure in order to function and therefore would be an active solution. This analogy does not apply so well to end-of-line flame arresters. However, there are many instances of flame arresters that have failed to function or otherwise contributed to hazardous incidents, due to neglect or lack of preventive maintenance.

5. GENERAL DESIGN

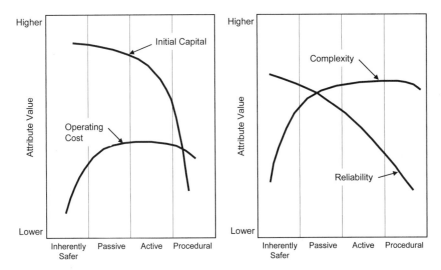

Figure 5.2 Comparison of Cost and Functional Attributes for Design Categories

5.1.6 Safety Factor

Often a safety factor is applied to critical design parameters to ensure that catastrophic failure of systems or components does not occur for unknown reasons. Examples of safety factors include ASME Code requirements for allowable stresses vs. yield stress of materials. Safe operating limits should be set to prevent system operation in this safety factor zone.

Safety factors should be provided for expected degradation and deviations (e.g., corrosion or surges / operating deviations), as well as conditions such as contamination in process streams.

Over-designing a process is not considered a safety factor. Over-design most often involves planning for future conditions and equipment, for example, providing extra capacity in a pump because there is the potential for unit expansion.

Safety factors address variations, uncertainties, and potentially erroneous assumptions and decisions regarding operating conditions, operating rates, exact chemistry and composition of the process material, future expansions, type and frequency of mechanical integrity programs, and effectiveness of change management programs.

5.1.7 Safeguard Stewardship

Safeguard stewardship involves two very important concepts:
- The safeguard being installed will work under existing process conditions.
- The safeguard does not create new hazards.

In the first case, it is important that safeguards are installed so that they can be tested and maintained. Facilities may have a false sense of trust, particularly management, believing that everything is safe because there are many safeguards. An incident in the

Buncefield gasoline terminal (Ref. 5-4) in the UK resulted in a major fire and explosion because the level alarm on a gasoline storage tank was inoperative.

In the second case, a poorly planned safeguard can create hazards. An example is a Safety Instrumented System (SIS) installed to stop the flow to a process vessel to prevent overfilling. When the SIS actuates an isolation valve, the pump providing feed to the vessel could deadhead, resulting in seal failure and loss of containment. To avoid this situation, the SIS should have activated shutdown of the feed pump or an automatic spillback (recycle) could be provided on the pump to satisfy minimum flow requirements. In the Buncefield incident (Ref. 5-4), the fire water pump was suspected as being the source of ignition for the explosion and resulting fire.

5.2 INHERENTLY SAFER DESIGN

Inherently Safer Technology (IST), also known as Inherently Safer Design (ISD), permanently eliminates or reduces hazards to avoid or reduce the consequences of incidents. IST is a philosophy, applied to the design and operation life cycle, including manufacture, transport, storage, use, and disposal. IST is an iterative process that considers such options, including eliminating a hazard, reducing a hazard, substituting a less hazardous material, using less hazardous process conditions, and designing a process to reduce the potential for, or consequences of, human error, equipment failure, or intentional harm. Overall safe design and operation options cover a spectrum from inherent through passive, active, and procedural risk management strategies. There is no clear boundary between IST and other strategies (Ref. 5-5):

- *ISTs are relative:* A technology can only be described as inherently safer when compared to a different technology, including a description of the hazard or set of hazards being considered, their location, and the potentially affected population. A technology may be inherently safer than another with respect to some hazards but inherently less safe with respect to others and may not be safe enough to meet societal expectations.

- *ISTs are based on an informed decision process:* Because an option may be inherently safer with regard to some hazards and inherently less safe with regard to others, decisions about the optimum strategy for managing risks from all hazards are required. The decision process should consider the entire life cycle, the full spectrum of hazards and risks, and the potential for transfer of risk from one impacted population to another. Technical and economic feasibility of options should also be considered.

A chemical manufacturing process is inherently safer if it reduces or eliminates the hazards associated with materials and operations used in the process *and* this reduction or elimination is permanent and inseparable. A process with reduced hazards is described as inherently safer compared to a process with only passive, active, and procedural controls. A note of caution, sometimes eliminating a hazard in one place may increase hazards elsewhere.

An inherently safer process should not, however, be considered "inherently safe" or "absolutely safe." While implementing inherently safer concepts will move a process in the direction of reduced risk, it will not remove all risks. No chemical process is without risk, but all chemical processes can be made safer by applying inherently safer concepts (Ref. 5-6).

5. GENERAL DESIGN

Inherently safer design should be an essential aspect of any process safety program. If hazards can be eliminated or reduced, extensive layers of protection to control those hazards may not be required or may be less robust. However, inherently safer concepts are not the only process risk management strategy available and may not always be the most effective. A system of strategies that includes both inherently safer design and additional layers of protection may be needed to reduce risks to an acceptable level.

An inherently safer process can offer greater safety potential, often at a lower cost. However, selection of an inherently safer approach does not guarantee that the actual implementation of those approaches will result in a safer operation than an alternate process that is safer due to multiple layers of protection. The traditional strategy of providing layers of protection for a hazardous process can be quite effective, although the expenditure of resources to install and maintain the layers of protection may be very large. In some cases, benefits of the inherently more hazardous technology will be sufficient to justify the costs needed to provide the layers of protection required to reduce its risk to a tolerable level.

Approaches to the design of inherently safer processes and plants have been grouped into four major strategies (Ref. 5-6):

Minimize   Reduce quantities of hazardous substances

Substitute   Replace a material with a less hazardous substance

Moderate   Use less hazardous conditions, a less hazardous form of a material, or facilities that minimize the impact of a release of hazardous material or energy

Simplify   Design facilities which eliminate unnecessary complexity and make operating errors less likely and which are forgiving of errors that are made

These four strategies form a protocol by which the risks associated with loss of containment of hazardous materials or energy can be significantly reduced and in some cases eliminated. The elimination of risk due to loss of containment is very difficult, if not impossible to achieve using other risk reduction measures, i.e., active or passive safeguards. These measures, while effective if installed and maintained properly, generally reduce the likelihood of release and sometimes will mitigate the consequences of a release. However, they cannot reduce the risk to zero. Kletz's statement "What you don't have can't leak" embodies the ultimate goal of inherently safer strategies and describes the elimination of the risk of hazardous materials releases. However, while they are highly effective techniques, it is usually not possible to eliminate all process-related risks since the properties that make a material hazardous are often the same properties that make it useful.

5.2.1   Minimize

In the context of inherently safer, minimize means to reduce the quantity of material or energy contained in a manufacturing process or plant. Process minimization is often thought of as resulting from the application of innovative new technology to a chemical process, for example, tubular reactors with static mixing elements, centrifugal distillation techniques, or innovative, high surface area heat exchangers. These types of minimization strategies are also discussed in this section. However, much can be accomplished in process inventory reduction simply by applying good engineering design principles to more conventional technology.

Often, the inventory of hazardous materials onsite is driven by operational and business considerations, particularly in the number of transportation containers that are stored or used onsite at any given time. Railroad dispatching schedules, trucking schedules, and other transportation-related issues, most of which are established independently of safety considerations, often influence the amount of hazardous materials present onsite. Sometimes inventories are determined by onsite purchasing considerations, such as timing of incoming or outgoing shipments related to price. Careful coordination with shippers and carriers is required in order to minimize inventories related to transportation scheduling.

When designing a process facility or unit, the dimensions of every item of process equipment should be specified as large enough to accomplish its intended purpose and no larger. Required surge capacities, either for normal operations or for emergency situations, sometimes demand larger equipment. They are part of the intended purpose of a process design and should be maintained. But, this extra space should be kept empty and unused, and the process should not be modified in the future to accommodate additional process capacity. Raw material and in-process intermediate storage tanks should be minimized, if feasible.

5.2.2 Substitute

In the context of inherently safer, substitution means the replacement of a hazardous material or process with an alternative that reduces or eliminates the hazard. Process designers, line managers, and plant technical staff should continually inquire if less hazardous alternatives can be effectively substituted for all hazardous materials used in a manufacturing process. However, the substitution concept of inherent safety is best applied during the initial design of a process. Substituting raw materials and intermediates after the process has been built, while possible in some cases, is usually very difficult.

Examples of substitution in two categories - reaction chemistry and solvent usage - are discussed below. However, there are many other areas where opportunities to substitute less hazardous materials can be found, including materials of construction, heat transfer media, insulation, and shipping containers.

Basic process chemistry that uses less hazardous materials and chemical reactions offers significant potential for improving inherent safety in the chemical / processing industry. Alternate chemistry may use less hazardous raw materials or intermediates or result in reduced inventories of hazardous materials (minimization) or less severe processing conditions (moderation). Identifying catalysts that can enhance reaction selectivity or allow desired reactions to be carried out at a lower temperature or pressure is often the key to developing inherently safer chemical synthesis routes.

Replacement of volatile organic solvents with aqueous systems or less hazardous organic materials improves the safety of many processing operations and final products. In evaluating the hazards of a solvent, or any other process chemical, it is essential to consider the properties of the material at the processing conditions. For example, a combustible solvent is a major fire hazard if handled above its flash point or boiling point.

5. GENERAL DESIGN

5.2.3 Moderate

In the context of inherent safety, moderate means using materials under less hazardous conditions. Moderation of conditions can be accomplished by strategies that are either physical (e.g., lower temperatures, dilution) or chemical (e.g., development of a reaction chemistry which operates at less severe conditions).

5.2.4 Dilution

Dilution reduces the hazards associated with the storage and use of a low boiling hazardous material in two ways:

1. By reducing the storage pressure
2. By reducing the initial atmospheric concentration if a release occurs

Materials that boil below normal ambient temperature are often stored in pressurized systems under their vapor pressure at the ambient temperature. The pressure in such a storage system can be lowered by diluting the material with a higher boiling solvent. This reduces the pressure imposed on the storage container, as well as the pressure difference between the storage system and the outside environment, thereby reducing the rate of release in case of a leak in the system. If there is a loss of containment incident, the atmospheric concentration of the hazardous material at the spill location and the downwind atmospheric concentration and hazard zone are reduced.

5.2.5 Simplify

In the context of inherently safer, simplify means designing the process to eliminate unnecessary complexity, thereby reducing the opportunities for error and misoperation. A simpler process is generally safer and more cost-effective than a complex one. For example, it is often cheaper to spend a relatively small amount of money to build a higher pressure reactor, rather than spend a large amount of money for an elaborate system to collect and treat the discharge from the emergency relief system of a reactor designed for a lower maximum pressure. *Inherently Safer Chemical Processes, A Life Cycle Approach* (Ref. 5-6) offered a few reasons why process designs are unnecessarily complex:

- *The Need to Control Hazards* - Instead of avoiding hazard using inherently safer design principles, most designers choose to control them actively using controls, alarms, and safety instrumented systems.
- *The Desire for Technical Elegance* - To some designers, simple equates to crude or primitive, whereas, if carefully designed, a simple process can achieve what it needs to do without excess equipment. A simple process design that contains only the essential elements to safely carry out its intended task(s) is actually more elegant than a complicated process that does the same thing.
- *The Failure to Conduct Hazard Analyses Until Late in the Design* - PHAs and similar studies performed late in the design usually result in more active controls and equipment rather than more inherently safer solutions.
- *Following Standards and Specifications That Are No Longer Appropriate or Not Completely Applicable* - Active solutions to potential hazards that are sometimes contained in design / engineering standards and specifications can accumulate in a design and create an over-complicated process.

- *Flexibility and Redundancy* - While some level of redundancy is necessary and desirable with basic process equipment, particularly where the failure of the component will have serious effects, this should be limited to what carefully performed PHAs and other studies reveal as the correct level. For every extra pump, heat exchanger, or other basic component, additional controls, utility requirements, piping / valves, and other mechanical equipment will follow, thereby greatly expanding the complexity of the process. Additionally, not every risk can or should be solved by specifying some piece of equipment to deal with it. Only those risks that have been identified in the PHA process that exceed a pre-determined value should be addressed using active equipment or where law or regulation specifies such a solution.

For more information on inherently safer design, see *Inherently Safer Chemical Processes, A Life Cycle Approach* (Ref. 5-6).

5.3 BASIC PROCESS CONTROL SYSTEMS

The Basic Process Control System (BPCS) responds to input signals from the process and its associated equipment, other programmable systems, and / or an operator and generates output signals causing the process and its associated equipment to operate in the desired manner.

The primary function of a BPCS is business and production goals, especially uptime, production, and quality. The BPCS consists of many hardware and software components and relies heavily on communication equipment to access and display process information. The BCPS logic solver is often referred to as the "controller" and can utilize pneumatic, hydraulic, electrical, electronic, or Programmable Electronic (PE) technology. A modern PE-based BPCS provides nearly seamless integration of controllers and operator displays. PE technology enables complex control algorithms, such as advanced process controls, sequencing, predictive controls, and batch reactor recipe management.

For more information, refer to *Guidelines for Safe Automation of Chemical Processes* (Ref. 5-7) and *Guidelines for Safe and Reliable Instrumented Protective Systems* (Ref. 5-8).

If active controls are needed to prevent the process operating parameter from reaching a hazardous condition, the control system design should specify the desired operating conditions to provide adequate time for controls to function before reaching the safe operating limits. Calculating the adequate time requires knowing the speed of the process transient, the response time of the device being controlled, and the lag time of the control sensing element and the final control element (e.g., a control valve). For example, an upset in feed to a tank could lead to an overflow on high level. A proper design provides enough time for the tank level control to sense the upset and to take corrective action on the flow into or out of the tank before it overflows. For such a tank, the maximum set point for the level should be reduced to allow adequate response time.

There are few chemical plants that are so robust that an active control system is not required. Using both active and passive controls can assure product yield and quality, and maintain safe operating conditions. The BPCS may initiate alarms or automatically act to moderate a high or low operating condition within the never-exceed limits. A

5. GENERAL DESIGN

Safety Instrumented System (SIS) may be required to rapidly shut down or otherwise place the process in a safe state if the BPCS fails to maintain safe operating conditions (Ref. 5-8). **A BPCS may not be adequate as the sole source of a process safety shutdown.** Many of the following guidance items related to the design, operation, and testing of BPCSs are not inherently safer technology in a strict sense, because they relate to active safeguards. However, much of this guidance can also be considered part of the inherently safer strategy to simplify systems.

5.3.1 Alarm Management

The need for an alarm is usually specified by process design and good engineering principles. Alarms may originate from Operations personnel, from a process hazards analysis, or as a result of a team's investigation of an incident. Typical alarms include:

- *A Warning Against Operational Error* - An alarm can be justified if an operational error will lead to a plant upset or equipment damage. The upset will be such that the control scheme will not be able to bring the plant back to normal condition.
- *Equipment Malfunction* - Malfunction of equipment can lead to plant upset which the plant control scheme may not be able to correct. For example, a pressure control valve in an overhead vapor line which gets stuck in the closed position may cause the pressure in the system to rise and result in the lifting of a relief valve.
- *Equipment Protection* - The malfunction of a system which can lead to damage to the associated (or downstream) equipment, for example, high temperatures on a product rundown line that may exceed a tank design limit.
- *Signal a Shutdown of Major Equipment* - The shutdown of a certain piece of equipment will cause major plant upset and will require substantial operator intervention to mitigate the effect of the shutdown.
- *High Furnace Tube Skin Temperature* - Refinery furnace tubes may be provided with skin temperature indicators. Skin temperature indicators should have high temperature alarms and should be set at the Maximum Allowable Skin Temperature (MAST).
- *Minimum Flow for Rotating Equipment* - For centrifugal pumps, an alarm should be provided to warn of an operation with less than minimum safe flow.
- *Flammable and Toxic Gas Detectors* - Flammable and toxic gas detectors or those devices which indicate immediately dangerous to life and health should be configured with an alarm to warn personnel in the affected area.

With the ability to make every signal into an alarm in a BPCS, operator information overload is a genuine safety concern. ANSI / ISA 18.2-2009, *Management of Alarm Systems for the Process Industries, Instrumentation, Systems, and Automation Society* (Ref. 5-9), is a standard that provides guidance and requirements for the design and implementation of an effective alarm system. Proper alarm design will follow a rationalization and prioritization process to determine the need for the alarm, the required response for the alarm, and the priority of the response. Operators should be trained to understand the importance of the safe operating limit alarms. BPCS digital and analog alarm displays should be grouped to be readily identifiable by color, physical position, and distinctive sound annunciation. An on-screen list of potential action alternatives should be displayed. The navigation of digital BPCSs should be intuitive and user friendly, particularly with respect to alarm screens.

Alarm priority assignment is determined according to how fast an operator should respond to a situation. The most important alarms, at any given time, should be obvious to the operator. Alarms are typically prioritized considering the following two factors:
1. *Severity of the Consequences* - The expected outcome that the operator can prevent by taking the corrective action associated with the alarm.
2. *Time Available* - Compared with the time required for the corrective action to be performed and its desired effect.

Alarm prioritization makes it easier for the operator to identify important alarms when a number of them occur together. Alarms can be prioritized as:
- *Critical* - Operator action is required to avoid a serious incident (e.g., safety and environmental impact; or may be initiated by a safety shutdown system).
- *High* - Timely operator action required (e.g., to avoid severe equipment damage or unit shutdown).
- *Medium* - Operator action required (e.g., to avoid off-spec product and equipment level management).
- *Low* - Operator action required, but unit is still in steady state operation.

5.3.2 Testing Instrumentation

Process control and safety shutdowns should be provided during all modes of operation, not only in the normal, steady state operating mode. Properly designed automated systems include provisions for functional testing of the entire function, as well as for calibrations of individual devices, such as sensors. Functional tests may be performed online or offline depending on the function, test facility design, and mechanical integrity plan (Ref. 5-7 and Ref. 5-8).

Periodic testing of the complete function should include the final element(s). To do so without creating transient or frequent unit / plant shutdowns, it is often necessary to perform functional tests of the final elements during shutdown periods when the process is offline. Depending on the nature of the plant response, it may be possible to initiate a planned shutdown by tripping the final elements. This should be attempted only if the resulting shutdown will be orderly and stable and will not cause transients in other process parameters that are outside their normal limits.

Periodic testing is essential for ensuring that automated systems have adequate reliability and dependability. Test records should be maintained to support reliability analysis, tracking, and auditing. The importance of testing and documentation is illustrated in the following example: a 15-year-old heater was designed to automatically shut down and provide an alert in the central control room in response to high heat transfer oil pressure, high tube wall temperature, low fuel gas pressure, and flame-out. After a fire destroyed the heater, it was determined that there were no records documenting initial validation (acceptance test), periodic proof testing, or preventive maintenance. It was further determined that there was no systemic program in place to periodically test the instrumented functions (Ref. 5-10).

5.4 INSTRUMENTED SAFETY SYSTEMS

An Instrumented Safety System (ISS) is comprised of instrumentation and controls that implement safety functions identified as safeguards for process safety hazards in the PHA. ISSs include many types of Instrumented Safety Functions (ISFs), such as safety controls, safety alarms, safety interlocks, safety permissives, detection or suppression equipment, and Safety Instrumented Functions (SIFs) (Refs. 5-11 and 5-12).

While some ISFs may be implemented in the BPCS, the SIFs must be independent from the BPCS to ensure their effectiveness in preventing hazards due to BPCS failure (Refs. 5-8, 5-13, and 5-14). SIFs are implemented to detect the existence of unacceptable process conditions and to take action on the process to achieve or maintain a safe state. The automation systems that implement SIFs are now called Safety Instrumented Systems (SISs), but in the past, these systems have also been referred to as emergency shutdown systems, safety interlock systems, and safety critical systems (Ref. 5-12).

See Chapter 7, Section 7.2, for a detailed discussion of ISS.

5.5 PROCESS DESIGN / PROCESS CHEMISTRY

5.5.1 Process Equipment Safe Operating Limits

The zones of operation are defined as:

- *Normal Operating Zone* - The minimum or maximum values of a critical operating parameter that define the boundaries of normal operations.
- *Troubleshooting Zone* - An area that provides time for troubleshooting so that operations personnel can make adjustments in time to return critical operating parameters to the normal operations zone. Human factors and process response time generally indicates zone size. Immediate actions and in some cases predetermined actions to avoid Safe Operating Limit (SOL) deviation are taken in this zone.
- *Buffer Zone* - The upper and lower area of the known safe zone provides a buffer to ensure no critical operating parameter can reach the unknown / unacceptable operation zone. Factors that influence buffer zone size may include engineering judgment, reliability of instrumentation, operating experience, probability and consequence of human error, etc. A process will not be intentionally operated in this zone.
- *Safe Operating Limit (SOL)*- A value for a critical operating parameter that defines the equipment or process unit safe operating envelope beyond which a process will not intentionally be operated due to the risk of imminent catastrophic equipment failure or loss of containment. Operational or mechanical corrective action ceases and immediate predetermined actions are taken at these critical operating parameter values in order to bring equipment and process units to a safe state.
- *Unacceptable or Unknown Operation Zone* - An area beyond the Safe Operating Limit (SOL). A process will not be intentionally operated in this zone.

Examples of operating parameters might include:
- High or low pressure
- High and low level
- High and low temperature
- High and low pH
- High and low flow

Figure 5.3 illustrates a typical zone of operation for processes.

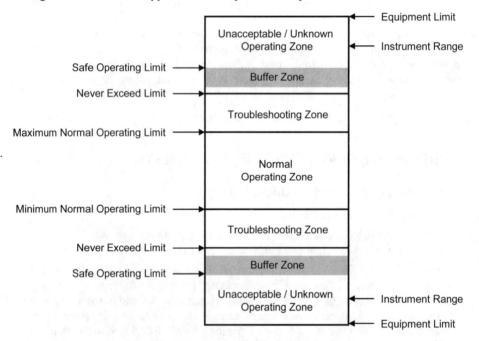

Figure 5.3 Illustration of Zones of Operation

A Safe Operating Limit (SOL) pre-alarm is acceptable provided there is sufficient time after the pre-alarm to perform effective corrective action prior to exceeding an SOL. The setpoint for the pre-alarm is established by considering the process dynamics, required operator response time to ensure effectiveness, and the instrumentation detection and response lag. Refer to ISA TR84.00.04 (Ref. 5-14) for more guidance on safety setpoints.

Each hazardous process unit should have SOLs identified. Typical SOLs should meet the following requirements:
- SOLs are established on critical operating parameters only if it is physically possible to exceed the limit and if exceeding the limit could lead to a catastrophic failure of process equipment or catastrophic loss of containment.
- SOLs are determined by identifying design limits of equipment within a system. The most limiting elements will establish the SOLs for the system.
- Safety and environmental consequences of pressure relief system activation to the atmosphere (relief valve, rupture disc, etc.) should be considered in the determination of SOL settings.

5. GENERAL DESIGN

Each SOL should be documented in plant Process Safety Information (PSI). Typically, SOL information is presented in table layout and includes the following:
- Description of critical operating parameter that provides instrument tag name
- Minimum and / or maximum normal operating limits (i.e., pre-SOL alarm point) and units of measure
- Minimum and / or maximum SOLs and units of measure
- Technical basis for SOL

5.5.2 Consequences of Deviation

Consequences of deviation are generally associated with SOLs. The consequence would be the impact of fire, explosion, and loss of containment if immediate action was not taken.

For each SOL, there will be immediate actions required (manual or automatic) to correct deviation within a predetermined time limit. This could be as simple as an open bypass valve around the control valve to divert feed and shut down the unit.

Normally, the steps to correct are actions that operation personnel perform. The actions can be taken by either the board operator or an outside operator.

5.6 PLANT SITING AND LAYOUT

Siting and layout appear to be synonymous; however, they are slightly different. Siting is the process of locating a complex, site, plant, or unit. Layout is the relative arrangement of equipment or buildings within a given site.

The arrangement of process units and buildings are crucial factors in the safety and economics of a chemical plant. The plant layout (plot plan) should incorporate safety while providing access for operations and maintenance. Some of the safety benefits of a good layout are:

- Minimization of:
  - Explosion damage, since explosion overpressure falls off rapidly with distance from the center of the explosion
  - Thermal radiation damage, as the intensity of thermal radiation also falls off with the distance
- Easier access:
  - For emergency services, such as firefighting
  - To equipment for maintenance and inspection
- Efficient and safe construction
- Reduction of onsite and offsite personnel exposure to incident consequences

Plant layout can have a large impact on plant economics. Additional space increases the initial investment due to higher capital costs (more land, piping, cabling, etc.) and operating costs. However, additional space also tends to enhance safety. Overall life cycle costs may actually be lower due to reduced consequence damage gained by spacing and potentially lower maintenance and turn-around costs. It is important, therefore, to carefully weigh these issues to optimize the plant layout.

5.6.1 Site Layout

Preliminary identification of various hazards during early planning stages of the project will help establish proper layout at the beginning of the project and prevent design rework later. Good layout can reduce the effects of some of the controllable factors, such as liquid spills, and uncontrollable factors, such as exposure to natural hazards, site slope, and wind direction and force, that contribute to losses.

It is not unusual for separation distances to be compromised as a result of subsequent plant expansions, process changes, or other modifications. For this reason, it is essential that minimum separation distances be clearly defined and maintained. If future plant modifications are anticipated which might impact separation distances, consideration should be given to employing larger initial separation distances and / or other protective means.

As a general guideline, the layout of the units is based on the flow principle so that the material flow follows the process flow diagram. The goal is to minimize the transfer of materials for both economic and safety reasons and allow a release to be contained at its source. Plant layout is largely constrained by the need to observe minimum safe separation distances.

Adequate separation is often achieved by dividing up a plant into process blocks of similar hazards, e.g., process units, tank farms, loading / unloading operations, utilities, waste treatment, and support areas and then separating individual operations or hazards within each block. The block approach also serves to reduce the loss potential from catastrophic events, such as unconfined vapor cloud explosions, and to improve accessibility for emergency operations. References for safe separation distances include:

- API RP 752, *Management of Hazards Associated with Location of Process Plant Permanent Buildings* (Ref. 5-15), and API RP 753, *Management of Hazards Associated with Location of Process Plant Portable Buildings* (Ref. 5-16)
- NFPA 30, *Flammable and Combustible Liquids Code* (Ref. 5-17)
- CCPS, *Guidelines for Facility Siting and Layout* (Ref. 5-18)
- Insurance and internal company guidelines

Design considerations for layout and spacing include:

- A maximum block size limitation with adequate spacing between the blocks allows access for firefighting
- Adequate overhead and lateral clearance for pipeways, pipe racks to prevent possible damage by large moving vehicles, cranes, and trucks

Two methods exist for determining minimum separation distances within chemical process plants. The first method is to use recommended separation distances for generic plant hazards. These distances are generally conservative and will cover most situations (Ref. 5-18).

The second method for determining minimum separation distances is calculating the amount of heat received by an object from a fire involving the actual hazards in question. While this method generally results in more realistic separation distances, the calculations are often complex and should only be performed by persons familiar with the concepts involved. In addition, the calculations should consider all possible scenarios, and selection of endpoint values used is very important and can make a

5. GENERAL DESIGN

substantial difference in results. Space does not permit complete discussion of this subject here; however, additional information can be found in CCPS, *Guidelines for Fire Protection in Chemical, Petrochemical, and Hydrocarbon Processing Facilities* (Ref. 5-19).

In addition to radiant heat exposure, other factors which should be considered in determining separation distances and plant layout include topography, prevailing winds for normal and accidental vapor / gas releases, liquid drainage paths for accidental liquid spills, location of fire protection equipment, and accessibility for emergency vehicles. Specifically for toxics, dispersion modeling can be used to assist in the location of buildings and the need for shelter-in-place (Ref. 5-20).

5.6.2 Unit Layout

Unit layout is the arrangement of equipment within a particular block on the site. Process units are usually grouped because they are generally more hazardous than central services. The unit layout also depends on whether the unit uses single- or multi-stream operation. Space for future expansion of plant equipment or pipe work as well as access for installation is another factor to consider. Large vessels and equipment needing frequent maintenance or cleaning should be located close to unit boundaries for ease of access by cranes. Plant items such as heat exchangers and reactors that need removal of internals should be provided with necessary space and lifting arrangements. An incident occurred in Texas City when a heat exchanger was being lifted over a storage tank of toxic material. The heat exchanger fell, resulting in a significant offsite release of toxic material.

Some further considerations in unit layout are:
- Location of fired heaters in relation to units with flammable materials.
- Separation of equipment that is a potential source of explosions, such as chemical reactors, by blast-resistant walls, if increased spacing is not practical.
- Location of pumps handling flammable material. These items are frequent sources of releases and should not be grouped in one single area. They should not be located under vessels, air-cooled heat exchangers, or pipe racks.
- A model review (built or 3d CAD model) is generally conducted to review the layout and spacing.

The design should consider the effects of congestion and confinement on the potential for aggravating an explosion event. To this end, it is preferable to space equipment as far apart as possible and to avoid confinement where possible (e.g., use grated decks rather than solids decks).

The greater spacing also improves access for emergency responders, personnel trying to escape from a hazard, etc. - although greater spacing also costs more in piping (and, in principle, provides more surface area from which a leak could occur in the first place) and costs more in land. When determining unit layout, consideration should be given to potential spacing needs during periodic and major shutdowns and turnarounds. Additional equipment laydown, staging, and cleaning are often needed to reduce need for excessive movement of equipment and materials during these periodic events.

5.6.3 Storage Layout

Layout of hazardous materials storage areas requires careful attention. Typically a far larger quantity of material is held in storage than in process. Siting, design, fabrication, and operation of storage facilities are thoroughly addressed in *Guidelines for Safe Handling and Storage of High Toxic Hazard Materials* (Ref. 5-21) and *Guidelines for Facility Siting and Layout* (Ref. 5-18). Some of the important aspects of storage layout are:

- Storage tanks should be arranged in groups so that common dike and firefighting equipment can be used for each group.
- It is essential to keep storage tanks away from process areas since a fire or explosion in a process unit may endanger the large inventory of the storage tank.
- Storage tanks should be diked in accordance with NFPA 30 (Ref. 5-17). Piping, valves, and flanges should be kept to a minimum when located within dikes. Valves, manifolds, and piping should be installed outside dikes or impounding areas.
- The effect of intensity of thermal radiation from an adjacent tank on fire should be considered in spacing the tanks. Tolerance of tanks to thermal radiation can be increased by insulating or fireproofing the tank shell and providing water cooling arrangements.
- Secondary containment systems are considered passive protective systems. They do not eliminate or prevent a spill or leak, but they can significantly moderate the impact without the need for any active device. Containment systems can actually be defeated by manual or active design features. For example, a dike may have a drain valve to remove rainwater, and the valve could leak or be left open. Another example is a door in a containment building that could be left open.

5.6.4 Occupied Building Location

Many of the fatalities and serious injuries resulting from process safety incidents are caused not by direct consequences of the actual incident but by damage to buildings where personnel work or congregate (occupied buildings). Therefore placement of occupied buildings within a facility is critical to minimizing the consequences of an incident and the overall risk a facility presents to personnel.

Occupied buildings should be evaluated using the methodologies of API RP 752 (Ref. 5-15), API RP 753 (Ref. 5-16), and CCPS, *Guidelines for Vapor Cloud Explosions, Pressure Vessel Burst, Flash Fires, and BLEVEs* (Ref. 5-22).

5.7 MATERIALS OF CONSTRUCTION

Equipment service life is influenced by many factors, such as materials of construction, design details, fabrication techniques, operating conditions, and inspection and maintenance procedures. Material failures, while relatively infrequent, can be extremely severe, resulting in catastrophic accidents. The best way to reduce the risk of material failure is to:

- Fully understand the internal process, the exterior environment, and failure modes
- Select materials for the intended application
- Apply proper fabrication techniques and controls
- Follow good maintenance, inspection, and repair techniques

Corrosion refers to the degradation or breakdown of materials due to chemical attack. Corrosion is one of the most important process factors in material selection and yet the most difficult to predict. In general, equipment service life can be predicted from well-established general corrosion data for specific materials in specific environments. However, localized corrosion is unpredictable, difficult to detect, and can greatly reduce service life. Even more insidious are subsurface corrosion phenomena. API RP 571 *Damage Mechanisms Affecting Fixed Equipment in the Refinery Industry* provides a detailed discussion on corrosion mechanisms (Ref. 5-23).

5.7.1 Properties of Materials

The basis for selection is performance under design conditions, that is, how the material will function in the process environment, not only at standard operating conditions but also under startup, shutdown, and upset conditions. The behavior of a material in a process environment is determined by its physical, chemical, and mechanical properties. These properties determine how the material will be affected by process chemicals as well as how the material will affect the process.

Chemical and physical properties are important in determining corrosion reactions which could affect system integrity and appropriate corrosion prevention measures. Some properties and their effects are:

- Thermal expansion (especially differences in expansion of different components; e.g., vessel and cladding)
- Melting point or range (affects weldability, hot forming; e.g., hot-short cracks may become focal points for corrosive attack and mechanical failure)
- Brittle fracture of carbon steel when exposed to low temperatures
- Acid / base resistance
- Resistance to solvents
- Susceptibility to various types of corrosion

Metals and alloys are often subjected to heat treatment to improve mechanical properties and corrosion resistance or to bring about thermal stress relief. Heat treatment can be done before fabrication to get better mechanical properties (e.g., increase ductility and impact strength) and corrosion resistance or done after cold / hot work to reduce the residual stress.

5.7.2 Corrosive Environments

If the range of process conditions is accurately specified by the process engineer, the materials engineer can generally select suitable materials of construction without additional testing. However, upsets and impurities, trace elements, and contaminants are likely to cause most of the problems; therefore, any potential contact with impurities, in all process fluids, ambient environment, utilities, etc., and for *all* operating scenarios, should be identified to the materials engineer.

Both the external (ambient) and internal (process) conditions in contact with materials need to be examined. The external environment, that is, the ambient conditions in the plant, may be corrosive. Atmospheric pollutants include corrosive species as well as those which may have adverse catalytic effects upon other pollutants (e.g., coal dust). Contaminants in soil or groundwater as well as naturally occurring variations in groundwater composition and pH should be considered for equipment or pipelines in contact with the ground.

The internal environment is defined by the process, its chemistry, and its conditions. The process engineer should provide the materials engineer with sufficient information about the process, ambient conditions and utilities, for startup and shutdown as well as routine operations, to ensure adequate selection, especially for corrosive service. Preliminary materials selection is usually based on process conditions, such as:

- Process chemicals, including the major and minor constituents of each process stream, trace contaminants, pH, and oxidizing or reducing agents and water content. For example, styrene will leach copper, and thus materials in contact with styrene are generally specified to not contain copper. Additionally, chlorine can lead to stress corrosion cracking in stainless steel.
- Operating conditions, including temperature, pressure, velocity, and solids content.
- Process variations, including operational excursions in process chemistry, temperature, or pressure; excursions associated with startup or shutdown conditions. The order in which the conditions occur can be important (Ref. 5-24), e.g., purging / cleaning with steam may constitute a temperature excursion.
- Contaminants in feedstock, process intermediate, product, or utility. Contaminants introduced by small or midsized internal leaks in heat exchanger tubes or other internals. Impact of contaminants on gaskets and packing and seals.
- Catalysts. Metal ions in the material may affect either the chemistry of the process itself or the product quality. For example, nickel is known to catalyze many synthesis reactions and its inclusion can result in unwanted side reactions.
- Utilities, including trace elements in cooling water, hydrotest water, steam, etc.

5.7.3 Pitfalls in Material Selection

Process criteria often determine materials of construction for pressure vessels, heat exchangers, valves, piping, pumps, tanks, and instrumentation. These requirements should be adequately documented in complete equipment or instrument data sheets. Fabrication and corrosion control techniques should also be specified.

Specific types of process equipment have characteristic corrosion problems. Bimetallic heat exchangers are frequently subject to electrolytic corrosion, particularly where the two metals are in contact. Distillation or extraction columns have corrosion problems associated with the presence of distinctly different environments at different locations in the same vessel. Pumps, some piping configurations, and valves are subject to a higher incidence of velocity effects (erosion).

If corrosion testing is performed to provide a basis for material selection or fabrication techniques, the test conditions should be as close as possible to the actual (design) service environment. Velocity of process fluids, for example, may be overlooked, but it is just as important to test as composition, concentration, temperature, pressure, and time factors.

If operating or failure conditions differ from design conditions, the original material selection might not be valid. Design bases should be fully and clearly documented and communicated to the operators (through procedures, training, etc.); inadequate documentation frequently causes confusion and can invalidate any management of change procedure.

Requirements may be imposed upon the manufacturer and the supplier to ensure that the materials are accurately represented. A big problem is traceability of materials. Manufacturers may be required to attest that the material is in accordance with the material specification. Materials certification or a certificate of conformance may be required to provide the paperwork certifying the materials are as specified. If further work is done on the material, the manufacturer may also have to provide a certified material test report, verifying the quality of welding or other treatments. Some means of identification, for example, lot number, weld number, or heat number, is required to trace the material to the manufacturer. The *ASME Boiler and Pressure Vessel Code* contains guidance on material segregation, traceability, and alloy verification.

5.8 CORROSION

Corrosion is chemical attack on a metal. Corrosion may occur at a uniform, predictable rate or it may be localized, on the surface or as a subsurface phenomenon. The following discussion of corrosion, although normally thought of in terms of the internal, i.e., process environment, also applies to external surfaces of equipment and piping.

5.8.1 General Corrosion and Metallurgical Changes

General corrosion means the entire surface loses metal uniformly due to attack by chemical or electrochemical reaction. Reaction with gases present in the process may cause oxidation, sulfidation, reactions with halogens and hydrohalides, and various other types of corrosion. The corrosion rate is predictable, based on previous experience and can be compensated for by adding a corrosion allowance to the wall thickness of piping and equipment. For example, for carbon steel 1/16 inch or more is added for typical project life. The National Board Inspection Code (Ref. 5-25) provides an explanation and formula for determining corrosion allowance. Decarburization and carburization are other metallurgical changes, although there is no metal loss or surface change.

5.8.2 Stress-Related Corrosion

5.8.2.1 Stress Corrosion Cracking

Every alloy is subject to Stress Corrosion Cracking (SCC) in some environment; however, chloride stress corrosion cracking is commonly associated with stainless steel. The majority of SCC problems are associated with stainless steels and aqueous chloride salts, but both sulfide and chloride stress cracking are common in the process industry. It occurs when material has been under tensile stress in an environment containing sulfide compounds or chloride salts for a period of time. For example, salt water, brackish water, and chlorinated city water have chlorides and, in most cases, are not compatible with stainless steel.

5.8.2.2 Corrosion Fatigue

Corrosion fatigue can be defined as a combination of normal fatigue and corrosion that causes failure at stress levels far below the design endurance limit of the metal involved. Corrosion fatigue resistance is remarkably decreased by an increase in the stress cycle frequency, even in the low frequency ranges. Compressive stresses will not cause corrosion fatigue. Corrosion fatigue is influenced by:

- Environmental factors, such as temperature, pH, oxygen content, and composition of process fluids
- Mechanical factors, such as vibration

5.8.2.3 Pitting

Pitting results from electrochemical potential set-up by differences in oxygen concentration inside and outside the pit (Ref. 5-26). Pitting is also used as a generic term to refer to other types of localized corrosion.

Because of its localized and deeply penetrating nature, pitting is one of the more damaging types of corrosion in the process industry. Pits can extend through the material within a short period of time. Pitting is difficult to detect by online monitoring or field testing. Addition of corrosion inhibitors (e.g., oxygen scavengers) can prevent this type of corrosion. Pitting often occurs or is accelerated when vessels / piping are opened for inspection or other reasons.

5.8.2.4 Intergranular Corrosion

Intergranular Corrosion (IGC) is a severe corrosion problem for austenitic stainless steels. IGC is caused by impurities (in the case of nickel alloys) or alloying elements (for stainless steels) that migrate from the surrounding areas to the grain boundaries and then precipitate between the grains. These precipitated materials have a different corrosion potential than adjacent grains and become either cathodic or anodic. If the precipitate is anodic, it will be corroded. If the precipitate is cathodic, a narrow zone next to the grain boundary will be corroded. Then a fine crack will form along the grain boundary and degrade the mechanical properties of the metal. Many unstablized austenitic steels are susceptible to IGC.

5.8.2.5 Galvanic Corrosion

Accelerated corrosion may occur when two dissimilar metals are joined. The metal with the lower position in the galvanic series may be corroded. Proper electrical isolation can protect the metal from galvanic corrosion. Also, coating the cathodic member of the couple can be effective in reducing galvanic corrosion.

5.8.2.6 Hydrogen-Induced Attack

Hydrogen is commonly encountered in process environments, for example, in hydrocarbon reforming operations and hydrogenation and dehydrogenation reactions. Some of the problems associated with use of hydrogen in chemical and refining

5. GENERAL DESIGN

processes are discussed in API RP 941, *Steels for Hydrogen Service at Elevated Temperatures and Pressures in Petroleum Refineries and Petrochemical Plants* (Ref. 5-27), commonly known as the "Nelson curves."

Hydrogen sulfide (H_2S) in refinery operations significantly increases corrosion in carbon steel. Guidance on materials for use in H_2S service can be found in NACE MR0175/ISO 15156, *Petroleum and Natural Gas Industries - Materials for Use in H_2S-Containing Environments in Oil and Gas Production* (Ref. 5-28), and MR0103-2007, *Materials Resistant to Sulfide Stress Cracking in Corrosive Petroleum Refining Environments* (Ref. 5-29).

5.8.3 Design Considerations

5.8.3.1 Crevice Corrosion

Corrosion often occurs where corrosive fluids are trapped in a cavity, such as a gasket surface or welded lap joint. The following considerations may help minimize this type of corrosion:

- Minimize the use of threaded joints and socket weld connections.
- Minimize flanged connections and try to use welded joints. When using a single butt joint, a permanent backing strip should not be used.
- Specify "solid" non-absorbent gaskets.
- Use continuous seal welds in corrosive environments.
- Seal weld the tube to tube sheet joint in heat exchangers when practical.
- Use a full weld around the top side of tray support rings in distillation columns.

5.8.3.2 Trapped Liquids

Providing free drainage (via a sloped floor under storage tanks, proper drain line for pressure vessels, sloped tube for condensers, point drain for piping systems, etc.) will eliminate the possibility of liquid trapped inside a tank, equipment, or piping and thus avoid the aggressive corrosion caused by stagnant fluid in dead pockets.

5.8.3.3 Corrosion Under Insulation

Various types of corrosion may occur hidden under insulation, including general corrosion, pitting, crevice corrosion, and external stress corrosion cracking. Selection of insulation systems, including materials which do not absorb moisture or process chemicals, as well as vapor barriers and weatherproof covers can minimize the risk of external corrosion under insulation systems. See Section 5.10.3 for further discussion of corrosion under insulation.

5.8.3.4 Cathodic Protection and Anodic Protection

There are two types of galvanic protection: cathodic and anodic. Cathodic protection is a process in which electrons are transferred from an external source to the metal, suppressing dissolution of the metal. Cathodic protection supplies electrons from an external power supply or a sacrificial anode. Cathodic protection is only good for moderately corrosive environments. This method is widely used in oil fields, in cooling water service, and for underground piping or structures.

Anodic protection is based on formation of a protective film on a metal by externally applied anodic currents. Thus, the anodic protection can be applied to passive metal only.

5.8.3.5 Corrosion Allowance

Although technically not a way to control corrosion, specification of a corrosion allowance is a commonly used method to address the problem of general (uniform) corrosion. A corrosion allowance is added to the wall thickness based on the general corrosion rate predicted by previous experience and the design life of the equipment or piping. Corrosion allowance cannot be used to compensate for pitting or localized corrosion. Periodic inspection and wall thickness determinations should be made and monitored to determine when the equipment or piping should be derated or replaced.

The addition of increased corrosion allowance does not work for external corrosive environments. The proper protection measure in such cases is proper surface protection (painting and coating).

5.8.4 Erosion

Erosion is a mechanical effect and therefore not technically within the scope of this section, but it is a significant factor in material selection. Erosion is wearing away of a material by mechanical energy that can lead to loss of containment. Erosion occurs by impingement of solid particles or liquid drops on a surface. Erosion is seen very frequently in high velocity slurry and pneumatic solids transport services, but it can also occur in more common scenarios, such as particles in steam, bubbles in a liquid, or where restrictions in flow exist.

Erosion can typically be found at inlet and outlet nozzles, on internal piping, on grid or tray sections, on vessel walls opposite inlet nozzles, on internal support beams, on piping elbows, and on impingement baffles. Impingement protection, smoother curvature, and higher corrosion allowances are generally used to combat erosion. Materials selected for equipment construction should consider the potential for erosion from the process stream based on the highest anticipated process stream velocity. Higher velocities will accelerate erosion rates. Harder faced materials are more resistant to erosion. Erosion can also result from cavitation in a flowing fluid, usually in or downstream of throttling service. Erosion may remove the protective passive layer, resulting in accelerated corrosion.

5.9 CIVIL / STRUCTURAL / SUPPORT DESIGN

The safety of the plant can depend on the civil, structural, and architectural design. Failures of foundations, walls, or supporting structures can rupture piping and vessels and lead to release of hazardous materials. As long as the structural loads are below or at design limits, failures are usually not a problem, because "structural failure probabilities under such conditions are usually one to three orders of magnitude smaller than mechanical, electrical and equipment failure probabilities" (Ref. 5-30). In rare situations, like natural hazards and explosions, these structural failure probabilities should be incorporated into the risk assessment (Ref. 5-31).

5. GENERAL DESIGN

5.9.1 Site Preparation and Analysis

Preparation of the site, governed by plot plans and grading and paving drawings, will establish the safe placement of the plant, provide for drainage and runoff containment, and define environmental considerations to be addressed.

5.9.1.1 Geotechnical Studies

Geotechnical investigations will establish excavation requirements, types of foundations required, and site drainage requirements. Any existing hazardous conditions discovered during site selection, such as contaminated soil, buried waste pits, etc., should be addressed in accordance with environmental regulations.

5.9.1.2 Surface Drainage

There are two key process safety considerations with respect to surface drainage. One is the potential for hazardous flammable, explosive, or toxic materials to enter the normal surface water drainage and collection system; another is adequate collection, treatment, and disposal of firefighting water.

Each facility should have a well-drained working surface and a drainage system to carry off storm water and / or spills to a holding area or treatment facility. Local, state, and federal regulations should be consulted to determine drainage or treatment required. Drain lines for these systems should be adequately sized not only for the chemicals involved but also for rainwater and runoff fire water that might be introduced. Drains should be sized to carry firewater flows as required by NFPA guidelines (Ref. 5-17).

5.9.1.3 Foundations

Foundations should be designed to transmit all loads and forces from the equipment or structures to the soils or rock beneath the foundations. Loads should be calculated using actual density of liquids and solids used in the process if heavier than water. Seismic and explosion or blast loads also should be considered. Foundation design of facilities related to the containment of hazardous material should address internal and external pressures, equipment loads, dynamic forces from vibrating equipment, and hydraulic uplift pressure from groundwater.

The geotechnical report will specify flood design considerations, such as reduced lateral pressure factor or lower shear resistance for foundation designs. For any large-volume underground chambers, such as buried drainage lines, below-grade storage tanks, or "basement" levels used for maintenance or storage, flotation should be considered in the design to assure anchorage. Similarly, open concrete pits or reservoirs have to be designed with this problem in mind. An American Petroleum Institute (API) separator or other concrete chamber, even a manhole, should be investigated to ensure that the weight of the item, plus its normally expected contents, will not float out of the ground or otherwise be dislodged from its designed location due to hydrostatic buoyancy forces.

Foundation design is determined by bearing pressure geotechnical investigation and testing. In situ pile testing (test piles) should include not only bearing tests but uplift resistance tests as well.

Good engineering practice or regulatory criteria may require that foundation designs for vessels containing hazardous materials also provide for containment and detection of leaks. For example, a ring foundation may not be appropriate for a tank storing hazardous material because it provides an undetected path for leaks to migrate to groundwater. For corrosive fluids, the design should include protection against seepage of the fluid into soil areas around the foundation.

Similar to the impact of environmental contaminants on piping and equipment selection, consideration should be given to selecting proper materials of construction for foundations, dikes, and containment structures. For example, a bare concrete containment dike or tank foundation can be rapidly degraded by even small leaks of strong acids. Coatings, linings, or alternate materials of construction may be required to ensure long-term integrity of foundation systems.

5.9.1.4 Underground Piping

Two explosions and fires within one week in the Houston, Texas, area in early 1992 involving underground pipelines point out the necessity of being absolutely sure, before the start of excavation or piling, that a seemingly clear site is free of hazardous obstacles. Many heavily industrialized areas rely upon underground pipelines as a vital part of the product transportation infrastructure. Where products are potentially hazardous, it is wise to consider protected aboveground, rather than underground, transfer. Protected aboveground transport makes leak detection and correction easier and will generally result in a safer operation. In many areas pipeline "easements" have been granted by individual real estate owners to allow this type of product transport. Where major easements exist, real estate title documents are generally amended to assure that a purchaser is aware of these restrictions on use. Therefore a scan of title documents may reveal nearby underground pipelines.

Pipeline easements generally restrict aboveground use in the easement. Process plant erection will not be allowed, and possibly more important, site access will be severely restricted. Vehicular crossings may be prohibited, except on established roads that usually have limited bearing loads. New crossings will have to be carefully constructed and supported, in effect being "bridges" across the easement though constructed at grade. Other crossings, such as pipe bridges and power lines, will similarly require careful consideration and design. Underground crossings may require special permission and documentation.

Along the sides of the easement branch take-offs may run through the proposed site. These may be more insidious than larger lines as they may not have the documentation that the easement does. An abandoned branch could be the most dangerous, as it may be capped or sealed at only the user's end and could be live from the supply underground end. An undocumented line also could exist within the boundaries of a single site where development occurred at distant locations. It is likely that "isolated" units were once connected to other units or to a central utilities center. Though most interconnections are aboveground, there is a high probability of underground lines as well.

The most likely existence of underground lines, but fortunately the most easily anticipated, is in the reuse of an old site where a unit was demolished. It may have been razed *to* the ground but not *below* the ground. Foundations, tanks, sumps, and diversion boxes, some of which may be connected to process lines containing toxic or explosive chemicals, may be encountered. Therefore, it is as important, if not more important, to conduct an underground survey as well as an aboveground survey for any proposed site.

5. GENERAL DESIGN

For older plants, it is not a good practice to rely only on underground piping drawing. In many instances these drawings contain significant errors and omissions.

Underground piping in process plants is generally utility piping, including services such as sewers and drains, city and service water, fire protection, and cooling water supply and return. Electrical power lines and pressure piping also may be underground. Special elements of design should be considered for safety, such as anchoring and thrust blocks to prevent movement of pressured lines, use of cathodic protection to prevent corrosion, and avoidance of process water tie-ins to fire water supply or sanitary water. Points where lines either enter the ground or come out of the ground should be protected from vehicular traffic.

Headers or mains for these services are normally located in open corridors outside plant operational areas for maintenance and modification accessibility. Elevations of lines containing liquids should be below any nearby underground electrical conduits.

Underground process drains should be evaluated for creation or transportation of hazardous or flammable vapors. In normal operation, an open area above the fluid in the drains allows vapors to migrate beyond the areas where they are generated. Such vapors could enter an area where an open flame or electrical sparks could cause combustion. Therefore, oily water sewer systems should be designed with P-traps, submerged outlets, vent tubes, and vapor sealed manholes to prevent flammable vapors from migrating to sources of combustion. Monitoring of the concentration of flammable materials may be necessary.

In transporting hazardous liquids, particularly hazardous wastes, double-walled piping has become the preferred or required method of transport, to prevent the release of the transported materials to the environment. Double-walled piping is also used for transporting highly toxic gases. Double-walled piping normally consists of an inner pipe, an outer pipe, a spacer system which suspends the inner pipe within the outer, and a leak detection system. This type of system is normally used where any release of the material would create a major health hazard. In designing this system, certain elements need to be addressed:

- Both pipe walls and the piping supports should be compatible with the material being transported.
- The supports should be spaced so that the inner pipe will not sag, and potentially rupture, between supports.
- For long pipe runs it may be desirable to zone the leak detection system to pinpoint the location of the leak.

5.9.1.5 Below-Grade Structures

Process or support structures below grade include items such as API separators, pump pits, spill ponds, water treatment facilities, and sumps. Structural failure of pump pits may damage the pumps and associated piping causing uncontrolled release of process fluids.

There may occasionally be a requirement for a hot or cold liquid "dump" system to an isolated underground tank to conserve or isolate expensive or hazardous liquids. The dump piping will be installed and stay at ambient temperature until actually used. Introduction of the process fluid will cause the underground lines to expand or contract.

As with aboveground lines, this movement should be considered in the design. The lines generally run in trenches, with solid or open grating covers, with expansion room at turns. If for some reason (generally, the depth of the lines) it is not practical to trench, the lines should be sleeved, usually with larger bore piping, to allow free movement during growth or shrinkage.

5.9.1.6 Grade Level Structures

The primary plant layout determines the location of roads and other structures that affect excavations and underground piping. For example, road bases can produce heavy loading on underground piping; ruptured piping could lead to process spills or washouts involving dislocation of other plant piping or equipment. An envelope is normally established to ensure proper clearance between pipe racks and any plant roads. Small piping lines can deliver flammable, toxic, or corrosive products as well as large pipes; small piping and electrical lines, only shown diagrammatically, should be kept out of the roadway envelope. Encroachment could lead to an electrical fire or explosion or a power outage. Sometimes this potential hazard is identified only during review of a request for changes to structural steel or revisions to the radius of a road curve.

The ground (concrete, paving, etc.) should be sufficiently sloped to avoid pooling of released materials under process equipment. Creation of low curbs to control runoff can result in pool fires under process equipment and therefore significantly increase fire damage.

5.10 THERMAL INSULATION

Insulation may be applied to a surface to perform one or more functions, such as temperature control (heat conservation or freeze protection), personnel protection, condensation prevention, or sound attenuation. The major process safety issues related to thermal insulation are:

- Fire exposure protection of equipment and piping
- Corrosion under wet insulation
- Spontaneous ignition of insulation wet with flammable or combustible liquids

5.10.1 Properties of Thermal Insulation

This section discusses how these process safety considerations are affected by the properties of insulation, such as thermal performance, moisture absorption, and fire resistance.

5.10.1.1 Thermal Performance

Insulation is used to prevent heat loss or gain for process control and it is often necessary for the protected process system to function properly. For example, if a process fluid condenses or freezes or vaporizes in a line, a hazardous condition may exist, such as overpressurization, loss of process control, or runaway reaction. For calculating heat transfer rates and determining simple heat loss or gain, guidelines are published (Ref. 5-32). Computer programs are available to aid the engineer in selecting the optimum thickness based on a pre-determined set of parameters such as energy costs, local usage rates, and capital costs.

5. GENERAL DESIGN

Insulation is also applied to protect workers from injury. Equipment and piping are generally insulated for personnel protection when the exterior temperature exceeds 140°F (60°C).

5.10.1.2 Absorption of Liquids

Absorption of moisture or process liquids can lead to a hazardous condition, such as lowered thermal performance, corrosion under wet insulation, or a fire if the absorbed liquid is flammable or combustible.

Thermal performance is impaired when the insulation material is wet. Moisture can enter insulation material through a break in the weather barrier, by a leak in steam trace tubing, or by a process leak in the insulated system. When the air spaces in insulation become filled with water or other liquid, the insulation's conductivity approaches that of the liquid. (Ref. 5-33).

5.10.1.3 Fire Safety

Fire safety is related to three major properties of insulation:
- Combustibility of the insulation itself
- Combustibility of absorbed liquids
- Integrity during fire

For maximum safety, insulation should be non-combustible, non-absorptive, non-melting, and well maintained throughout the life of the facility. Insulation materials that increase the facility's combustibility should be avoided. Avoid using plastic foam insulation materials of the polyisocyanurate type. Some plastic foam insulation materials that emit toxic gases when subjected to fire are prohibited in some locations. Insulation materials are tested in accordance with ASTM E 84-10 (Ref. 5-34) for flame spread and smoke development.

Absorption of flammable material creates a fire hazard even though the insulation itself might be non-combustible. Spontaneous insulation fires may occur when a combustible liquid leaks into porous insulation and reaches a temperature where runaway self-heating occurs (Ref. 5-35). There have been numerous serious fires caused by hydrocarbon saturation of open cellulite insulation material. An example is leaking Dowtherm into insulation and spontaneous combustion.

The ability to withstand high temperature exposure, combustion, and smoke development is a desirable quality in an insulation system. Fire-resistant insulation material will not only be fire safe, it will also provide fire protection for the insulated component. In this role, the insulation minimizes the heat transfer to the protected surface and minimizes the potential for failure of the equipment and subsequent release of fuel or hazardous materials. Fire resistance is an alternative to the use of other protective systems such as sprinklers or physical barriers to protect critical systems in the plant. Drips, leaks, and spills from above onto hot process surfaces can result in fire when hydrocarbon contacts a surface that is above the auto-ignition temperature of the hydrocarbon that was released.

5.10.1.4 Fabrication

Some insulation materials perform well thermally but are difficult to fabricate; they do not form well to the substrate or to adjoining insulation sections or shrink after application and leave gaps in the system. These gaps cause "hot spots" on the jacketing surface or cold spots on hot process temperature systems. Poor insulation fit-up and the resulting problems can be reduced if the chosen insulation material is fabricated to standard dimensions and is tested for linear shrinkage and dimensional stability at the conditions for which it is being specified. In addition, allowances should be made for the differential expansion between the pipe and the insulation.

5.10.1.5 Durability

If the insulation does not hold up well in service, the thermal performance and ultimately the safety of the whole system can be affected. Insulation which is crushed or torn may allow a heat flow path or expose the equipment or piping surface to outside elements such as fire, moisture, or corrosive atmospheres. For example, if insulation is damaged on a high temperature line where cabling or instrument tubing runs in close proximity, the tubing could become overheated and fail. Also, insulation should not rip off when hit by fire water.

5.10.2 Selection of Insulation Materials

For optimum thermal performance, the selection of material is the key factor. However, the choice is not as simple as selecting the material with the lowest thermal conductivity. After materials engineers and piping designers have made preliminary choices of materials, the process engineer / safety engineer should look at safety issues of the system as a whole.

Thermal insulation, usually as blocks or batts, provides for thermal efficiency as well as fire protection. Cementitious materials, usually applied wet and activated by fire exposure, can be used for fire protection when thermal efficiency in normal operations is not important. Wright and Fryer (Ref. 5-36) present a good summary of fire protection materials options.

Insulation systems (including jacket, banding, and supports) commonly installed on piping and equipment for reactive chemical service for the purpose of fire protection should incorporate the following features:

- A non-combustible inorganic insulation material such as calcium silicate or cellular glass
- Double-layer construction with all joints staggered
- High melting point jacketing
- Well-secured jacketing, typically by stainless steel bands

High melting point jacketing may be stainless steel or other lower cost jacketing materials developed as alternates. One such material is a sheet steel product with a coating of corrosion-resistant aluminum-zinc alloy applied by a continuous hot dipping process. Aluminum covering should not be used if the insulation is for fire protection.

Some insulation materials may contain trace contaminants such as chlorides which can induce stress corrosion cracking problems in austenitic stainless steel materials.

5. GENERAL DESIGN

ASTM C795 identifies requirements for insulation materials acceptable for use over austenitic stainless steel including corrosion testing and chemical analysis (Ref. 5-37).

5.10.3 Corrosion Under Insulation

Corrosion under thermal insulation, both wet and dry, is recognized as a potential problem. Corrosion is often the initiating event for loss of containment, fire, or explosion. Because the corrosion is hidden, it is usually not discovered until it's too late. Ironically, both the causes and methods of prevention are relatively simple and have been known for years. Selection of thermal insulation has become routine, but potential for deficiencies in fabrication and installation still occur. For example, a serious problem occurred on a multi-storied column subjected to monthly testing of the firewater high pressure spray. The metal weather-jacketing system was not designed to be impenetrable to the upward spray of the system, resulting in water under the insulation (Ref. 5-38).

5.10.3.1 Contributing Factors

Materials of construction for piping and equipment are usually selected on the basis of the internal environment, that is, the process fluids contained. Selection of insulation also should consider the external environment, that is, vapors or fluids, such as rainfall, process fluids, and corrosive gases that may be absorbed by the insulation. The combination of physical and chemical factors in the environment will accelerate corrosion.

5.10.3.1.1 Service Temperature

The temperature range for corrosion under insulation is 140-250°F (60-121°C). At higher temperatures, the corrosion rate is higher even though water is driven off faster. High temperatures can cause localized, very aggressive corrosion at points of evaporation. Corrosion occurs even at lower temperatures; therefore it needs to be considered at all service temperatures.

5.10.3.1.2 Intermittent and Cyclic Service (Temperature Transition)

In high temperature systems when the water is driven from the insulation, salts collect and may result in very aggressive corrosion when the location is rewetted. In low temperature systems, thawing locations exist that typically stay wet, creating localized corrosion. Both thawing and vaporizing transition zones exist on vessel and pipe nozzles, clips, and skirts. Even on the body of a single piece of equipment, the temperature may range from below to above freezing, creating a temperature transition zone.

Corrosion problems are intensified by the cyclical nature of process operations. Service cycles cause temperature cycles and temperature transition zones. Many insulated items spend time in a down cycle for maintenance or for other reasons.

5.10.3.1.3 Equipment Design

In the past, equipment design typically assumed that vapor barriers would remain intact; they do not. New designs can include vapor barrier improvements to keep water out and methods such as drains and vents to let moisture escape. Attachment of nozzles, clips, and insulation should be designed to control moisture into and out of the insulation.

Certain designs contribute to especially corrosive situations. The location of vents and drains, along with faulty sealing methods, allows water entry (and often retention). Size reductions in towers create water trap potentials. Low temperature refrigerated systems can condense and freeze atmospheric moisture resulting in ice buildup which, once begun, further damages vapor barriers and insulation materials.

5.10.3.1.4 Climate

Proximity to airborne salt is a significant problem; plants on the sea coast are more prone to problems. The facility itself may provide a source of moisture and contaminants (such as cooling tower fall-out areas). Olefin plants with sub-ambient conditions can result in condensation dripping which creates an unfavorable climate, especially when airborne salts can be washed from adjacent equipment into insulation.

5.10.3.2 Material Stress Conditions

Residual stresses from fabrication are typically relieved by some sort of thermal stress relief. However, certain fabrication techniques leave steel in as-fabricated conditions. The cold bending of pipe for non-corrosive service (as defined by process material contained) reduces initial fabrication costs significantly for smaller (8 inch and under) diameter pipe but leaves residual stresses that can cause galvanic attack of the outer diameter of the stressed part. Stainless steels typically have sufficient residual stresses from fabrication so that chlorides will cause severe cracking above 140°F (60°C).

5.10.3.3 Prevention of Corrosion

The primary methods of preventing corrosion under wet insulation are preventing the entry of water into the system and protecting the surface of the piping or equipment. Since no insulation system can be presumed to be entirely waterproof, protective coatings are extremely important in preventing corrosion. Methods to reduce corrosion under insulation are:

- Avoid direct contact between dissimilar metals by coating the parts with insulating coatings or petrolatum tape to minimize galvanic corrosion.
- Avoid primary reliance on mastic seals and caulking as a weather barrier, both of which tend to dry with age and exposure to elevated temperatures.
- Design nozzles, manways, ladder and lifting lug clips, platform angle iron mounts, bleeder valves, fittings, valves, etc., for all connections to be outside the insulation.
- Design weather-proofing jacketing such that natural runoff will occur.
- Ensure inspection ports which are designed for water-tight construction are available to allow for corrosion inspection of the substrate.
- Prime and paint carbon steel lines prior to insulation and sealing.
- Use insulating materials which contain low concentrations of chlorides or other contaminants that might induce SCC of stainless steel piping and equipment.

5. GENERAL DESIGN

Most insulating materials contain or can absorb moisture in storage and installation. If a tight, impermeable weather barrier is installed over such insulation and then placed in hot service, the moisture should be allowed to evaporate through release vents.

Installing and maintaining flashing and caulking at structural or piping penetrations of the insulation can prevent water ingress at these locations. The condition of the insulation sealant can determine whether or not corrosion occurs under the insulation. Hydroscopic insulation should be carefully maintained at joints. Although keeping water out is effective in preventing corrosion, it is very difficult to do consistently.

Corrosion problems are most prevalent on insulated steel surfaces operating in the temperature range of 140-250°F (60-121°C). For this service, external protective coatings are especially important. Immersion grade epoxy-phenolics and amine-cured coal tar epoxies are frequently used, depending on the operating temperature. Proper preparation of the surface is critical in determining how well the protective coating works.

For protecting insulated surfaces at 270-1,000°F (130-540°C) a NACE publication (Ref. 5-39) describes coating systems and tapes which are chemically resistant to humid environments containing chlorides and sulfides. Although corrosion may be reduced at very low temperatures, it can be appreciable at intermediate temperatures in the range of -50-35°F (-45-2°C). For these temperatures, NACE provides recommendations for suitable coating materials as well as surface preparation and application methods required for reliable performance.

5.11 HUMAN FACTORS IN DESIGN

Appropriate consideration and implementation of human factors in process design will improve process safety by:
- Making the process and its intended operation easier to understand
- Making procedures clearer and easier for operators to do what is intended
- Limiting potential deviations from intended operations

New facilities should be reviewed for ergonomics and human factors issues during design, construction, and startup. Existing facilities should be reviewed periodically for opportunities to improve human factors in an inherently safer way, including through process hazards analyses. Such reviews are usually performed both periodically for an entire process as well as for significant modifications. The use of human factors checklists, such as the one provided in the CCPS book, *Human Factors Methods for Improving Performance in the Process Industries* (Ref. 5-40), can help improve the application of inherently safer design in existing and modified processes.

Human Factors Methods for Improving Performance in the Process Industries (Ref. 5-40) also describes human factors as the discipline of addressing interactions in the work environment between people, a facility, and its management systems. Reference is made to an International Association of Oil and Gas Producers (Ref. 5-41) model for human factors, which is applicable to the process industries. This model is based on three major areas, each with a number of sub-topics:

1. Facilities and equipment
2. People
3. Management systems

Culture is a factor overriding all of these issues, as it defines the norms in which a system operates, both socially and technically.

Table 5.1 Culture and Working Environment

Facilities & Equipment	People	Management Systems
Work space design	Human characteristics & behavior (physical and mental)	Management commitment
Maintenance	Fitness	Safety culture
Physical characteristics	Stress	Procedures
Reliability	Fatigue	Training
		Hazard identification
		Risk assessment

Simply put, human factors involves working to make the environment function in a way that seems as natural as possible to people. The goal of human factors is to fit the task and environment to the person, rather than forcing the person to significantly adapt in order to perform the work. This reduces the potential for human error that can cause or contribute to process safety and other types of incidents.

Human factors has its origins in the Industrial Revolution and emerged as a full-fledged discipline during World War II when it was recognized that aircraft cockpit designs needed to consider the human interface for controls and displays to ensure safety and reliability of operations. Likewise, human factors has an essential role in the application of inherently safer design. A system or procedure that is designed with human factors as a core focus will be less prone to human error, resulting in reduced risk of safety, process safety, or environment-related incidents.

The subject of human factors in the process industries is treated in depth in *Human Factors Methods for Improving Performance in the Process Industries* (Ref. 5-40), which includes approaches for implementation of such strategies in the designs of plants and their management systems. A Human Factors Tool Kit is also provided.

Designing for human factors minimizes the potential for these types of errors and improves the potential for identification and corrective action in order to minimize the consequences of the error.

The guiding premises for making systems inherently safer against human error are:

- *Humans and the systems designed and built by them are susceptible to error.* Human factors design reviews of new and existing facilities and modifications, such as through process hazard analyses (PHAs) or separate human factors evaluations, as well as reviews of human factors-related root causes or contributing factors in incident investigations (particularly near-misses), can help identify means to reduce the potential for human error.

5. GENERAL DESIGN 157

- *Existing facilities can contain many traps to cause human error.* It is important to identify these potential traps based on operator input, as they alone may be aware of them. Input from both experienced and newer operators should be sought because newer operators may be more aware of the traps that more experienced operators have become used to and found ways to routinely avoid. Elimination of such traps is inherently safer than training and expecting people to avoid them. Input from operators and maintenance personnel can also be valuable in identifying other human factors-related issues. Human factors training often helps personnel identify issues that they may have previously recognized but were unable to understand and express in terms of human factors and the potential for error that could lead to adverse safety consequences.
- *Designers can provide systems to facilitate operator involvement in the process and ensure an appropriate workload.* In modern highly automated chemical plants it is possible for the operators to become too removed from the process such that, should an unexpected event occur, they do not have the knowledge to respond appropriately. Operator workload also has a significant influence on their reliability. Operators that are too busy or not busy enough have both been shown to have an increased likelihood of error. Including operator involvement and workload as parameters in the process design can reduce operator error and facilitate better performance from the operators in responding to unplanned events. (Ref. 5-40).

CCPS (Ref. 5-40), Lorenzo (Ref. 5-42), and Attwood (Ref. 5-43) discuss human error in detail.

The tools in *Human Factors Methods for Improving Performance in the Process Industries* (Ref. 5-40) can be used in each stage of the chemical process life cycle to help evaluate the tradeoffs involving human factors between various options. In many cases, low cost options in design can make the operations inherently safer from a human factors perspective.

Well-designed human systems can produce inherently safer plant designs and operating procedures. Plants and processes that are designed and constructed with careful attention to human factors are inherently safer than those that are not. If we understand how humans work and how human errors occur, we can design better systems for managing, supervising, designing, reviewing, training, auditing, and monitoring. Human factors consideration is an integral part of an inherent safety effort in a company.

5.11.1 Human Factors Tools for Project Management

Figure 5.4 illustrates the human factors tools that have been designed for use during each phase of the planning, development, detailed design and construction, and startup process. Some of the tools are stand-alone, others are integrated into existing key project tools, such as Hazard and Operability (HAZOP) studies. Refer to Attwood, Chapter 9 (Ref. 5-43), for a complete description of each tool in Figure 5.4.

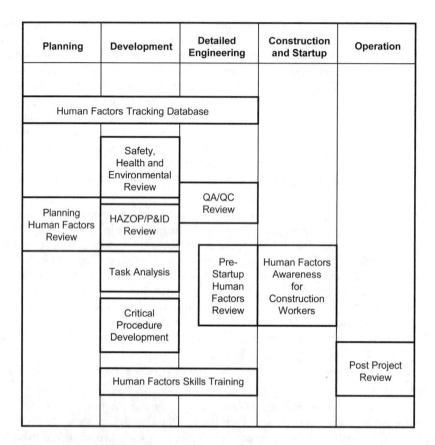

Figure 5.4   Human Factors Tools

5.12   SITE SECURITY ISSUES

Security efforts, like process safety efforts, protect the community and company employees and assets while keeping a facility operational and profitable. A large incident, such as a release of hazardous materials, can injure people, harm the environment, and seriously damage a company by disrupting operations, inviting multimillion-dollar lawsuits, requiring costly remediation, upsetting employees, and injuring the company's reputation.

Security plays a small part in the process safety aspects of design. However, that small part is very important.

The CCPS book *Guidelines for Analyzing and Managing the Security Vulnerabilities of Fixed Chemical Sites* (Ref. 5-44) is a resource for determining the potential vulnerabilities of a processing facility to security events.

Key concepts explained in the book include the following:
- *Layers of Protection* - A concept whereby several different devices, systems, or actions are provided to reduce the likelihood and severity of an undesirable event.
- *Security Layers of Protection* - Also known as concentric *Rings of Protection,* a concept of providing multiple independent and overlapping layers of protection in depth with prevention and mitigation to both increase the reliability of the safeguards as well as to lessen the likelihood of an event escalating to extreme consequence. For security purposes, this may include various layers of protection such as counter-surveillance, counter-intelligence, physical security, and cyber security.
- *Delay* - A security strategy to provide various barriers to slow the progress of a perpetrator in penetrating a site to prevent an attack or theft or in leaving a restricted area to assist in apprehension and prevention of theft.
- *Detect* - A security strategy to identify a perpetrator attempting to commit a chemical security event or other criminal activity in order to provide real-time observation as well as post-incident analysis of the activities and identity of the perpetrator.
- *Deter* - A security strategy to prevent or discourage the occurrence of a breach of security by means of fear or doubt. Physical security systems, such as warning signs, lights, uniformed guards, cameras, and bars, are examples of systems that provide deterrence.

5.12.1 Physical Security

The term physical security refers to equipment, building and grounds design, and security practices designed to prevent physical attacks against a facility's people, property, or information.

Some commonly used physical security measures for processing facilities include:
- Perimeter fences with anti-climbing features
- Adequate illumination of perimeter and key areas at night
- Locked gates at road and railroad entrances
- Bollards to protect process equipment and piping from vehicle impact
- Guard / security personnel sufficient to staff a central station and provide routine checks at key points in the facility
- An electronic access control system that requires the use of key cards at main entrances and on other appropriate doors and that provides an audit trail of ingress and egress
- A closed-circuit television system to monitor key areas of the facility. Where appropriate, employ motion sensors that mark the video recording and alert security staff when someone enters a restricted area
- A system of parcel inspection (using magnetometers, X-ray screening, or explosives detectors). Require the use of property passes for removal of property from the site.

A primary key consideration for new design is to locate the most vulnerable or important locations so that it is the hardest for adversaries to reach.

Facility management should assess its unique security needs and establish an appropriate level of security protection service.

5.12.2 Cyber / Electronic Security

In a process facility, protecting information and computer networks means more than safeguarding a company's proprietary information and keeping the business running, as important as those goals are. It also means protecting chemical processes from hazardous disruptions and preventing unwanted chemical releases. To an adversary, information and network access can equal the power to harm the company, its employees, and the community at large (Ref. 5-44).

The chemical industry well understands the importance of protecting its trade secrets. However, it is also vital to protect information that could be useful to criminals, demonstrators, and terrorists who wish to plan attacks on a chemical site or obtain hazardous materials for weapon building. Examples of such information include:

- Process flow diagrams
- Piping and instrumentation diagrams
- Formulations
- Recipes
- Client and supplier lists
- Site maps
- Other information that describes the workings of a chemical facility

Important concepts of measures for enhancing computer and network security at their facilities include the following:

- Physically secure computer rooms, motor control centers, rack rooms, server rooms, telecommunications rooms, and control rooms, ideally with electronic or biometric access control systems that record ingress and egress and if possible place the computer room above the first floor of the building to reduce the likelihood of theft and water damage. The computer room should not be adjacent to an exterior building wall.
- Employ firewalls, virus protection, encryption, user identification, and message and user authentication to protect both the main computer network and any subsidiary networks, such as access control systems, that are connected to it or to the outside.
- Allow the principles of "least access," "need to know," and "separation of functions" to guide the determination of user authorizations, rather than position or precedent.

5.13 REFERENCES

5-1. API RP 2003. *Protection Against Ignitions Arising out of Static, Lightning, and Stray Currents*, Seventh Edition. American Petroleum Institute. Washington, D.C. 2008.

5-2. CCPS. *Guidelines for Preventing Human Error in Process Safety*, Center for Chemical Process Safety of the American Institute of Chemical Engineers. New York, NY. 1994.

5-3. EPA. *Risk Management Program (RMP)*. 40 CFR 68. U.S. Environmental Protection Agency. Washington, D.C. 1996.

5-4. MIIB. The Buncefield Incident, 11 December 2005. The final report of the Major Incident Investigation Board. 2008. http://www.buncefieldinvestigation.gov.uk/reports/

5-5. CCPS. Final Report: *Definition for Inherently Safer Technology in Production, Transportation, Storage, and Use*. Center for Chemical Process Safety of the American Institute of Chemical Engineers. New York, NY. 2010.

5-6. CCPS. *Inherently Safer Chemical Processes, A Life Cycle Approach*. Center for Chemical Process Safety of the American Institute of Chemical Engineers. New York, NY. 2009.

5-7. CCPS. *Guidelines for Safe Automation of Chemical Processes*. Center for Chemical Process Safety of the American Institute of Chemical Engineers. New York, NY. 1993.

5-8. CCPS. *Guidelines for Safe and Reliable Instrumented Protective Systems*. Center for Chemical Process Safety of the American Institute of Chemical Engineers. New York, NY. 2007.

5-9. ANSI / ISA 18.2-2009, *Management of Alarm Systems for the Process Industries*. International Society of Automation, Research Triangle Park, NC. 2009.

5-10. Sanders, R. *Chemical Process Safety: Learning from Case Histories*, 3rd *Edition*. Elsevier. Oxford, UK. 2005.

5-11. CCPS. *Guidelines for Independent Protection Layers and Initiating Events*. Center for Chemical Process Safety of the American Institute of Chemical Engineers. New York, NY. 2011.

5-12. ISA 84.91.01. *Identification and Mechanical Integrity of Instrumented Safety Functions in the Process Industry*. International Society of Automation, Research Triangle Park, NC. 2011.

5-13. ANSI / ISA 84.00.01-2004 (IEC 61511 modified). *Functional Safety: Safety Instrumented Systems for the Process Industry Sector*. International Society of Automation, Research Triangle Park, NC. 2004.

5-14. ISA TR84.00.04. *Guidelines on the Implementation of ANSI/ISA 84.00.01-2004 (ISA 61511 Modified)*. International Society of Automation, Research Triangle Park, NC. 2006.

5-15. API RP 752. *Management of Hazards Associated with Location of Process Plant Permanent Buildings, Third Edition*. American Petroleum Institute. Washington, D.C. 2009.

5-16. API RP 753. *Management of Hazards Associated with Location of Process Plant Portable Buildings, First Edition*. American Petroleum Institute. Washington, D.C. 2007.

5-17. NFPA 30. *Flammable and Combustible Liquids Code, 2008 Edition*. National Fire Protection Association. Quincy, MA. 2008.

5-18. CCPS. *Guidelines for Facility Siting and Layout*. Center for Chemical Process Safety of the American Institute of Chemical Engineers. New York, NY. 2003.

5-19. CCPS. *Guidelines for Fire Protection in Chemical, Petrochemical, and Hydrocarbon Processing Facilities*. Center for Chemical Process Safety of the American Institute of Chemical Engineers. New York, NY. 2003.

5-20. Hanna, S.R. and Britter, R.E. *Wind Flow and Vapor Cloud Dispersion at Industrial and Urban Sites*. Center for Chemical Process Safety of the American Institute of Chemical Engineers. New York, NY. 2002.

5-21. CCPS. *Guidelines for Safe Storage and Handling of High Toxic Hazard Materials*. Center for Chemical Process Safety of the American Institute of Chemical Engineers. New York, NY. 1988.

5-22. CCPS. *Guidelines for Vapor Cloud Explosions, Pressure Vessel Burst, BLEVE and Flash Fires Hazards*. Center for Chemical Process Safety of the American Institute of Chemical Engineers. New York, NY. 2010.

5-23. API RP 571. *Damage Mechanisms Affecting Fixed Equipment in the Refining Industry*. American Petroleum Institute. Washington, D.C. 2007.

5-24. Hurst, L.R. *Brittle Fracture of a Brick-lined Pressure Vessel*. Materials Performance, Volume 25, No. 3, pp 24-26. 1986.

5-25. National Board Inspection Code. *The National Board of Boiler and Pressure Vessels Inspectors*. Columbus, OH. 2010.

5-26. Lees, F.P. *Loss Prevention in the Process Industries, Third Edition*. Elsevier, Inc. Oxford, UK. 2005.

5-27. API RP 941. *Steels for Hydrogen Service at Elevated Temperatures and Pressures in Petroleum Refineries and Petrochemical Plants, Seventh Edition*. American Petroleum Institute. Washington, D.C. 2008.

5-28. ANSI / NACE MR0175 / ISO 15156. *Petroleum and Natural Gas Industries - Materials for Use in H2S-containing Environments in Oil and Gas Production - Parts 1, 2 and 3*. National Association of Corrosion Engineers. Houston, TX. 2008.

5-29. NACE. MR0103-2007, *Materials Resistant to Sulfide Stress Cracking in Corrosive Petroleum Refining Environments*, National Association of Corrosion Engineers. Houston, TX. 2007.

5-30. Sundararajan, C. *Guide to Reliability Engineering: Data, Analysis, Applications, Implementation and Management*. Van Nostrand Reinhold, New York, NY. 1991.

5. GENERAL DESIGN 163

5-31. Sundararajan, C. *Structural Engineering Aspects of Plant Risk Assessment.* AIChE Process Plant Safety Symposium, p. 940. American Institute for Chemical Engineers. Houston, TX. 1992.

5-32. ASTM C680-08. *Standard Practice for Estimate of the Heat Gain or Loss and the Surface Temperatures of Insulated Flat, Cylindrical, and Spherical Systems by Use of Computer Programs.* American Society for Testing Materials Philadelphia, PA. 2008.

5-33. Malloy, J.F., and Turner, W.C. *Thermal Insulation Handbook*, McGraw-Hill. New York, NY. 1981.

5-34. ASTM E84-10. *Standard Test Method for Surface Burning Characteristics of Building Materials.* American Society for Testing Materials. Philadelphia, PA. 2010.

5-35. Britton, L.G. *Spontaneous Fires in Insulation.* Plant / Operations Progress, Vol. 10, No.1. 1991.

5-36. Wright, J.M., and K.C. Fryer. *Alternative Fire Protection Systems for LPG Vessels.* GASTECH 81 LNG / LPG Conference, Gastech Ltd., Herts, U.K. 1981.

5-37. ASTM C795-08. *Standard Specification for Thermal Insulation for Use in Contact with Austenitic Stainless Steel.* American Society for Testing Materials. Philadelphia, PA. 2008.

5-38. Pollock, W.I., and C.N. Steely Eds. CORROSION / 89 Symposium: *Corrosion Under Wet Thermal Insulation: New Techniques for Solving Old Problems.* National Association of Corrosion Engineers. Houston, TX. 1990.

5-39. NACE. *A State-of-the-Art Report of Protection Coatings for Carbon Steel and Austenitic Stainless Steel Surfaces Under Thermal Insulation and Cementitious Fireproofing.* National Association of Corrosion Engineers. Houston, TX. 1989.

5-40. CCPS. *Human Factors Methods for Improving Performance in the Process Industries.* Center for Chemical Process Safety of the American Institute of Chemical Engineers. New York, NY. 2007.

5-41. International Association of Oil and Gas Producers. *Human Factors*. London, UK. 2005. (www.info.ogp.org.uk/hf/).

5-42. Lorenzo, D.K. *A Manager's Guide to Reducing Human Errors: Improving Human Performance in the Chemical Industry.* Chemical Manufacturers Association. Washington, D.C. 1990.

5-43. Attwood, D.A., Deeb, J.M., and Danz-Reece, M.E. *Ergonomic Solutions for the Process Industries.* Elsevier, Inc. Oxford, UK. 2004.

5-44. CCPS. *Guidelines for Analyzing and Managing the Security Vulnerabilities for Fixed Chemical Sites*, Center for Process Safety of the American Institute of Chemical Engineers. New York, NY. 2003.

6

EQUIPMENT DESIGN

The design solutions presented in the tables in this chapter are established and offer well proven approaches for mitigating the failure scenarios. However, a potential design solution is false protection if it is not reliably engineered and maintained. Active solutions in particular may need redundancy (i.e., dual sensors, separation of control and interlock functions) to provide the required level of reliability and risk reduction. True redundancy must include the absence of common mode failures by providing independence and functional diversity (e.g., independent power supplies, sensors operating on different principles). The advantage of a risk based approach to design selection is that it provides the means for determining how much redundancy is enough.

The design should also take into account the need for periodic inspection and proof testing of systems. For example, Pressure Safety Valves (PSVs) may need testing at intervals that are shorter than scheduled plant turnarounds. A good engineering design solution is the installation of dual PSVs to allow testing at prescribed intervals without interfering with production.

Safety design solutions can contribute to hazards if not properly maintained. While system maintenance is not specifically addressed, this book assumes the safety equipment will be subjected to a maintenance and inspection program once installed. Material of construction should be specified and selected to minimize corrosion, because external visual inspection would be difficult and interior visual inspection would be expensive and would increase downtime.

The importance of a documented Design Engineering Package cannot be emphasized enough. This documentation is not only critical during the design phase, but is essential for operations and maintenance throughout the life cycle of the facility. Design Engineering packages are further discussed in Chapter 8, Documentation.

It should also be recognized that the failure scenarios presented in the tables focus on process-related hazards rather than maintenance-initiated incidents. It is further assumed that the facility has adequate safe work practices, which encompass hot work permits, confined space entry, ignition control, lockout / tagout, etc.

Information on equipment failure scenarios and associated design solutions is introduced in table format. The organization of the tables is the same in each section. The table headings are described below.

- *Events* - Specific failure mechanism / cause (e.g., control system failure).
- *Consequence* - Potential outcome if the cause were to occur and no intervention happens. In many cases, loss of containment is the final consequence.

Depending on the material released, the ultimate outcome could be fire, explosion, toxic exposure, environmental impact, or no impact at all.

- *Potential Design Solution* - Potential design solutions that could be implemented to reduce the risk of failure scenario. The design solutions are grouped into the following three categories: inherently safer / passive, active, and procedural. Design solutions identified could be considered protection layers in a hazard assessment (e.g., HAZOP or LOPA).

Particular attention should be given to the "Generally Applicable' row of each individual equipment table that begins each "Event / Consequence" category. These potential design solutions are relevant to each event / consequence scenario for a given process deviation. For example, under the "Generally Applicable - High Pressure" category, pressure relief devices are likely listed as an "active" potential design solution and, therefore, they are an "active" design solution for all subsequent "Event / Consequence" scenarios in the high pressure category.

It should be recognized that the design solutions presented are only some of the possible approaches for reducing the risk of the associated failure scenario. The authors of this book could not anticipate all the possible applications or conditions that may pertain to a specific design situation. Individual company risk tolerability criteria also vary, which could affect the chosen design solutions. The design solutions are also not equal in cost / benefit ratio. Therefore, it is intended that the table be used in conjunction with the design basis selection methodology presented in Chapter 2 to arrive at the optimal design solution for a given application.

Use of the design solutions presented in the tables for each equipment section should be combined with sound engineering judgment and consideration of all relevant factors. For example, assume that it is decided that a nitrogen blanketing system will be installed on an atmospheric storage tank to reduce the risk of forming a flammable mixture that could result in an internal explosion. Typically, nitrogen supply pressures are significantly higher than the design pressure of a storage tank (Ref. 6-1). Consequently, the total system design also needs to address the hazard of overpressure due to uncontrolled opening of a high pressure utility system.

This example illustrates an important aspect of the intended use of the equipment failure tables. The design and installation of safety systems, especially active solutions, can also introduce potential hazards that were not originally present. Therefore, it is necessary to use the table in the context of the total design concept to ensure that all hazards have been considered. As shown in the example, this may involve combining several scenario design solutions to arrive at a final acceptable design. Consequently, the table should be consulted at various stages of the design to reaffirm that failure mechanisms are considered.

Utilizing several design solutions for the same scenario is also possible and often desirable. Again referring to the design of a flammable liquid storage tank, employing ignition source controls (e.g., non-splash filling, grounding) as well as vapor space inerting may be desirable based on the consequences of catastrophic tank failure.

6. EQUIPMENT DESIGN

In addition to providing the required degree of reliability for any one failure scenario, multiple safeguards may be the optimum approach to process deviations caused by very different failure scenarios. The LOPA analysis is one technique that determines if sufficient layers of protection are available. The LOPA analysis is discussed further in Chapter 4.

> For example, suppose a vessel can be overpressured by deflagration in the vapor space in one scenario and by runaway reaction in another scenario. The deflagration event may be characterized by a high pressure rise rate but a modest maximum pressure rise. The runaway reaction may be characterized by a very high pressure rise but a modest maximum reaction rate early in the runaway. With this disparity in the scenarios, the optimum safeguard design might be pressure containment for the deflagration and emergency pressure relief for the runaway reaction. In this situation, these safeguards are not redundant.
>
> The tables contain numerous design solutions derived from a variety of sources and actual situations. Many of the solutions are readily understood. In some instances, additional explanation is warranted to fully appreciate the approach. The failure scenarios and design solutions section contains additional information on selected design solutions. The information is organized and cross referenced by the scenario number in the table.

6.1 VESSELS

This section presents potential failure mechanisms for vessels and suggests design alternatives for reducing the risks associated with such failures. The types of vessels covered in this section include:

- In-process vessels (surge drums, accumulators, separators, etc.)
- Pressurized tanks (spheres, bullets)
- Atmospheric, fixed roof storage tanks (cone / dome roof)
- Atmospheric storage tanks (cone, cone with internal floating roof, floating roof tanks)

Reactors and mass transfer equipment are a unique subset of vessels, in that they are specifically intended to process chemical reactions. Because reactors have unique failure scenarios specifically attributable to the reaction (e.g., reactant accumulation), Section 6.2 is devoted to this class of equipment. However, many of the generic vessel failure modes discussed in this section, such as corrosion-related failures or auto-polymerization, may also apply to reactors.

6.1.1 Past Incidents

Important lessons can be learned from prior mistakes. Several case histories of incidents involving vessel failures are provided to reinforce the need for the safe design and operating practices presented in this chapter.

6.1.1.1 Storage Tank Stratification Incident (Scenarios 5 and 24)

Acetic anhydride is used as an acetylating agent for many compounds. When it reacts with a hydroxyl group, acetic acid is formed as a by-product. Pure acetic anhydride will react energetically with water to form acetic acid. In typical acetylation reactions, an excess of anhydride is used to drive the reaction to completion. This excess is then reacted in the receiver tank with water to convert the excess anhydride to acid. The acid is then refined and reconverted into anhydride. This operation can be performed safely, since the presence of acetic acid makes water and acetic anhydride miscible, and therefore the rate of reaction can be controlled by the rate of water addition.

In this case, the acetylation reaction did not proceed as designed, due to an inadvertent omission of the strong mineral acid catalyst needed to initiate the reaction at low temperatures [-10°F (-23°C)]. Thus, the receiver tank did not contain a mixture of acetic anhydride and acetic acid, but contained only very cold, pure anhydride. The operator in charge of the water addition did not realize the change in composition and additionally failed to turn the tank agitator on prior to beginning the water addition. After several minutes of water addition, he realized his mistake with the agitator and hit the start button. Immediately, the water, which had layered out on top of the cold anhydride, mixed and reacted violently. This caused a partial vaporization in the tank and eruption through an open manway, resulting in fatal burning of the operator.

Lessons learned include the importance of verification that the agitator is turned on prior to beginning the water addition. If this had occurred, the reaction rate would have again been controlled by the water addition rate. However, the water was added at near-stoichiometric concentrations virtually instantaneously, resulting in an uncontrolled exothermic reaction.

Design solutions for a safer process include:
- A design that does not require an opening (Inherent).
- A design with a much smaller opening and / or located away from the operator (Passive).
- Interlock water addition with agitator operation (Active).
- Interlock a surrogate indication of acetylation "non-reaction" (temperature, or cooling duty) prohibiting transfer to receiver vessel (Active).

6.1.2 Failure Scenarios and Design Solutions

Table 6.1 presents information on equipment failure scenarios and associated design solutions specific to vessels.

6.1.2.1 Ignition of Flammable Atmosphere (Scenario 2)

When applying vapor space inerting, there are some special circumstances that need to be recognized; namely, the presence of oxygen is needed for some hazard mitigation measures. For example, the corrosion inhibiting mechanism of certain metals (e.g., stainless steel) depends on the presence of some oxygen. Likewise, some polymer formation inhibitors that are added to reactive materials need oxygen to stay active. In such situations, a limiting oxygen atmosphere may achieve the desired balance between inhibitor activity and flammability protection.

6. EQUIPMENT DESIGN

Flame arresters are often implicated in vessel incidents, not because they are ineffective, but because they are misapplied or improperly maintained. Flame arresters that are not routinely inspected can become plugged (e.g., condensation / corrosion by stored fluids, foreign debris). Eventually, the protected vessel can be subjected to overpressure or vacuum conditions if the vessel is not protected by a relief device (Ref. 6-2).

Table 6.1   Common Failure Scenarios and Design Solutions for Vessels

No.	Event	Consequence	Potential Design Solutions		
			Inherently Safer / Passive	Active	Procedural
Pressure					
	Generally Applicable – High Pressure (Applicable to all high pressure scenarios)		Vessel designed for maximum utility pressure, supply pressure, upstream pressure	Pressure relief device BPCS control loop to vent pressure to safe location High pressure interlocked to isolate source Interlock to isolate vessel inlet or trip feed pump on high pressure Pressure control valve to open to safe location (flare)	Operator response to high pressure alarm
	Generally Applicable – Low Pressure (Applicable to all low pressure scenarios)		Vessel designed for maximum vacuum (full vacuum rating)	Vacuum relief system Automatic blanketing pressure control to minimize vacuum Low pressure interlocked to isolate vessel	Operator response to low pressure alarm
1	Opening of high pressure utility system	Potential increased pressure	Incompatible utility couplings to prevent connections of high pressure utilities No utility connections above pressure rating of vessel		Labeling of utility connections Written procedures and training to verify pressure before operating

Table 6.1   Common Failure Scenarios and Design Solutions for Vessels

No.	Event	Consequence	Potential Design Solutions		
			Inherently Safer / Passive	Active	Procedural
2	Flammable atmosphere in vessel vapor space	Potential ignition in vapor space resulting in fire / explosion	Floating-roof tank instead of fixed roof (see procedural) Ignition source controls (e.g., lightning protection, permanent grounding / bonding, non-splash filling including dip pipe, fill line flow restriction, or bottom inlet) Vessel designed for deflagration pressure	Explosion venting (e.g., frangible roof for fixed-roof tank) Vapor space combustible concentration control Vapor space inerting Emergency purge and / or isolation activated by detection of flammable atmosphere	Oxygen analyzer with alarm Written procedures and training for no transfers during electrical storms Written procedures and training to feed empty tanks at low rate until fill line submerged, avoiding splash filling
3	Inadequate or obstructed vent path	Potential increased pressure	Outlet block valve minimization Outlet sized to eliminate or reduce likelihood of plugging Vent screen to avoid entrance of foreign objects	Heat tracing of vent to avoid condensation and solidification	Written procedures and training for securing valves open via seals or locks Written procedures and training to periodically examine vent opening for obstructions Written procedures and training to verify open vent path before initiating fill operation
4	Contamination with high vapor pressure material	Potential increased pressure	Incompatible couplings to prevent unintended mixing of materials	Explosion venting (e.g., frangible roof for fixed-roof tank)	Written procedures and training for isolation of volatile materials by blinding, removable spool, disconnection, etc.
5	Roll-over or collapse of stratified layers	Potential increased pressure or rapid uncontrolled reaction of stratified layers	In-line mixer external to vessel to premix feeds Tank filling system design that avoids tank stratification (e.g., top splash filling)	Mechanically agitate or recirculate tank contents Provide recycle loop to mix vessel contents	Written procedures and training on filling procedure to avoid stratification
6	Failure of upstream process controls, resulting in vapor or flashing liquid feed	Potential increased pressure	Control valves properly sized Restriction orifice to limit pressure rise	Interlock to isolate vessel on high or low level Redundant level measurements on upstream vessels interlocked on low level	Operator response to high pressure alarm Written procedures and training on how to respond to low level event

6. EQUIPMENT DESIGN

Table 6.1 Common Failure Scenarios and Design Solutions for Vessels

No.	Event	Consequence	Potential Design Solutions		
			Inherently Safer / Passive	Active	Procedural
7	Uncontrolled condensation / absorption of vapor phase component	Potential increased vacuum	Insulation on vessel Open vent to atmosphere	Seal pots Temperature controllers with alarms and interlocks	Written procedures and training for monitoring temperature and addition rate of materials Operator response to low pressure alarm
8	Control or equipment failure in vapor recovery system on refrigerated / chilled storage	Potential increased vacuum	Additional insulation to prolong acceptable refrigeration outage	High pressure interlock to automatically start spare compressor	Written procedures and training for operator startup of spare compressor on high pressure indication
9	Vent / seal freezing – in high humidity, low pressure (near atmospheric tanks)	Potential high pressure on level increase, potential vacuum on level decrease		Freeze protection for overflow seals and tank vents	Written procedures and training for visual inspection by operator, especially during cold weather
Flow					
	Generally Applicable – More Flow *(Applicable to all more flow scenarios)*			Feed interlock activated by high flow Automated flow control loop on fill line based on vessel level	Operator response to high flow alarm Operator response to high level alarm Written procedures and training to limit flow to a maximum safe value Written procedures and training to monitor filling rate and prevent excessive fill rate
10	Excessive fill rate	Potential increased level and pressure in vessel	Flow restriction orifice in fill line Grounding and bonding on vessel and transfer lines Non-static producing material	Pressure controllers with alarms and interlocks	Written procedures and training to feed empty tanks at low rate until fill line submerged, avoiding splash filling

Table 6.1   Common Failure Scenarios and Design Solutions for Vessels

No.	Event	Consequence	Potential Design Solutions		
			Inherently Safer / Passive	Active	Procedural
		Potential static accumulation, potential fire / explosion	Flow restriction orifice in fill line Grounding and bonding on vessel and transfer lines Non-static producing material	Pressure controllers with alarms and interlocks	Written procedures and training to feed empty tanks at low rate until fill line submerged, avoiding splash filling
11	Internal heating / cooling coil leak or rupture	Potential reaction with vessel contents	Electrical heating External heater / cooler (jacket) Heating / cooling medium that is not reactive with vessel contents Lower pressure heating or cooling medium	Back pressure control with external heating / cooling circulation to avoid leak into vessel High temperature and / or pressure alarm and automatic addition of quench / diluent fluid or inhibitor	
12	Excessive emptying rate	Potential vacuum, potential loss of containment	Vessel designed for vacuum Restriction orifice on line	Vacuum relief	Operator response to high flow alarm
13	Electrostatic spark discharge during charging of liquids	Potential fire / explosion	Dip leg to minimize static accumulation Ground and bonding on vessel Non-static producing material Bottom filling of vessel	Automatic inerting of vessel prior to addition	Written procedures and training for manual grounding and bonding of container to vessel Written procedures and training for manual inerting of vessel prior to liquid addition Written procedures and training to avoid use of non-conductive containers
Temperature					
Generally Applicable – High Temperature (Applicable to all high temperature scenarios)			Vessel designed for maximum expected temperature	High temperature alarm and interlock that isolates the heating medium	Operator response to high temperature alarm
Generally Applicable – Low Temperature (Applicable to all low temperature scenarios)			Vessel designed for minimum expected temperature	Low temperature alarm and interlock	Operator response to low temperature alarm

6. EQUIPMENT DESIGN

Table 6.1   Common Failure Scenarios and Design Solutions for Vessels

No.	Event	Consequence	Potential Design Solutions		
			Inherently Safer / Passive	Active	Procedural
14	External fire	Potential increased temperature and pressure in vessel	Buried (underground or bermed) tank (consider environmental issues) Fire safe valves Fireproof insulation (limits heat input) Locate outside fire affected zone Slope-away with remote impounding of spills Tank-to-tank separation to minimize escalation	Automatic closure of isolation valves on fire detection Fixed water spray,(deluge) and / or foam systems activated by flammable gas, flame, and / or smoke detection devices Relief valves sized for external fire scenario	Emergency response plan Emergency response team Written procedures and training for manual activation of fixed water spray (deluge) and / or foam systems Written procedures and training preventing flammable materials (including insulation, enclosures, etc.) in area of the vessel or tank
15	Insulation fires	Potential increased temperature and pressure in vessel	Closed cell insulation provided Liquid tight seal provided where there is likelihood for liquid hydrocarbon soaking into the insulation	Fixed water spray,(deluge) and / or foam systems activated by flammable gas, flame, and / or smoke detection devices Relief valves sized for external fire scenario	Emergency response plan Emergency response team Written procedures and training for manual activation of fixed water spray (deluge) and / or foam systems
16	Excessive heat input or loss of cooling	Potential initiation of an uncontrolled thermal runaway reaction resulting in increased pressure	Temperature control of heating medium (e.g., use hot water instead of steam)	Addition of quench on high temperature	
17	Excessive mechanical agitation	Potential increased temperature resulting in unexpected chemical reaction	Limit agitator motor or re-circulating pump power Uninsulated vessel to allow heat loss	Motor shutdown on high temperature detection	Written procedures and training to turn off motor on high temperature indication

Table 6.1   Common Failure Scenarios and Design Solutions for Vessels

No.	Event	Consequence	Potential Design Solutions		
			Inherently Safer / Passive	Active	Procedural
Level					
	Generally Applicable – High Level *(Applicable to all high level scenarios)*		Diking or drainage to remote impounding Overfill line to safe location	High level alarm and automatic feed cutoff / isolation	Operator response to high level alarm Written procedures and training to monitor level during transfer Written procedures and training to stop feed when level reaches a certain point Written procedures and training to verify tank has sufficient free board prior to transfer
	Generally Applicable – Low Level *(Applicable to all low level scenarios)*		Gravity feed or run-dry type pump	Low level alarm with interlock to automatically shut down the transfer pump	Operator response to low level alarm
18	Liquid overfill	Potential contamination of common vent headers, utility headers, and other connected equipment	Independent vent paths		
19	Level control valve fails closed	Potential increased level in vessel	Closed-loop filling		
20	Low level (floating-roof tank)	Potential for floating roof sitting on its internal legs, possible ignition of flammable atmosphere in tank vapor space	Underflow nozzle located to maintain a minimum liquid level in the tank	Electrical bonding of floating roof to tank	Written procedures and training to monitor tank level periodically

6. EQUIPMENT DESIGN

Table 6.1   Common Failure Scenarios and Design Solutions for Vessels

No.	Event	Consequence	Potential Design Solutions		
			Inherently Safer / Passive	Active	Procedural
21	Fill rate exceeds overflow line / vent capacity	Potential to overpressure tank on high level (back pressure on overflow line floods vent and exceeds tank pressure rating)	Overflow sized for maximum fill rate (vent line and overflow lines are separate)	High level interlocks below overflow point	
Equipment Failure					
22	Subsidence of soil below vessel	Potential vessel damage	Tank foundation design and construction (piling and soil compaction)		Written procedures and training for operator response to indication of tank subsidence Leak detection testing
23	Floating roof sinks from snow or water on top of roof or corrosion of roof / pontoons	Potential seal failure	Corrosion-resistant material selection for floating roof Double deck or pontoon floating roof Fixed roof to protect the floating roof Internal legs or downward limiting stop devices		Emergency response procedures Written procedures and training for periodic draining of roof Written procedures and training for periodic inspection and repair of pontoons
24	Failure of agitator	Potential stratification of immiscible layers resulting in poor product quality	Compatible / mutually soluble materials External, inline mixing of feeds before entering tank	Agitator monitor interlocked to stop feed stream Automatic backup pump around system In tank sensor to monitor agitator blade movement	Written procedures and training for manual activation of back-up pump around system Written procedures and training for manual shut off of feed on detection of loss of agitation

Table 6.1   Common Failure Scenarios and Design Solutions for Vessels

No.	Event	Consequence	Potential Design Solutions		
			Inherently Safer / Passive	Active	Procedural
25	Tank lining failure	Potential rapid corrosion of tank wall / floor, potential tank failure	Materials of construction compatible with material stored (including temperature considerations)		
			Corrosionresistant secondary containment, including complete resistant foundation under tank as part of secondary containment		

6.1.3   Design Considerations

6.1.3.1   Process Vessels

The process conditions of a vessel will influence all activities that contribute to the safe operation of the vessel. The reliability and integrity of process vessels begin with the definition of the process requirements followed by mechanical design activities including material selection and continue with the fabrication techniques and quality assurance practices. After the vessel is in operation, the service requirements, maintenance practices, and inspection techniques will influence the length of time that the vessel can remain in service.

During the design phase, special attention is required to properly define vessel design parameters. Codes and standard practices are available to address design pressures and temperatures but attention to less obvious design factors must also be made, including the need for internal or external corrosion allowance, fluid-specific gravity, thermal stress, and external loads such as wind, snow, and earthquake.

Process conditions (including commissioning, startup, normal operation, shutdown, and upset conditions) must be accurately defined before the mechanical design efforts are started. Issues such as cyclic pressure and temperature, the potential for auto-refrigeration, and very high and low operating temperature will affect the vessel design. Sulfur- and hydrogen-containing environments are significant and measures must be employed to prevent hydrogen embrittlement, delaminations, and stress cracking in the vessel (Ref. 6-3).

An accurate definition of the vessel operating conditions, maximum and minimum excursions as well as normal, is required. Rapid cyclic heating of vessels is not desirable since this may cause local cracking of material. Minimum design metal temperatures dictate impact test requirements for materials in cold service. Auto-refrigeration upon depressurizing should be addressed.

6. EQUIPMENT DESIGN

The fabrication techniques and inspections conducted during fabrication will greatly influence the quality of the finished vessel. Faulty fabrication, for example, poor welding, improper heat treatment, dimensions outside tolerances allowed, or improper assembly, may cause problems to develop in pressure vessels. Vessel fabrication should be independently verified to ensure the vessel is fabricated per the specification.

Mechanical forces can cause a vessel to fail unless adequate provision has been made for such forces, e.g., thermal shock, cyclic temperature changes, vibration, excessive pressure surges, thrust from relief devices, and other external loads.

Internal components such as baffles, agitators, and trays should be installed in such a manner that liquid and vapors are not trapped, which might prevent them from being drained or vented from the vessel. Although intermittent tack welding may provide sufficient mechanical strength for baffles or tray support rings, complete fillet welds are preferred so that crevices and pockets are not created that could produce hidden locations for corrosion.

Agitators present a different set of challenges for pressure vessels. They not only bring with them the usual hazards of leaking seals, vibration, and alignment, agitators also apply additional loads beyond static and dynamic (torque) to the vessel head. Normal torque loads are in the same plane as the nozzle face and determined from the horsepower required for the agitator motor.

6.1.3.2 Gas / Liquid Separators

Gas-liquid separators are commonly used to disengage liquid from a two-phase mixture of gas and liquid by gravity or centrifugal force. Typical applications for gas-liquid separators include, natural gas-crude oil separators, compressor suction liquid knockout drums, and distillation tower reflux drums. All of these applications share the same design basis and concerns as process vessels. Gas-liquid separators are frequently equipped with a demisting pad to prevent the carryover of liquids into the exiting vapor and a vortex breaker located above the bottom outlet nozzle of the separator to prevent vapor entrainment in the liquid (gas blowby). Gas entrainment in the liquid stream can damage control valves, overpressure downstream vessels, and lead to product contamination.

Process variables and parameters to be considered include vapor and liquid velocity, liquid level, and vapor and liquid density. Liquid carryover may occur when vapor velocities are far in excess of design velocities or when the liquid level in the separator rises past the elevation at which the gas-liquid stream enters the vessel. If the separator is used as a compressor suction drum, liquid carryover can cause serious damage to the compressor. Liquid carryover can be prevented by maintaining good level control of liquid in the vessel. High level instrumentation can be used to alert the operator and shut down critical equipment (compressors) if necessary.

6.1.3.3 Storage Tanks and Vessels

The first approach for safer storage tanks and vessels is a good passive design, including:
- Foundations, fabrication techniques, and anchorages
- Design of related pipework and fittings to consider stresses due to movement, expansion / contraction, vibration, connections, valves, and layout

- Selection of ancillary equipment including pumps, compressors, vaporizers, etc.
- Consideration of the range of operations as well as non-operational periods such as commissioning, decommissioning, unit shutdowns, and tank cleaning

Detailed information on mechanical design, fabrication, and non-destructive examination of storage vessels is found in many standard references. Design of storage vessels and related piping is addressed in API, ASME, and UL standards. For additional information, refer to the references and suggested readings in Section 6.1.4.

Whether intended for use at atmospheric, low pressure, or high pressure conditions, the primary considerations of tank design are stresses, both pressure and thermal, including fire exposure. The objective is to maintain working pressure within permissible limits by providing adequate pressure and vacuum relief.

The two main types of large tanks used for storing liquids at near-atmospheric conditions are the welded vertical flat bottom tank with a fixed roof (cone, flat, or domed) and the welded vertical tank with a floating roof in place of the fixed roof. Both types can be used to store hazardous materials. The fixed-roof tank is normally preferred in applications where it is desirable to collect and treat all emissions from the tank or where an inert gas is used to reduce the possibility of fire, explosion, or chemical reaction. Floating-roof tanks are typically used where the vapor pressure of the stored fluid would be excessive for a cone roof tank or where control of emissions from the tank is not required but still desirable. It should be recognized that a drain should be provided for removing water from external floating-roof tanks, but drains can plug up and the roof could then flood and sink.

For environmental emission controls, domed or cone roof tanks with internal floaters are sometimes used. For many types of materials, particularly for organics, the type of tank that may be used will be governed by the EPA or by state environmental authorities. The material's vapor pressure is the main determining factor. Most materials with a vapor pressure below 1.5 psia can be stored in fixed-roof tanks. Materials with a vapor pressure between 1.5 and 11 psia should be stored in at least a floating-roof tank. Tank emissions must be recovered for reuse or destruction for materials with a vapor pressure over 11 psia.

For materials such as butane or ammonia that are normally stored as pressurized liquids, pressure vessels are used. For liquids or gases requiring high pressure storage, horizontal tanks on saddles are used. These tanks are cylindrical with elliptical or domed pressure heads.

6.1.3.4 Atmospheric Storage Tanks

Hazards associated with atmospheric tanks (ambient pressure to 0.5 psig) include overpressure and vacuum, vapor generation, spills, tank rupture, fire, and product contamination. In addition, differential settlements, and seismic wind loadings are important concerns (Refs. 6-1 and 6-4).

Internal deflagration is a concern because of the presence of a flammable / air mixture in the presence of an ignition source. Static is a common ignition source and will be impacted by the conductivity of the fluid, the manner in which the vessel is filling (e.g., splash filling), or the contents mixed and the grounding of the vessel. This mixture can occur during filling, emptying, or mixing in tanks that contain organic vapors near their flash point. Air ingress can occur from daily breathing caused by daily

6. EQUIPMENT DESIGN

temperature changes. A flammable mixture may also occur in stored products containing impurities or light gases such as hydrogen in petroleum fractions as a result of an upset in an upstream process unit.

Fixed-roof tanks can be constructed as "weak-seam roof tanks" which are designed so that the roof-to-shell connection will fail preferentially to any other joint and the excess pressure will be safely relieved if the normal venting capacity should prove inadequate (Ref. 6-5). Weak-seam tanks for storing toxic materials are discouraged since a tank rupture would release the material to the atmosphere.

Vacuum in fixed-roof tanks can be caused when material is rapidly withdrawn or when a sudden drop in temperature or pressure, usually caused by weather conditions or steaming out, reduces the volume of the vapor in the tank. Pressure / vacuum (PV) vent protection should be sized to handle the maximum withdrawal rate plus the maximum temperature / volume reduction occurring simultaneously (Ref. 6-5 and Ref. 6-6). The vacuum relief device should be located at or near the highest point in the tank.

Excessive vapor generation may be the result of a deviation of temperature or routing of products more volatile than the design fluid. For tanks provided with internal heaters, adequate liquid level above the surface of the heater should be maintained so as not to overheat the tank contents and cause vapor generation or reach the auto-ignition temperature. Adequate venting capacity should be provided for excess vapor generation or coil rupture.

Polymerization of materials in a tank can lead to high overpressure combined with elevated temperatures in the tank. In this situation standard pressure relief valves may not be enough, both because very large two-phase flows may be involved, and because solid, polymerized materials may plug the relief valve. In these cases rupture discs with ducting leading to the atmosphere may be used, with the relief effluent being directed to a safe area of the plant. Discharge of the vent stream to a blowdown drum should be considered if the stream contains large amounts of liquid.

Common causes of loss of primary containment are:
- Overfilling due to operator error or high level alarm failure (vehicular as well as stationary tanks)
- Backflow from tank vent header
- Withdrawal of fluid from the tank bottom without operator attention
- Mechanical failure of tank
- Accumulation of a large volume of water, snow, or ice on the tank roof causing collapse and subsequent exposure of liquid surface

An additional cause of spills is specific to floating-roof tanks. It is possible for the roof platform to tilt slightly and become wedged into one position. Withdrawal of material from the tank, leaving the roof unsupported, or the addition of material to the tank, forcing fluid up over the roof, may cause the collapse of the floating roof.

Strategies to avoid spills and minimize damage may include:
- Instrumentation for tank high level and flow total alarms and shutoffs should be completely separate from the normal level and flow measurement with separate sensors and control units. In some cases, a mass balance alarm may be useful.

Additionally, permissive interlocks can be used to ensure proper line-up. Also, blinding of infrequently used lines should be considered.
- Overflow lines routed to a safe location and secondary containment. Overflow lines should be sized to allow full flow in case of a tank overflow. A general rule of thumb for estimating the size of overflow piping is that it should be sized at least one standard pipe size larger than the inlet pipe.
- Provision of safe method of water withdrawal from tanks storing organics and water drainage from the roof of the tank.
- Provision of secondary containment around tanks to prevent spills from spreading to other areas. This can take the form of dikes, double-walled tanks, or tanks in a concrete vault.

The contamination of material in tanks caused by the introduction of incompatible materials or material of the wrong temperature may cause runaway reactions, polymerization, high temperature excursions, or vacuum in the tank. To avoid potential contamination of products or routing wrong materials to tanks, piping valves and manifolds to the tank should be clearly marked, operating procedures should be simple and well defined, and periodic operator training should be provided.

Tanks containing hazardous materials can be placed above ground or under ground. Underground tanks offer increased safety for flammable and explosive materials and they require a smaller buffer zone between the tanks and other plant processes. The underground placement, however, increases the potential for soil and groundwater contamination due to the difficulty of inspecting the underground tanks. To reduce the chances of leakage, the tanks should be double walled or contained in concrete vaults. The space between the primary tank and the secondary containment should be equipped with some form of leak detection system. Double-walled piping with a leak detection system is also recommended for underground installations.

For vessels containing flammable liquids, where the vessel design pressure is insufficient to contain a deflagration or open loading is performed, consideration should be given to providing an inert gas blanket (e.g., nitrogen) to reduce the oxygen concentration and prevent fires or explosions and documented in the design basis. For vessels containing flammable liquids, the design should be arranged to prevent or minimize free falling of liquid. One approach is bottom loading. Another approach is use of an internal dip tube that terminates near the bottom of the vessel.

6.1.3.5 Pressurized Storage Tanks

Pressurized storage tanks for gases, generally sphere or bullet, should meet all requirements under the ASME boiler and pressure vessel codes as well as the applicable NFPA codes, such as NFPA 58 for LPG storage (Ref. 6-7). Many of the safety considerations that apply to atmospheric tanks also apply to these tanks. However, there are design differences. For example:
- Overpressure is handled through pressure relief valves (Ref. 6-5). When the tank contents are flammable the tank often relieves to the plant's flare system. Vacuum is not normally a problem as many pressurized storage tanks are also designed for full or partial vacuum, but some types, such as large butane storage spheres, can collapse under certain conditions.

- Pressurized tanks are designed to relieve overpressure due to flame impingement or heat radiation from nearby fires. Protective water sprays for the tank are designed to cool and protect the exposed tank faces, but not to extinguish any flame coming from the tank. A depressurization valve may be provided to prevent a boiling liquid expanding vapor explosion (BLEVE) from occurring.

6.1.4 References

6-1. API STD 650. *Welded Steel Tanks for Oil Storage*, 11th Edition. American Petroleum Institute. Washington, D.C. 2008.

6-2. CCPS. *Deflagration and Detonation Flame Arresters,* Center for Chemical Process Safety of the American Institute of Chemical Engineers. New York, New York. 2002.

6-3. API RP 941. *Steels for Hydrogen Service at Elevated Temperatures and Pressures in Petroleum Refineries and Petrochemical Plants,* American Petroleum Institute. Washington, D.C. 2008.

6-4. API STD 620. *Design and Construction of Large, Welded, Low-Pressure Storage Tanks,* American Petroleum Institute. Washington, D.C. 2008.

6-5. API STD 2000. *Venting Atmospheric and Low-pressure Storage Tanks*, 6th Edition, American Petroleum Institute. Washington, D.C. 2008.

6-6. NFPA 30. *Flammable and Combustible Liquids Code,* National Fire Protection Association. Quincy, Massachusetts. 2008.

6-7. NFPA 58. *Liquefied Petroleum Gas Code*, 2008 Edition National Fire Protection Association. Quincy, Massachusetts. 2008.

6.1.4.1 Suggested Additional Reading

API Publication 2210. *Flame Arresters for Vents of Tank Storing Petroleum Products,* 3rd Edition. American Petroleum Institute. Washington, D.C. 2000.

UL 525. *Flame Arresters for Use on Vents of Storage Tanks for Petroleum Oil and Gasoline,* Underwriter's Laboratory. Camas, Washington. 2008.

ANSI / API Spec 12B. *Specification for Bolted Tanks for Storage of Production Liquids,* 15th Edition. American Petroleum Institute. Washington, D.C. 2008.

API Spec 12P. *Specification for Fiberglass Reinforced Plastic Tanks*, 3rd Edition. American Petroleum Institute. Washington, D.C. 2008.

API Spec 12F. *Specification for Shop Welded Tanks for Storage of Production Liquids,* 12th Edition. American Petroleum Institute. Washington, D.C. 2008.

API Spec 12D *Specification for Field Welded Tanks for Storage of Production Liquids* American Petroleum Institute. Washington, D.C. 2008.

ASME. *Boiler and Pressure Vessel Code*, Section VIII American Society of Mechanical Engineers. New York, New York 2010.

API STD 2510. *Design and Construction of Liquefied Petroleum Gas (LPG) Installations*, American Petroleum Institute. Washington, D.C.

ASME Code for Pressure Piping B31.3. *Chemical Plant and Petroleum Refinery Piping*, American Society of Mechanical Engineers. New York, New York.

UL 58. *Steel Underground Tanks for Flammable and Combustible Liquids*, Underwriter's Laboratory. Camas, Washington.

UL 142. *Steel Aboveground Tanks for Flammable and Combustible Liquids*, Underwriter's Laboratory. Camas, Washington.

Myers, P. *Above Ground Storage Tanks,* McGraw-Hill. New York, New York. 1997.

6.2 REACTORS

This section presents potential failure mechanisms for reactors and suggests design alternatives for reducing the risks associated with such failures. The types of reactors covered in this section include:

- Batch reactors
- Semi-batch reactors
- Continuous-flow stirred tank reactors (CSTR)
- Plug flow tubular reactors (PFR)
- Packed-bed reactors (continuous)
- Packed-tube reactors (continuous)
- Fluid-bed reactors

This section presents only those failure modes that are unique to reaction systems. A number of the generic failure scenarios pertaining to vessels and heat exchangers may also be applicable to reactors. Consequently, this section should be used in conjunction with Section 6.1, Vessels, and Section 6.4, Heat Transfer Equipment. Unless specifically noted, the failure scenarios apply to more than one type of reactor.

Choosing a reactor design pressure high enough to contain the maximum pressure resulting from a worst case runaway reaction eliminates the need to size the emergency relief system for this scenario. It is essential that the reaction mechanisms, thermodynamics, and kinetics under *runaway conditions* be thoroughly understood to be confident that the design pressure is sufficiently high for all credible reaction scenarios. All causes of a runaway reaction must be understood, and any side reactions, decompositions, and shifts in reaction paths at the elevated temperatures and pressures experienced under runaway reaction conditions must be evaluated. Many laboratory test devices and procedures are available for evaluating the consequences of runaway reactions (Refs. 6-8, 6-9, and 6-10).

6.2.1 Past Incidents

Reactors are a major source of serious process safety incidents. Several case histories are presented to reinforce the need for safe design and operating practices for reactors.

6.2.1.1 Reactive Chemical Explosion

A powerful explosion and subsequent chemical fire killed four employees and destroyed a chemical manufacturer in Jacksonville, Florida. It injured 32, including 4 employees

6. EQUIPMENT DESIGN

and 28 members of the public who were working in surrounding businesses. Debris from the reactor was found up to one mile away, and the explosion damaged buildings within one quarter mile of the facility.

The facility was producing its 175th batch of Methylcyclopentadienyl Manganese Tricarbonyl (MCMT). The process operator had an outside operator call the owners to report a cooling problem and request they return to the site. Upon their return, one of the two owners went to the control room to assist. A few minutes later, the reactor burst and its contents exploded, killing the owner and process operator who were in the control room and two outside operators who were exiting the reactor area.

A loss of sufficient cooling during the process likely resulted in the runaway reaction, leading to an uncontrollable pressure and temperature rise in the reactor. The pressure burst the reactor; the reactor's contents ignited, creating an explosion equivalent to 1,400 pounds of TNT.

Lessons learned include not recognize the runaway reaction hazard associated with the MCMT it was producing. Additionally, the cooling system employed was susceptible to single-point failures due to a lack of design redundancy and the MCMT reactor relief system was incapable of relieving the pressure from a runaway reaction.

6.2.1.2 Hydroxylamine Explosion

A process vessel containing several hundred pounds of Hydroxylamine (HA) exploded at a manufacturing facility near Allentown, Pennsylvania. Employees were distilling an aqueous solution of hydroxylamine and potassium sulfate, the first commercial batch to be processed at the facility. After the distillation process was shut down, the HA in the process tank and associated piping explosively decomposed, most likely due to high concentration and temperature. Four employees and a manager of an adjacent business were killed. Two employees survived the blast with moderate-to-serious injuries. Four people in nearby buildings were injured. The explosion also caused significant damage to other buildings in an adjacent industrial park and shattered windows in several nearby homes.

Lessons Learned include:
- Process safety management systems were insufficient to properly address the hazards inherent in its HA manufacturing process and to determine whether these hazards presented substantial risks.
- Inadequate collection and analysis of process safety information contributed to failure in recognizing specific explosion hazards.
- Basic process safety and chemical engineering practices -such as process design reviews, hazard analyses, plant siting, corrective actions, and reviews by appropriate technical experts-were not adequately implemented.

6.2.1.3 Seveso Runaway Reaction (Scenario 9)

On July 10, 1976 an incident occurred at a chemical plant in Seveso, Italy, which had far-reaching effects on the process safety regulations of many countries, especially in Europe. An atmospheric reactor containing an uncompleted batch of 2,4,5-trichlorophenol (TCP) was left for the weekend. Its temperature was 316°F (158°C), well below the temperature at which a runaway reaction could start [believed at the time to be 446°F (230°C), but possibly as low as 365°F (185°C)]. The reaction was carried out under vacuum, and the reactor was heated by steam in an external jacket, supplied by exhaust steam from a turbine at 374°F (190°C) and a pressure of 174 psig (12- bar gauge). The turbine was on reduced load, as various other plants were also shutting down for the weekend (as required by Italian law), and the temperature of the steam rose to about 572°F (300°C). There was a temperature gradient through the walls of the reactor [572°F (300°C) on the outside and 320°F (160°C) on the inside] below the liquid level because the temperature of the liquid in the reactor could not exceed its boiling point. Above the liquid level, the walls were at a temperature of 572°F (300°C) throughout.

When the steam was shut off and, 15 minutes later, the agitator was switched off, heat transferred from the hot wall above the liquid level to the top part of the liquid, which became hot enough for a runaway reaction to start. This resulted in a release of TCDD (dioxin), which killed a number of nearby animals, caused dermatitis (chloracne) in about 250 people, damaged vegetation near the site, and required the evacuation of about 600 people (Ref. 6-11).

The lesson learned from this incident is that provision should have been made to limit the vessel wall temperature from reaching the known onset temperature at which a runaway reaction could occur. Additionally, transient conditions such as startup and shutdown should be considered adequately in the design.

6.2.2 Failure Scenarios and Design Solutions

Table 6.2 presents information on equipment failure scenarios and associated design solutions specific to reactors derived from a variety of sources and actual situations.

6.2.2.1 Loss of Agitation / Circulation (Scenario 9)

Runaway reactions are often caused by loss of agitation in stirred reactors (batch, semi-batch, and CSTR) due to motor failure, coupling failure, or loss of the impeller. Agitation can be monitored by measuring the amperage or power drawn by the agitator drive. Nevertheless, this has its drawbacks as the "measurement" of agitation takes place outside of the reactor, and sometimes, if the reactor contents are not viscous enough, the amperage or power draw will not detect that the agitator impeller has fallen off or corroded away. The loss of the impeller can be detected by using an internal flow sensor. The flow sensor, or a similar in-vessel detection device, can be interlocked to cut off feed or catalyst being added to a semi-batch reactor or CSTR.

If agitation is critical to the operation of a batch, semi-batch, or CSTR reactor, then an independent, uninterrupted power supply backup for the agitator motor should be provided. Alternatively, some degree of mixing can be provided by sparging the reactor liquid with inert gas or through the use of an external pumped loop to circulate material through the vessel.

6. EQUIPMENT DESIGN

Table 6.2 Common Failure Scenarios and Design Solutions for Reactors

No.	Event	Consequence	Potential Design Solutions		
			Inherently Safer / Passive	Active	Procedural
Runaway Reactions					
	Generally Applicable *(Applicable to all runaway reaction scenarios)*		Reactor designed for maximum expected temperature and pressure	Emergency relief device Automatic activation of bottom discharge valve to drop batch into a dump tank with diluent, poison or short-stopping agent, or to an emergency containment area Automatic addition of diluent, poison, or short-stopping agent directly to reactor (with effective mixing) Automatic feed shutdown based on detection of unexpected reaction progress (e.g., abnormal heat balance, high / low pressure, high / low temperature) Automatic venting of pressure to effluent system	Operator response to high temperature or pressure alarm Written procedures and training for manual activation of bottom discharge valve to drop batch into dump tank with diluent, poison, or short-stopping agent, or to an emergency containment area Written procedures and training for manual addition of diluent, poison, or short-stopping agent directly to reactor Written procedures and training for manual closure of isolation valve(s) in feed line on detection of unexpected reaction progress (i.e., abnormal heat balance)
1	Overcharge of catalyst (batch, semi-batch, and plug flow reactors)	Potential runaway reaction	Dedicated catalyst charge tank sized to hold only the amount of catalyst needed Reactor type selected that is less sensitive to catalyst change issues	Quantity of catalyst added limited by flow totalizer High level interlock / permissive to limit quantity of catalyst	Written procedures and training regarding the amount or concentration of catalyst to be added (might consider one person to stage the required catalyst amount and a second person to add the required amount, serving as a double check on type and quantity) Written procedures and training to establish an intermediate location for pre-weighed catalyst charges

Table 6.2   Common Failure Scenarios and Design Solutions for Reactors

No.	Event	Consequence	Potential Design Solutions		
			Inherently Safer / Passive	Active	Procedural
2	Addition of a reactant too rapidly (batch and semi-batch reactors)	Potential runaway reaction	Feed system capacity limited to within safe feed rate limitations (e.g., screw feeder for solids or flow orifice for liquids)	Automatically controlled feed system to limit feed rate within safe limitations	Written procedures and training for slowing or discontinuing charge if rate is exceeded or thermal event occurs
3	Addition of incorrect reactant or off-spec feed mixture	Potential runaway reaction	Dedicated feed tank and reactor for production of one product. Dedicated hoses and incompatible couplings for reactants where hose connections are used. Elimination of cross-connections	Control software preventing charge valve or pump operation until correct material bar code has been scanned	Dedicated storage areas / unloading facilities for reactants. Written procedures and training for double checking reactant identification and quality. Written procedures and training to verify material (scan bar code) prior to their addition
4	Overactive and / or wrong catalyst	Potential runaway reaction	Catalyst pre-diluted or pre-tempered		Written procedures and training to passivate fresh catalyst prior to use. Written procedures and training for testing and verification of catalyst activity and identification
5	Inactive and / or wrong catalyst	Potential delayed runaway reaction			Written procedures and training for testing and verification of catalyst activity and identification
6	Reactants added in incorrect order (batch & semi-batch)	Potential runaway reaction		Interlock shutdown of reactant addition based on detection of mis-sequencing. Sequence control via programmable logic controller	Written procedures and training for manual isolation of feed based on indication of mis-sequencing

6. EQUIPMENT DESIGN

Table 6.2   Common Failure Scenarios and Design Solutions for Reactors

No.	Event	Consequence	Potential Design Solutions		
			Inherently Safer / Passive	Active	Procedural
7	Reactor contents flow backwards	Potential runaway reaction	Feed vessel elevated above reactor with emergency relief device on reactor set below feed vessel minimum operating pressure Positive displacement feed pump instead of centrifugal pump	Automatic closure of isolation valve(s) in feed line on detection of low or no flow, or reverse pressure differential in feed line Check valve(s) in feed line Emergency relief device on feed vessel or feed line	
8	Loss of cooling	Potential runaway reaction	Large inventory of naturally circulating, boiling coolant to accommodate exothermic reaction	Automatic actuation of secondary cooling medium on detection of low coolant flow or pressure or high reactor temperature (e.g., city water or fire water or venting to overhead condenser for ebullient cooling)	Written procedures and training for manual activation of secondary cooling system
9	Loss of agitation (batch, semi-batch, and CSTR reactors)	Potential runaway reaction	Vessel design accommodating maximum expected pressure Alternative agitation methods (e.g., external circulation eliminates shaft seal as a source of ignition in vapor space)	Agitator power consumption or rotation indication interlocked to cut off feed of reactants or catalyst or activate emergency cooling Emergency relief device Inerting of vapor space Provide nitrogen buffer zone around seal using enclosure around seal Pressure or temperature sensors actuating bottom discharge valve to drop batch into a dump tank with diluent, poison, or short-stopping agent or to an emergency containment area Uninterrupted power supply backup to motor	Written procedures and training to visually check mechanical seal fluid on regular basis Manual activation of bottom discharge valve to drop batch into dump tank with diluent, poison, or short-stopping agent or to an emergency containment area Manual activation of inert gas sparging of reactor liquid to effect mixing

Table 6.2   Common Failure Scenarios and Design Solutions for Reactors

No.	Event	Consequence	Potential Design Solutions		
			Inherently Safer / Passive	Active	Procedural
10	Incomplete mixing before entering reactor	Potential runaway reaction	Static mixer ahead of reactor		Written procedures and training for the Operator to sample the monomer emulsion feed and observes that sample is stable without agitation for a predetermined length of time before feed is begun
11	Excessive heating	Potential runaway reaction	Temperature of heating media limited	Automatic activation of emergency cooling	Written procedures and training to close manual back-up heat media valves in the event of a primary control valve failure as indicated by excessive or uncontrolled heating Written procedures and training to limit rate of rise
12	Hot spot develops in catalyst (continuous packed bed or packed tube reactors)	Potential runaway reaction	Alternative reactor design (e.g., fluid bed) Flow distribution trays provided to minimize channeling Multiple small diameter beds to reduce maldistribution Reactor head space volume minimized to reduce residence time (partial oxidation reactors) and mitigate autoignition	Automatic switch to diluent	Written procedures and training for monitoring of exterior wall temperature with infrared optical detection system or other detection method Written procedures and training for packing tubes to ensure uniformity of catalyst filling

6. EQUIPMENT DESIGN

Table 6.2 Common Failure Scenarios and Design Solutions for Reactors

No.	Event	Consequence	Potential Design Solutions		
			Inherently Safer / Passive	Active	Procedural
13	Leakage of heating / cooling media	Potential runaway reaction	Heat transfer fluid that does not react with process fluid Heat transfer loop pressure lower than process pressure Jacket design rather than internal coil for heat transfer Metallurgy upgraded		Written procedures and training for periodic testing of process fluid for contamination Written procedures and training for leak / pressure testing of jacket, coil or heat exchanger prior to operation Written procedures and training for testing liner with continuity meter
14	Insufficient residence time	Potential incomplete reaction, leading to unexpected reaction in subsequent processing steps (in reactor or downstream vessel)	Flow limiting orifice on feed lines mechanically limit maximum flow capability of feed pumps	Automatic feed isolation based on continuous online reactor composition monitoring	Written procedures and training for manual feed isolation based on continuous online reactor composition monitoring Written procedures and training for sampling before manual transfer of material
15	Impurities in catalyst (adsorber)	Potential for runaway reaction		Automatic control to vent or quench the reaction Automatic controls for switching beds on regeneration	Operator response to bed high temperature alarms Written procedures and training for testing of adsorbents prior to loading into vessel Written procedures and training for verification of adsorbent compatibility with process materials

6.2.2.2 *Addition of Incorrect Reactant (Scenario 3)*

The addition of a wrong reactant can result in a runaway reaction. To minimize this error, the following measures can be taken:

- Provide dedicated feed tanks (for liquids) or feed hoppers (for solids) for batch reactors.
- Ensure two operators check the drums or bags of reactants before they are added and then sign off on a log sheet.
- Color-code and label all process lines so the operators know what is in them.

- Dedicate feed lines for critical reactants such as ethylene oxide.

If the risk of adding an incorrect reactant is still determined to present an unacceptable risk, further protective measures can be implemented, such as providing a temperature sensor to monitor the reaction and shut off a valve in the feed line upon detection of an abnormal temperature rise or rate of temperature rise.

6.2.2.3 Overactive, Inactive / Semi-Active, or Incorrect Catalyst Addition (Scenarios 4 and 5)

The addition of a semi-active or incorrect catalyst to a reactor may result in a runaway reaction either in the reactor or in downstream equipment. If the catalyst is fed continuously or at a controlled rate to a semi-batch reactor, protection can be provided by installing a temperature sensor in the reactor, interlocked with an isolation valve in the reactant feed line, which will shut the valve when the sensor detects an abnormal temperature rise. The temperature sensor could also be interlocked with a valve to stop the catalyst feed. Administrative controls, such as procedures for verifying catalyst identity and activity, can also be applied in addition to active controls.

Use of DIERS relief technology for two-phase flow can be used to design relief protection systems. One approach is to use two-stage pressure relief: one for minor excursions that might occur during pressure testing and a second level for actual reaction overpressure scenarios where large volume pressure relief is needed (often provided with rupture disks).

6.2.3 Design Considerations

6.2.3.1 Design Pressure

For reactors fabricated of metal (not glass-lined), it is recommended that a minimum design pressure of 50 psig be specified, even if the operating pressure is essentially atmospheric. A 50 psig design pressure will also generally provide some vacuum rating. This provides a measure of inherent safety for unexpected pressure swing events (pressure spikes). If an explosive mixture might be encountered, a deflagration test is recommended to determine what internal blast pressure might occur and what the design pressure should be. For deflagration design pressure requirements see NFPA 69 (Ref. 6-12).

6.2.3.2 Overpressure Relief

Reactors should be provided with overpressure relief protection. The relief design basis should include a review all reaction paths (intentional or unintentional) for the possibility of a runaway reaction, which often results in the need for an appreciably larger relief device than other relief scenarios may require. Where runaway reactions are known to occur, the piping from the relief device may be handling a multi-phase stream (vapor-liquid or vapor-liquid-solid) and should be routed first to an effluent handling system (knockout drum / catch tank) to separate the vapors from the liquid. The vapor line from the knock-out drum / catch tank should then be routed directly to a scrubber or flare stack if environmental considerations require further treatment and collection.

The pressure rating of a vessel will decrease as the temperature increases past the maximum allowable operating temperature. An incident at the Avon Refinery in

6. EQUIPMENT DESIGN 191

Martinez, California is a good example. Additional information on this incident is available at www.epa.gov/osweroe1/docs/chem/tosco.pdf.

For reactors containing flammable liquids, where the reactor design pressure is insufficient to contain a deflagration, consideration should be given to providing an inert gas blanket (usually nitrogen).

6.2.3.3 Addition

All flammable liquids should be charged into a reactor via dip legs or elbows which cause the liquid to run down the reactor wall to minimize static electricity accumulation. Where the addition rate of a reactant or catalyst could result in a runaway reaction if added too quickly, a restriction orifice should be installed in the feed line to limit the flow rate. Where overcharging (adding too great a quantity) of a reactant or catalyst can cause a runaway reaction, the use of a gravity flow head tank sized to hold only the quantity needed should be considered.

Where solids have to be added to a batch reactor containing flammable or toxic liquids, they should be charged by means of a rotary valve, lock-hopper, or screw feeder so that the operator will not have to open the reactor and be exposed to hazardous conditions or chemicals. The hopper or screw feeder may also be inerted to provide an additional protection layer. There should be instruments and procedures to assure that the solids are being fed as intended. In addition, special attention should be given to methods of safely unplugging valves and lines.

6.2.3.4 Agitation

A runaway reaction could occur due to unrecognized cessation of agitation (the shaft is still rotating although the impeller has fallen off or corroded out, or the circulation pump providing agitation has stopped, failed, or encounters a blockage in the discharge piping). To prevent this, a malfunction detector could be installed in the reactor in the vicinity of the impeller. The malfunction detector should have an alarm and be interlocked to stop feed of reactants or catalysts. Back-up power supply should be supplied to the agitator motor for critical reactions, such as polymerization reactions.

6.2.3.5 Runaway Reactions

Where runaway reactions are known to occur and an excessively large relief device is needed, consideration should also be given to providing means to inhibit (kill or "short stop") the reaction or drown (quench) the batch. It is recommended that independent and redundant temperature instruments in the reactor be interlocked to actuate any of the following remedial actions at a specified high temperature reading:

- Add a considerable amount of coolant or diluent to reduce the reaction rate. This measure requires that process design and detailed design provide for:
  - Choice of an appropriate fluid which does not react exothermically with the reaction mixture
  - Sufficient free volume in the reactor
  - Piping, instrumentation, etc., to add the fluid in the time required
- Rapidly depressure the vessel if the reactor is under pressure:

- Add an inhibitor to stop the reaction. This measure requires intimate knowledge of how the reaction rate can be influenced and whether effective mixing / inhibition is possible.
- Dump the reactor contents into a vessel or dump pit which contains cold diluent. This option also requires particular care that the dumping line is not blocked or does not become blocked during the dumping procedure.

6.2.3.6 Heating and Cooling Systems

Heat removal systems should be designed with all anticipated abnormal operating conditions taken into account. For systems where runaway reactions are possible, the heat removal system should be capable of functioning at the temperatures achieved during the runaway reaction, though it may not have the capacity to stop the runaway reaction (e.g., water coolers should not become vapor bound).

6.2.3.7 Glass-Lined Reactors

Because of the fragile nature of glass, precautions should be taken to avoid causing damage to a glass-lined reactor by thermal shock, mechanical impact, and corrosion. When specifying a glass-lined reactor, the vessel manufacturer should be given complete details about the reactants, the reaction conditions, and the batch cycles so that the proper type of glass can be provided. Glass-lined reactors should be periodically inspected for the presence of holes.

When specifying a glass-lined reactor, careful thought must be given to what chemicals are in the reactor and what the temperatures are during the batch cycle. Glass is not completely inert and is always undergoing local chemical reactions at the glass surface. What allows glass-lined steel to be used with corrosive materials is the low rate of reaction (kinetics). The slower the corrosion rate, the longer the glass lining will last. Glass-lined vessels can accumulate static, resulting in arcing within the vessel.

Acids (particularly hydrofluoric acid), alkalis, and even water can corrode glass in varying forms and degrees. Strange as it may seem, water can cause severe corrosion, and the rate increases with water purity. The corrosion rate also increases with increasing temperature and becomes greatest when the boiling point is exceeded. A small amount of acid added to water will greatly retard corrosion caused by water vapor condensation in the vapor area. This type of corrosion can also be reduced or eliminated by the introduction of an inert gas, insulating the vapor area, or both. These are important factors to consider in steam distillation processes.

6.2.4 References

6-8. CCPS. *Guidelines for Chemical Reactivity Evaluation and Application to Process Design*. Center for Chemical Process Safety of the American Institute of Chemical Engineers. New York, New York. 1995.

6-9. CCPS. *Guidelines for Process Safety Fundamentals in General Plant Operations*, Center for Chemical Process Safety of the American Institute of Chemical Engineers. New York, New York. 1995.

6. EQUIPMENT DESIGN 193

6-10. CCPS, *Guidelines for Pressure Relief and Effluent Handling Systems, 2nd Edition* The American Institute of Chemical Engineers, New York, New York. 2011.

6-11. Kletz, T.A. *What Went Wrong: Case Histories of Process Plant Disasters,* 3rd Edition. Gulf Publishing Company. Houston, Texas 1994.

6-12. NFPA 69. *Standard of Explosion Prevention Systems,* National Fire Protection Association. Quincy, Massachusetts. 2008.

6.2.4.1 Suggested Additional Reading

CCPS, *Problem Set for Kinetics,* Problem 16, Prepared for SACHE. Center for Chemical Process Safety of the American Institute of Chemical Engineers. New York, New York. 1995.

Benuzzi, A. and Zaldivar, J.M. (eds.). *Safety of Chemical Batch Reactors and Storage Tanks,* Kluwer Academic Publishers. Norwell, Massachusetts. 1991.

EPA Chemical Accident Investigation Report, Tosco Avon Refinery. Martinez, CA. EPA 550-R-98-0094, www.epa.gov/osweroe1/docs/chem/tosco.pdf

Gygax, R.W. *Chemical Reaction Engineering for Safety Chemical Engineering Science* 43(8), 1759-1771. 1988.

Scaleup Principles for Assessing Thermal Runaway Risks, Chemical Engineering Progress. February 1990.

International Symposium on Runaway Reactions. *Cooling Capacities of Stirred Vessel, Unstirred Container, Insulated Storage Tank, Uninsulated 1 cu meter Silo, Uninsulated 25 cu meter Silo: 65.* Sponsored by CCPS, IChemE and AIChE. Cambridge, Massachusetts. 1989.

Maddison, N. and Rogers, R.L. *Chemical Runaways: Incidents and Their Causes,* Chemical Technology, Europe. November / December, 28-31. 1994.

Noronha, J. and Torres, A. *Runaway Risk Approach Addressing Many Issues-Matching the Potential Consequences with Risk Reduction Methods*, Proceedings of the 24th Loss Prevention Symposium, AIChE National Meeting. San Diego, CA. 1990.

Wier, E., Gravenstine, G and Hoppe, T. *Thermal Runaways - Problems with Agitation,* Loss Prevention Symposium. Paper 830: 142. 1986.

6.3 MASS TRANSFER EQUIPMENT

This section presents potential failure mechanisms of mass transfer equipment and suggests design alternatives for reducing the risks associated with such failures. The types of mass transfer operations covered in this section include:

- Absorption
- Adsorption
- Extraction
- Distillation
- Scrubbing

- Stripping
- Washing

This section presents only those failure modes unique to mass transfer equipment. Many of the generic failure modes presented in Section 6.1 may also apply to vessels used for mass transfer. Mass transfer equipment failure may also result from disturbances in heat transfer processes in associated ancillary equipment. Refer to Section 6.4, Heat Transfer Equipment, for failures associated with heat transfer equipment. Unless specifically noted, the failure scenarios apply to more than one class of mass transfer equipment.

6.3.1 Past Incidents

This section describes past incidents that illustrate hazard scenarios involving mass transfer equipment.

6.3.1.1 Distillation Column Critical Concentration

In 1969, an explosion occurred in a butadiene recovery unit in Texas City, Texas. The location of the center of the explosion was found to be the lower tray section of the butadiene refining (final purification) column. The butadiene unit recovered by-product butadiene from a crude C4 stream. The overhead of the refining column was a highpurity butadiene product. The heavy components of the feed stream, including Vinyl Acetylene (VA), were removed as a bottoms product. The bottoms vinyl acetylene concentration was normally maintained at about 35%. Explosibility tests had indicated that VA concentrations as high as 50% were stable at operating conditions. Highly concentrated VA decomposes rapidly on exposure to high temperature.

When the butadiene unit was shut down to undertake necessary repairs, the refining column was placed on total reflux. The refining column explosion occurred approximately 9 hours after it was placed on total reflux. This operation had been performed many times in the past without incident. The operators did not observe anything unusual about this particular switch over to total reflux. Subsequent examination of the records indicated that the column had been slowly losing material through a closed but leaking valve in the column overhead line. As a result, reflux and reboiler steam flow continued to fall slowly throughout the shutdown period.

Loss of butadiene through the leaking valve resulted in substantial changes in tray composition in the lower section of the column. The concentration of vinyl acetylene in the tray liquid in the vicinity of the tenth tray apparently doubled to an estimated 60%. The loss of liquid level in the base of the column uncovered the reboiler tubes, allowing the tube wall temperature to approach the temperature of the steam supply. The combination of increased vinyl acetylene concentration and high tube wall temperature led to the decomposition of VA and set the stage for the explosion that followed (Ref. 6-13, Ref. 6-14, and Ref. 6-15).

Lessons learned include the need to monitor critical parameters even when in a static or hold mode. Often, minor process variables that do not warrant monitoring under active run conditions (e.g., seal leaks and purges, air ingress under vacuum conditions) can cause significant problems when accumulation occurs over an extended hold period.

6. EQUIPMENT DESIGN

6.3.1.2 Ethylene Purifier Vessel Rupture

Ethylene was purified in a bed containing 13X molecular sieve. The bed was regenerated using hydrogen-methane gas at 500°F (260°C), then flow purged with nitrogen. The temperature was allowed to drop to 338°F (170°C), and then the bed was pressurized with nitrogen. Ethylene was then introduced into the bed, and nitrogen was displaced.

The temperature in the bed was not being measured, but a temperature sensor was located 20 inches above the bed. After 7 hours of operation (preloading) with the bed open to a line pressure of 280-295 psig, the bed temperature had dropped to 266°F (130°C). A small flow was then started off the top with ethylene going in at the bottom. The bed temperature rose to 356°F (180°C) in 3 1/2 hours and over the next 4 hours the flow was adjusted to maintain this temperature. Shortly afterwards the shell ruptured, creating a longitudinal 3/8-inch- by- 32- inch hole. The gas caught fire immediately and burned for 25-30 minutes. The fire was not controlled because high temperature prevented the inlet valves from being closed; all the gas up to the closed feed valve at the gas plant was burned.

The principal cause of this incident was the failure to measure temperatures in the bed during regeneration and preloading with ethylene. Sieve 13X is a polymerization catalyst. Due to its large pore size, 13X also adsorbs ethylene and releases heat. The temperature measured above the bed gave no indication of the temperature anywhere within the bed, where these exothermic processes would occur. Even though the pressure of ethylene involved in this incident was unusually low (280 psig), evidently there was enough potential (via adsorption and polymerization) to generate the temperature required to cause thermal failure of the vessel. Had the bed temperature been comprehensively measured, any shortcomings in the purging and preloading procedures would have become apparent in time to take action. Such temperature measurement should be done via fast-acting thermocouples distributed throughout the bed and not via thermocouples mounted in heavy thermowells located near the walls, since the sieves are effective thermal insulators (Ref. 6-16).

Lessons Learned include the use of emergency isolation in the event of a fire and the need for measuring and alarming the temperature in the bed.

6.3.2 Failure Scenarios and Design Solutions

Table 6.3 presents information on equipment failure scenarios and associated design solutions specific to mass transfer equipment.

Table 6.3   Common Failure Scenarios and Design Solutions for Mass Transfer Equipment

No.	Event	Consequence	Potential Design Solutions		
			Inherently Safer / Passive	Active	Procedural
Pressure					
	Generally Applicable – High Pressure *(Applicable to all high pressure scenarios)*		Vessel designed for maximum expected pressure	Emergency relief device Automatic high pressure shutdown of heat input	Operator response to differential pressure indication Operator response to high pressure alarm Written procedures and training for manual shutdown on abnormal power consumption
	Generally Applicable – Low Pressure *(Applicable to all low pressure scenarios)*		Vessel designed for maximum vacuum	Vacuum relief system Automatic isolation and purge of equipment with inert gas on loss of vacuum	Operator response to low pressure alarm Written procedures and training for manual addition of vacuum breaking gas
1	Plugging from internals failure	Potential increased pressure	Large surface area screens to avoid entrance of internals into lines Support grids and hold down grids designed to minimize internal migration		Operator response to high differential pressure alarm Written procedures and training for proper operation of equipment to prevent damage to equipment internals
2	Blockage of packing / trays	Potential increased pressure	Internals selected and designed to minimize blockage and fouling Vessel designed without internals (e.g., spray tower)		Written procedures and training for on-line wash to eliminate fouling material
3	Loss of vacuum	Potential increased temperature resulting in liquid / vapor decomposition		Continuous injection of reaction inhibitor Automatic supply of nitrogen to the column	Written procedures and training to periodically test for inhibitor concentration
4	Air leakage into equipment operating under vacuum	Potential overpressure, potential fire		Oxygen analyzer with automatic activation of inert gas addition on detection of high oxygen concentration	Oxygen analyzer with alarm and manual activation of inert gas addition on detection of high oxygen concentration Written procedures and training to pressure check for leaks before start-up

6. EQUIPMENT DESIGN

Table 6.3   Common Failure Scenarios and Design Solutions for Mass Transfer Equipment

No.	Event	Consequence	Potential Design Solutions		
			Inherently Safer / Passive	Active	Procedural
5	Uncontrolled condensation	Potential vacuum and loss of containment		Automatic addition of blanketing gas pressure control system to minimize vacuum	Written procedures and training on monitoring conditions and breaking vacuum with nitrogen or other method
Flow					
6	Poor vapor flow distribution through adsorbers	Potential for hot spots	Adsorber cross-sectional area minimized Vessel distributors designed to avoid regions of flow maldistribution in the bed	Continuous monitoring of bed temperatures or by-products at certain locations and interlock shutdown and / or inerting / flooding on high temperature	Written procedures and training to monitor bed temperature / by-products and take appropriate action (e.g., inerting / flooding) High / low flow limits to set the bounds of good distribution as calculated in the design
7	Excessive vapor flow	Potential carryover of liquid to undesired location	Vessel designed with proper vapor-liquid disengagement (e.g., low superficial vapor velocity) Liquid removal via demister, cyclone, or other device with open liquid discharge	Removal of liquid from the vapor stream, e.g., knock-out pots with automatic level control Differential pressure indication and automatic reduction of vapor flow	Differential pressure indication and written procedures and training to reduce vapor flow
8	Accumulation of reactive material in section of fractionator (distillation columns)	Potential for runaway reaction	Change in feedstock to avoid reactive material	Online measurement (e.g., level, temperature, composition) and automatic side draw-off of reactive material	Online measurement (e.g., level, temperature, composition) and written procedures and training for manual removal of reactive material
9	Insufficient or excessive fractionation	Potential compositions outside of metallurgical limits resulting in increased corrosion	Metallurgy suitable for worst case composition.	Online measurement (e.g., corrosion probes, stream analysis, temperature) and automatic operating adjustment	Online measurement (e.g., corrosion probes, stream analysis, temperature) and written procedures and training for manual operating adjustment Independent site glasses for level verification by operators

Table 6.3   Common Failure Scenarios and Design Solutions for Mass Transfer Equipment

No.	Event	Consequence	Potential Design Solutions		
			Inherently Safer / Passive	Active	Procedural
Temperature					
	Generally Applicable – High Temperature *(Applicable to all high temperature scenarios)*		Vessel designed for maximum expected temperature	Interlock to isolate feed on detection of high bed temperature	Written procedures and training for reinstating process flow after regeneration and cooling Operator response to high temperature alarms
	Generally Applicable – Low Temperature *(Applicable to all low temperature scenarios)*		Vessel designed for minimum expected temperature		Operator response to low temperature alarms
Level					
	Generally Applicable – High Level *(Applicable to all high level scenarios)*		High reliability level device	Interlock to isolate feed on detection of high level	Operator response to high level alarms
	Generally Applicable – Low Level *(Applicable to all low level scenarios)*		High reliability level device	Interlock to shutdown withdrawal on detection of low level	Operator response to low level alarms
10	Interfacial level control failure (extractor)	Potential carryover of unwanted material to downstream equipment	Interface level controlled with overflow leg or weir	High / low interfacial level alarm with shutoff preventing further liquid withdrawal from vessel	Written procedures and training for manual vessel interfacial level control
		Potential to exceed design pressure rating, potential loss of containment.	Downstream equipment designed for maximum pressure		

6. EQUIPMENT DESIGN

Table 6.3   Common Failure Scenarios and Design Solutions for Mass Transfer Equipment

No.	Event	Consequence	Potential Design Solutions		
			Inherently Safer / Passive	Active	Procedural
Composition					
	Generally Applicable *(Applicable to all composition scenarios)*			Automatic isolate of feed on detection of high bed temperature Automatic emergency depressuring and / or flooding / inerting on detection of high temperature	Written procedures and training to manually isolate feed on detection of high bed temperature Written procedures and training for manual emergency depressuring and / or flooding / inerting on detection of high temperature
11	Premature introduction of process stream containing air (adsorber)	Potential internal fire on packing	Adsorbent selected to minimize combustion potential	Oxygen analyzer with automatic activation of inert gas addition on detection of high oxygen concentration	Oxygen analyzer with alarm and manual activation of inert gas addition on detection of high oxygen concentration Written procedures and training for reinstituting process flow after regeneration
12	High concentration of flammables in the inlet stream to carbon bed adsorber	Potential for hot spots		Automatic control of inlet stream outside flammable limits Automatic isolation of feed on detection of high temperature Inerting of process stream	Written procedures and training for manual control of inlet stream outside flammable limits Written procedures and training for manual isolation on detection of high temperature
13	Low moisture content in carbon bed adsorber	Potential for hot spots		Automatic steam injection to rehydrate bed prior to feed start Automatic water deluge on detection of fire	Written procedures and training for verification of adsorbent moisture content prior to placing in service Written procedures and training for manual steam injection to rehydrate bed prior to feed start-up Written procedures and training for manual water deluge on detection of fire

Table 6.3   Common Failure Scenarios and Design Solutions for Mass Transfer Equipment

No.	Event	Consequence	Potential Design Solutions		
			Inherently Safer / Passive	Active	Procedural
Maintenance / Startup					
14	Exposing packing internals during changeout	Potential fire if material is pyrophoric	Vessel designed with non-stick internals (e.g., plastic packing)		Written procedures and training for maintenance under inert atmosphere if necessary
			Vessel designed without internals (e.g., spray tower)		Written procedures and training for proper vessel wash-out / cool-down prior to opening *(continued)*
14	(continued)				Written procedures and training to monitor temperature and take appropriate action (e.g., flooding)
					Written procedures and training to chemically clean vessel to remove hazard prior to opening vessel
15	Adsorber bed not preconditioned (adsorber)	Potential increased temperature resulting in inefficient transfer	Adsorbents selected to adsorb only trace contaminants and not carrier gas (e.g., olefin purification)	Automatic preconditioning sequence prior to feed startup	Written procedures and training for CO monitoring with manual shutdown (for carbon bed adsorbers)
				CO monitoring with automatic shutdown (for carbon bed adsorbers)	Written procedures and training for multi-point temperature monitoring with manual shutdown of feed (for high pressure adsorbers)
				Multi-point temperature monitoring with automatic shutdown of feed (for high pressure adsorbers)	Written procedures and training for preconditioning adsorber bed

6.3.2.1   Line Blockage by Internals (Scenario 1)

During process upsets, the internals in mass transfer vessels may dislodge and be displaced into process lines where they create blockages. Such blockages can cause vessel pressure to increase, possibly to the relief device set pressure. Of particular concern is the possibility of internals lodging in the inlet piping of the relief device, thus impairing overpressure protection. This may result in a pressure condition that exceeds acceptable limits.

The first level of protection is to design supports and hold down grids to withstand fluctuations in differential pressure. Screens can be installed to prevent large pieces of internals from entering lines. For packings that are susceptible to abrasion, duplex filters supplied with differential pressure indication can be employed.

6. EQUIPMENT DESIGN

Pressure relief devices should be located upstream of potential blockage points. For example, the inlet to a Pressure Safety Valve (PSV) should be placed below the mist eliminator in the top of a column if severe fouling of the mist eliminator is possible.

6.3.2.2 Packing / Tray Blockage (Scenario 2)

Mass transfer equipment internals are susceptible to blockage due to process pressure and flow fluctuations of fouling material. When fouling conditions are encountered, a possible solution is to place chevron-type baffles or large-hole sieve trays where the most severe fouling is expected. If there is a possibility of packing becoming plugged due to polymerization or the feed stream contains solid particles, then a pressure relief valve should be installed in the vapor space below the bottom packing support plate. Also, the differential pressure should be monitored and alarmed.

6.3.2.3 Adsorbers (Scenarios 6, 11, 12, 13, 15)

Adsorption systems, such as dehydrators and purifiers, often require periodic regeneration with high temperature steam or gas. Should the process stream be reintroduced before the system is sufficiently cold, a hazardous situation could result. For example, an ignition hazard would exist if air containing organic vapor was prematurely introduced to a hot activated carbon bed. Another possibility is that an exothermic reaction will be initiated. The use of Programmable Logic Controllers (PLCs) for automatically switching adsorption beds into and out of regeneration can reduce the risk of human error.

When the potential for exothermic reaction exists, it is possible to generate high localized vessel wall temperatures. This can result in a lower MAWP for the vessel than the set pressure of the pressure safety valve. In such cases, some means to reduce vessel wall stress or quench the reaction is needed. Options include automatic emergency depressurization, injecting inert gas, or flooding with a compatible liquid. The UK Health and Safety Executive (HSE) issued an informative document on fire and explosion hazards of activated carbon adsorbers (Ref. 6-17).

6.3.3 Design Considerations

Batch distillation equipment can range from a free-standing column with a reboiler, condenser, receiver, and vacuum system to the use of a jacketed reactor with a condenser. Distillation often involves the generation of combustible vapors in the process equipment. This necessitates the containment of the vapor within the equipment, and the exclusion of air (oxygen) from the equipment, to prevent the formation of combustible mixtures that could lead to fire or explosion. Since distillation is temperature, pressure, and composition dependent, special care must be taken to fully understand any potential thermal decomposition hazards of the chemicals involved. Other potential hazards can result from the freezing or plugging in condensers, or blocked vapor outlets, which may lead to vessel overpressure if the heat input to the system is not stopped. Emphasis should be placed upon the use of inherently safer design alternatives using concepts such as:

- Limiting the maximum heating medium temperature to safe levels
- Selecting solvents which do not require removal prior to the next process step
- Using heat transfer medium to prevent freezing in the condenser

- Locating the vessel temperature probe on the bottom head to ensure accurate measurement of temperatures, even at a low liquid level.
- Minimizing column internal inventory

6.3.3.1 Columns

Columns, like other pieces of equipment, are available in a variety of mechanical designs. All of these various types are covered by the standard design codes, such as ASME Section VIII, Rules for Construction of Pressure Vessels (Ref. 6-18).

Column inventory can be minimized by understanding the different types of internal components that have differing operability flexibility and internal inventory. Choices for internal components include:

- Trays (bubble cap, valve, sieve reciprocating, baffles)
- Packed beds

Distillation columns often contain a large inventory of flammable liquids at elevated pressure and temperature. Inventory reduction may be obtained by prudent reduction of operating flexibility to obtain minimum holdup. Various tray designs and packing options can affect holdup volumes and, of course, column efficiency. Improved feed distribution, preheat, column pressure or multiple columns may be used to improve efficiency. The turndown ratio must be considered, particularly for large columns that may be on standby.

Minimizing column bottom inventories may make a column more sensitive to upsets if the response time of the control instrumentation is not capable of making quick adjustment. The same is true with the reflux inventory. For example, if a level controller fails open, the designer should determine if there is adequate time for response before the reflux pump runs dry. Operational problems include flooding, fouling, excessive pressure drop, or inefficient liquid / vapor contact. There is a need to provide pressure relief caused by loss of coolant, excessive heating in a reboiler, or fire. Design of pressure relief systems should account for all cases determined to be credible for the specific application under consideration.

Some chemicals are temperature sensitive and the bottom of the column should be sized down to minimize residence time, e.g., butadiene, ethylene oxide, etc.

Internal supports should be designed to withstand deviations such as flooding or pressure surge, a sudden collapse of packing, or tray failure. Process conditions may be particularly severe in distillation columns. The materials of construction should be thoroughly reviewed to understand any corrosion mechanisms that could occur in the vapor or liquid phases and with the vaporization and condensation processes.

Adequate instrumentation should be provided for monitoring and controlling pressure, temperature, level, and composition. The location of sensing elements in relation to column internals must be considered so that they provide accurate and timely information and are in direct contact with the process streams.

For vacuum towers, consideration should be given to installation of emergency block valves in the vacuum line which would close at selected column pressure and the purging of the column with nitrogen to break the vacuum. Another hazard associated with loss of vacuum is a rapid increase in the column bottoms temperature which may

6. EQUIPMENT DESIGN

lead to undesirable decomposition reactions depending on the chemical species involved in the distillation.

Opening packed columns for maintenance when not sufficiently cooled can result in fires when the high surface area, which may be coated with organics or pyrophoric materials, is exposed to air.

6.3.4 References

6-13.   Jarvis, H.C. *Butadiene Explosion at Texas City-2*, Plant Safety & Loss Prevention, Vol. 5. 1971.

6-14.   Freeman, R.H. and McCready, M.P *Butadiene Explosion at Texas City-1*, Plant Safety & Loss Prevention, Vol. 5. 1971.

6-15.   Keister, R.G., et al. *Butadiene Explosion at Texas City-3*, Plant Safety & Loss Prevention, Vol. 5. 1971.

6-16.   Britton, L.G., *Loss Case Histories in Pressurized Ethylene Systems*. Process Safety Progress, Vol. 13, No. 3. 1994

6-17.   HSE. *Carbon Bed Adsorbers – Fire and Explosion Hazards Report*. DIN SI5/62. Health and Safety Executive. UK. 2009. www.hse.gov.uk/foi/internalops/hid/din/562.pdf

6-18.   ASME. *Boiler and Pressure Vessel Code*, Section VIII, Division 1: Rules for Construction of Pressure Vessels. American Society of Mechanical Engineers, New York, NY. 2010.

6.4 HEAT TRANSFER EQUIPMENT

This section presents potential failure mechanisms for heat transfer equipment and suggests design alternatives for reducing the risks associated with such failures. The types of heat exchangers covered in this section include:

- Shell-and-tube exchangers
- Air-cooled exchangers
- Direct contact exchangers
- Others types including helical, spiral, plate and frame, wiped film, and carbon block exchangers

This section presents only those failure modes that are unique to heat transfer equipment. Some of the generic failure scenarios pertaining to vessels may also be applicable to heat transfer equipment. Consequently, this section should be used in conjunction with Section 6.1, Vessels. Unless specifically noted, the failure scenarios apply to more than one class of heat transfer equipment.

6.4.1 Past Incidents

This section provides several case histories of incidents involving failure of heat transfer equipment to reinforce the need for the safe design practices presented in this section.

6.4.1.1 Brittle Fracture of a Heat Exchanger

An olefin plant was being restarted after repair work had been completed. Leaks developed on the inlet flange of one of the heat exchangers in the acetylene conversion pre-heat system. To eliminate the leak, the control valve supplying feed to the conversion system was shut off and the acetylene conversion pre-heat system was depressured. Despite the fact that the feed-control valve was given a signal to close, the valve allowed a small flow (control valves are not intended for tight shut-off). High liquid level in an upstream drum may have allowed liquid carryover which resulted in extremely low temperature upon depressurization to atmospheric pressure (Ref. 6-19).

The leaking heat exchanger was equipped with bypass and block valves to isolate the exchanger. After the leaking heat exchanger was bypassed, the acetylene conversion system was repressured and placed back in service. Shortly thereafter, the first exchanger in the feed stream (converter pre-heater) to the acetylene converter system failed in a brittle manner, releasing a large volume of flammable gas. The subsequent fire and explosion resulted in two fatalities, seven serious burn cases, and major damage to the olefins unit.

The immediate cause of the converter pre-heater failure was that it was not designed for the low temperature deviation caused by depressuring the acetylene converter system. The heat exchanger that failed was fabricated from ASTM A515 grade 70 carbon steel. After the accident, the design of all process equipment in the plant which could potentially operate at less than 20°F (-7°C) was reviewed for suitable low temperature toughness.

Lessons learned include equipment design basis that should consider a wide range of possible operating conditions. It often costs relatively little to extend the design range beyond the minimum required. In this particular instance, it should have been recognized when defining the consequences of deviation that upstream cryogenic conditions may have a negative effect on downstream equipment during normal and abnormal operations.

6.4.1.2 Cold Box Explosion

Ethylene plants utilize a series of heat exchangers to transfer heat between a number of low temperature plant streams and the plant refrigeration systems. This collection of heat exchangers is known collectively as the "cold box." In one operating ethylene plant, a heat exchanger in the cold box that handled a stream fed to the demethanizer column required periodic heating and back-flushing with methane to prevent excessive pressure drop due to the accumulation of nitrogen-containing compounds (Ref. 6-20).

During a plant upset which resulted in the shutdown of the plant refrigeration compressors, the temperature of the cold box began to increase. During this temperature transient an explosion occurred which destroyed the cold box and disabled the ethylene plant for about 5 months. An estimated 20 tons of hydrocarbon escaped. Fortunately, the hydrocarbon did not ignite.

An investigation revealed that the explosion was caused by the accumulation and subsequent violent decomposition of unstable organic compounds that formed at the low temperatures inside the cold box (an unidentified inherent hazard). The unstable "gums" were found to contain nitro and nitroso components on short hydrocarbon chains. The

6. EQUIPMENT DESIGN

source of the nitrogen was identified as Nitrogen Oxides (NOx) present in a feed stream from a catalytic cracking unit. Operating upsets could have promoted unstable gums by permitting higher than normal concentrations of 1,3-butadiene and 1,3-cyclopentadiene to enter the cold box. To prevent NOx from entering the cold box, the feed stream from the catalytic cracking unit was isolated from the ethylene plant.

These incidents demonstrate that a thorough understanding of the inherent hazards of the process and a comprehensive consequence of deviation assessment are necessary during the equipment design and hazard identification phases.

6.4.2 Failure Scenarios and Design Solutions

Table 6.4 presents information on equipment failure scenarios and associated design solutions specific to heat transfer equipment.

Table 6.4   Common Failure Scenarios and Design Solutions for Heat Transfer Equipment

No.	Event	Consequence	Potential Design Solutions		
			Inherently Safer / Passive	Active	Procedural
Flow					
	Generally Applicable – More Flow (Applicable to all more flow scenarios)		Cold and hot side designed for maximum expected pressure		Operator response to high flow or high / low temperature alarms. Written procedures and training for manual isolation or bypassing of heating medium on indication of no flow on cold side
	Generally Applicable – No / Less Flow (Applicable to all no / less flow scenarios)		Antifouling design, e.g., pitch, baffle design and placement, designing for capability to periodically clean, etc.		Operator response to low flow or high temperature alarm
1	Control system failure, cold-side blocked in	Potential excessive heat input resulting in overpressure of cold side	Heat exchanger designed with an air pocket. Temperature of the heating medium limited	Pressure relief device	Written procedures and training to ensure heat exchangers are not blocked in
2	Flow maldistribution	Potential to overheat resulting in hot spots	Exchanger design / type less sensitive to flow distribution issues selected		Written procedures and training to detect maldistribution via bed temperature profile

Table 6.4   Common Failure Scenarios and Design Solutions for Heat Transfer Equipment

No.	Event	Consequence	Potential Design Solutions		
			Inherently Safer / Passive	Active	Procedural
Temperature					
	Generally Applicable – High Temperature *(Applicable to all high temperature scenarios)*		Alternative exchanger design Exchanger located outside fire-affected zone Fireproof insulation (limits heat input)	Automatic isolation of input flow on detection of high vent temperature Automatic shutdown of heat source on high temperature Backup cooling medium supply with automatic switch-over Fixed water spray (deluge) and / or foam systems activated by flammable gas or flame detection devices High temperature indication with alarm and interlock which isolates the heating medium	Emergency response plan Operator response to high temperature alarm Written procedures and training for manual activation of fixed fire protection water spray (deluge) and / or foam systems Written procedures and training for manual activation of backup cooling
	Generally Applicable – Low Temperature *(Applicable to all low temperature scenarios)*		Mechanical design to accommodate minimum expected temperature	Interlock to isolate feed on detection of low temperature	Operator response to low temperature alarm
3	Differential thermal expansion / contraction (shell-and-tube exchanger)	Potential leak or rupture resulting in overpressure of the low pressure side	Shell expansion joint, internal floating head or U tubes Alternative exchanger design other than shell and tube (e.g., spiral, plate, and frame) Alternative flow arrangement to avoid thermal stress Low pressure side designed for 10/13 design pressure of high pressure side (ASME)	Automatic control of introduction of process fluids on startup and shutdown	Written procedures and training for control of introduction of process fluids on startup and shutdown to reduce cycling Written procedures and training for periodic inspection / analysis of low pressure fluid for high pressure fluid leakage

6. EQUIPMENT DESIGN

Table 6.4 Common Failure Scenarios and Design Solutions for Heat Transfer Equipment

No.	Event	Consequence	Potential Design Solutions		
			Inherently Safer / Passive	Active	Procedural
4	Sudden ambient temperature drop (air-cooled exchanger)	Potential excessive heat transfer rate resulting in freezing of material	Different type of exchanger selected to minimize or eliminate consequences of freezing	Automatic air inlet temperature control via air pre-heating with steam or air recirculation Air flow control (e.g., variable pitch / speed fans)	Written procedures and training for monitoring and manual adjustment of air inlet temperature
Level					
Generally Applicable – High Level (Reboilers) *(Applicable to all high level scenarios)*				Interlock to isolate feed on detection of high level	Operator response to high level alarm
Generally Applicable – Low Level (Reboilers) *(Applicable to all low level scenarios)*				Interlock to shut down heat source on detection of low level	Operator response to low level alarm
5	High level in kettle vaporizer	Potential carryover to downstream equipment	Kettle vaporizer designed for adequate height for disengaging liquids Demister installed in kettle vaporizer		
6	Low level in kettle vaporizer	Potential superheating	Kettle vaporizer designed with weir		
Equipment Failure					
Generally Applicable *(Applicable to all equipment failure scenarios)*			Alternative heat exchanger designs Mechanical design (e.g., proper baffle spacing) accommodating maximum anticipated inlet feed pressure / velocity Mechanical design to accommodate maximum expected temperature and pressure of a possible exothermic reaction	Emergency relief device Automatic shutdown on detection of high pressure on low pressure side	

Table 6.4   Common Failure Scenarios and Design Solutions for Heat Transfer Equipment

No.	Event	Consequence	Potential Design Solutions		
			Inherently Safer / Passive	Active	Procedural
7	Corrosion / erosion	Potential leak or rupture resulting in overpressure of the low pressure side	Corrosion-resistant materials Design changes to reduce erosion (e.g., lower velocities, inlet baffle) Double tube sheets Less corrosive heat transfer media Low pressure side designed for 10/13 design pressure of high pressure side (ASME) Open low pressure side return Seal welding of tube-to-tube sheet joints Seamless versus seam-welded tubes Impingement Protection	Pressure relief device	Corrosion detection device (e.g., coupons) Written procedures and training for periodic inspection / analysis of low pressure fluid for high pressure fluid leakage Shutdown for mechanical integrity inspections Inspection programs, e.g., RBI
8	Tube leak / rupture (shell-and-tube exchanger)	Potential leak or rupture resulting in overpressure of the low pressure side	Seamless versus seam-welded tubes Alternative exchanger design other than shell and tube (e.g., spiral, plate, and frame) Low pressure side designed for 10/13 design pressure of high pressure side (ASME) Seal welding of tube-to- tube sheet joints	Pressure relief device	Written procedures and training for periodic inspection / analysis of low pressure fluid for high pressure fluid leakage
		Potential for flammable material at cooling tower, potential fire			Operator response to gas detection alarm on top of cooling tower

6. EQUIPMENT DESIGN

Table 6.4   Common Failure Scenarios and Design Solutions for Heat Transfer Equipment

No.	Event	Consequence	Potential Design Solutions		
			Inherently Safer / Passive	Active	Procedural
9	Fouling, accumulation of non-condensables	Potential loss of heat transfer	Additional surface area in air cooler to transfer heat via natural convection Continuous open venting of non-condensables Exchanger designed for suitable velocity to minimize fouling Heat exchanger design less prone to fouling (e.g., direct contact)	Automatic tempering of cooling medium temperature to avoid low tube wall temperature resulting in solids deposition Automatic venting of non-condensables	Written procedures and training for manual adjustment of cooling medium tempering Written procedures and training for periodic exchanger cleaning Written procedures and training manual isolation of input flow on detection of high vent temperature
10	Corrosion / erosion, vibration or differential thermal expansion	Potential tube leak resulting in mixing of fluids resulting in exothermic reactions, phase changes, and / or fluid system contamination	Double tube sheet design Heat transfer media selected that is chemically compatible with process materials Seal welding of tube-to-tube sheet joints	Downstream fluid analyzers with concentration alarms interlocked with automatic shutdown	Operator response to downstream fluid analyzers with concentration alarms Written procedures and training for periodic sampling and analysis of fluids Written procedures and training on steps to reduce thermal stress on startup and shutdown. Written procedures and training to reduce need for thermal cycling
11	Fan blade failure (air-cooled exchanger)	Potential vibration resulting in tube rupture due to impact	Design of passively cooled system Machine guarding	Vibration monitoring with automatic fan shutdown	Written procedures and training for manual fan shutdown on indication of excessive vibration
12	Misalignment or entrance of foreign objects (scraped surface)	Potential for scraper punctures heat transfer surface resulting in equipment damage	Screens at entrance of heat exchanger to remove foreign objects	Automatic shutdown of motor on high amperage or power	Written procedures and training for manual shutdown of motor on high amperage or power

6.4.2.1 Leak / Rupture of the Heat Transfer Surface (Scenarios 3, 7, and 8)

This common failure scenario may result from corrosion, thermal stresses including cryogenic embrittlement, or mechanical stresses of heat exchanger internals. The leak / rupture of tubes leads to contamination or overpressure of the low pressure side. Failure to maintain separation between heat transfer and process fluids may lead to violent reaction in the heat transfer equipment or in the downstream processing equipment. To make the heat transfer process inherently safer, designers must look at possible interactions between heating / cooling fluids and process fluids.

For equipment with design pressures <1000 psig, a complete failure of tubes may not be a credible overpressure scenario if the design pressure of the low pressure side and associated equipment is greater than the ratio of the test pressure of the low side to the design pressure of the high pressure side (Ref. 6-21), or if the geometry of the tube layout is such that a complete break is not physically possible. For equipment with design pressures >1000 psig, however, a complete failure should be considered credible, regardless of pressure differential.

Double tube sheets or seal welding may be used for heat exchangers handling toxic chemicals. For heat transfer problems involving highly reactive / hazardous materials, a triple-wall heat exchanger may be used. This type of heat exchanger consists of three chambers and uses a neutral material to transfer heat between two highly reactive fluids. Alternatively two heat exchangers can be used with circulation of the neutral fluid between them.

There are known cases of cooling tower fires that have resulted from contamination of cooling water with hydrocarbons attributable to tube leakage. Gas detectors and separators may be installed on the cooling water return lines or in the cooling tower exhaust (air) stream.

Thermal stresses can be reduced by limiting the temperature differences between the inlet and outlet streams. In addition, alternate flow arrangements may be used to avoid high thermal stresses. Thermal cycling of heat transfer equipment should be kept to a minimum to reduce the likelihood of leaks and ruptures.

6.4.2.2 Fouling and Accumulation of Non-Condensable Gases (Scenario 9)

It is desirable to design heat exchangers to resist fouling. Sufficient tube side velocity may reduce fouling. However, higher tube-side velocities may also lead to erosion problems. In some cases fouling will cause higher tube wall temperatures, leading to overheating of reactive materials, loss of tube strength, or excessive differential thermal expansion.

Accumulation of non-condensable gases can result in loss of heat transfer capability. Heat exchangers in condensing service may need a vent nozzle or other means of removing non-condensable gases from the system.

6.4.2.3 External Fire

Emergency relief devices are often sized for external fire. Heat transfer equipment, such as air coolers, present a unique challenge when it comes to sizing relief devices. These exchangers are designed with large heat transfer areas. This large surface area may result in very large heat input when exposed to an external fire. Indeed, it may not be practical to install a relief device sized for an external fire case due to large relief area requirements. Other mitigation measures, such as siting outside the potential fire zone or diking with sloped drainage, may be used to reduce the likelihood and magnitude of external fire impinging on the heat exchanger. Alternative heat exchanger designs may also be used to reduce the surface area presented to an external fire.

6.4.3 Design Considerations

Heat transfer is one of the most widely used operations in the chemical process industries. Not only is heat transfer used in physical operations (distillation, drying), but it is a required component of most reactions. This category of equipment includes heat exchangers, vaporizers, reboilers, process heat recovery boilers, condensers, coolers, and chillers. Some design considerations are included in the following:

- ASME Code (Ref. 6-22)
- API Standard 520, *Sizing, Selection, and Installation of Pressure-Relieving Devices in Refineries* (Ref. 6-23)
- API Standard 660, *Shell and Tube Heat Exchangers* (Ref. 6-24)
- Tubular Exchanger Manufacturers Association (TEMA)
- Heat Exchanger Institute standards

Control of temperature is critically important in maintaining control of the process. Loss of temperature control has many adverse effects, including increase in pressure, increase in reaction rate, increase in corrosion rate, change in equilibrium conditions, destruction of products, and instability of products. Temperature excursions beyond normal operating limits may put excessive stress on the shell side, tube side or both. Startup, shutdown or maintenance procedures may present a situation where one side has no fluid in it while the other side is at an extreme.

Common problems of exchangers include tube rupture, leaking, fouling, tube vibration, and polymerization and solidification (Ref. 6-25). Failures in heat exchangers result in pressure changes (overpressure or vacuum) and contamination of the heat transfer fluid or process fluid. The primary hazard is failure to maintain separation of materials which might react violently upon contact.

Design considerations for exchanger include:

- Use of double tube sheets for heat exchangers handling toxic chemicals.
- Selection of which material is on the shell or tube side.
- Design for drainage to reduce corrosion by installing exchanger in a sloped orientation (avoid baffles, which allow fluids to be trapped).
- Design for periodic cleaning.
- Design for ability to drain, purge, wash, and prevent / minimize dead trapped liquid sections, particularly if heat continues.
- Provide a tube sheet vent nozzle and / or a means to vent noncondensable gases from the process system.
- Careful selection of materials to resist corrosion on both sides. The use of bimetallic tubes may create a new set of potential problems as each tube may respond in a different manner.

- Tube pitch and spacing, flow distribution, fluid velocity, and ΔT should be considered to prevent fouling.
- The bending of exchanger tubes to form U-bends introduces residual stresses in the tube material which may make it more susceptible to stress corrosion cracking. Stress relief of U-bend exchanger tubes depends on the alloy and service conditions (temperature and constituents); in fact, stress relief may introduce undesirable metallurgical effects.
- External stress corrosion cracking from chlorides in cooling water must be addressed; for example, the designer may consider using alloys more resistant to chloride attack.
- Selection, installation, and maintenance of insulation to avoid corrosion under thermal insulation.
- Design to prevent ice plugging in cold condensers when inadvertent moisture gets in the system or the system temperature control goes colder than intended.

The minimization strategy of inherent safety can be applied in some instances by using several smaller exchangers rather than one large one. Besides the reduction in hazardous material retained, more corrosion resistant materials can be used in the first exchanger, which experiences the greatest temperature differential. This first exchanger could either be a sacrificial type under continuous corrosion monitoring or be fabricated from a more corrosion resistant alloy.

One safeguard strategy to protect leaking exchanger tubes that contaminate the cooling water is to provide gas detectors or gas separators for the cooling water return. In addition to analyzing the compounds exchanging heat, the designer should consider the potential effects of inhibitors (or other water treatment chemicals) in the cooling water or heat transfer fluid.

Another safeguard strategy may be to protect against leaking tubes by considering potential interaction between the materials exchanging heat in the event of a leak. The decision as to which is the high pressure side may depend on the potential reactions between process chemicals and the heating medium. If a small amount of chemical "A" is introduced through a tube leak into large amounts of chemical "B" without a considerable reaction, then try to design the process so that "A" is slightly higher in pressure than "B". In case corrosion or tube failure occurred, then the only hazard would be poor product quality and heat exchange. Other hazardous conditions may exist if water can poison a catalyst or react with an acid.

Consideration must be given to possible tube rupture and an adequately sized relief device must be provided.

6.4.4 References

6-19. Viera, G.A., Simpson L. L., and Ream B. C. *Lessons Learned from the Ethylene Oxide Explosion at Seadrift, Texas*, Chemical Engineering Progress. August 1993.

6-20. Price, J. H. *Cold Box Explosion at Shell Steam Cracker in Berre, France.* Paper presented at AIChE Spring National Meeting, Houston, Texas. 1989.

6-21. API STD 521. *Guide for Pressure Relieving and Depressuring Systems, Fifth Edition.* American Petroleum Institute, Washington D.C.. 2007.

6. EQUIPMENT DESIGN

6-22. ASME Section VIII-DIV 1. *ASME Boiler and Pressure Vessel Code,* Section VIII, Division 1: Rules for Construction of Pressure Vessels. American Society of Mechanical Engineers. New York, New York. 2010.

6-23. API STD 520. *Sizing, Selection, and Installation of Pressure Relieving Devices in Refineries*, Part I - Sizing and Selection, Eighth Edition. American Petroleum Institute. Washington D C. 2008.

6-24. API STD 660. *Shell-and-Tube Heat Exchangers*, Eighth Edition. American Petroleum Institute. Washington D. C. 2007.

6-25. Lees, F.P. *Loss Prevention in the Process Industries,* Third Edition. Elsevier, Inc. Oxford, UK. 2005.

6.4.4.1 Suggested Additional Reading

Kletz, T. A. *Learning from Accidents,* Oxford: Butterworth-Heinemann Ltd. 1994.

Kuppan, T. *Heat Exchanger Design Handbook*, CRC Press, Boca Raton, Florida. 2000.

McCarthy, A. J., and Smith, B. R.. *Reboiler System Design - The Tricks of the Trade*, Process Plant Safety Symposium, February 28-March 2, 1994, Houston, Texas. 1994.

Yokell, S. *A Working Guide to Shell-and-Tube Heat Exchangers.* McGraw-Hill, New York, New York. 1990.

6.5 DRYERS

This section presents potential failure mechanisms for dryers, drying systems and suggests design alternatives for reducing the risks associated with such failures. The types of equipment covered in this section include:

- Spray dryers
- Tray dryers
- Fluid-bed dryers
- Conveying (flash, mechanical, and pneumatic) dryers
- Rotary dryers

This section presents only those failure modes that are unique to dryers. Some of the generic failure scenarios pertaining to vessels and heat transfer equipment may also be applicable to dryers. Consequently, this section should be used in conjunction with Section 6.1, Vessels and Section 6.4, Heat Transfer Equipment. Also, since drying equipment is often associated with solid-fluid separators and solids handling and processing equipment, refer to Section 6.7 for additional information. Unless specifically noted, the failure scenarios apply to more than one class of dryers.

6.5.1 Past Incidents

This section presents case histories involving fires and explosions (deflagrations) to reinforce the need for safe design and operating practices for dryers and drying systems.

6.5.1.1 Drying of Compound Fertilizers

A fire and explosion occurred in a dryer handling a blended fertilizer that contained single and triple super-phosphates and a mixture of nitrogen-phosphorous-potassium fertilizers. The blend was prone to self-sustained decompositions, and began decomposing while passing through the dryer. When the temperature of the blend rose to about 130°C, the operator intervened and shut down the dryer. Subsequently, a rapid exothermic reaction occurred within the dryer that resulted in a fire and explosion. One person was killed and 18 were injured (Ref. 6-26).

Lessons learned include the consideration of a different type of dryer for this application that better controls the temperature.

6.5.1.2 Fires in Cellulose Acetate Dryer

A continuous belt dryer used to dry cellulose acetate powder had experienced repeated small internal fires over a two-year period. After performing a basket (self-heating) test to determine if exothermic behavior was present under various solids depths, investigators discovered that an exothermic reaction was detected at 433°F (223°C) under process conditions. Because the dryer was heated with 100-psig steam [saturation temperature of 342°F (172°C)] it was initially thought that this exothermic behavior was not the cause of the fires. Further examination revealed that the 100-psig steam at this particular location was superheated to 455°F (235°C), well above the exotherm initiation temperature. After a steam desuperheater was installed immediately upstream of the dryer, the fire problem disappeared.

Lessons learned include the need to understand the temperature sensitivity of the material being dried as well as knowing the actual characteristics of the heating medium being used.

6.5.1.3 Pharmaceutical Powder Dryer Fire and Explosion

An operator had tested dryer samples on a number of occasions. After the last sampling, he closed the manhole cover, put the dryer under vacuum, and started rotation of the dryer. A few minutes later an explosion and flash fire occurred, which self-extinguished. No one was injured. Investigations revealed that after the last sampling, the dryer manhole cover had not been securely fastened. This allowed the vacuum within the dryer to draw air into the rotating dryer and create a flammable mixture. The ignition source was probably an electrostatic discharge on the internal lining of the dryer. No nitrogen inerting had been used (Ref. 6-26).

Lessons learned include the following precautions to prevent similar incidents from occurring in the future:

- Nitrogen purging is carried out before charging or sampling of the dryer.
- If the absolute pressure rises to about 4 psia, the rotation stops, an alarm sounds, and a nitrogen purge starts automatically.

6.5.2 Failure Scenarios and Design Solutions

Table 6.5 presents information on equipment failure scenarios and associated design solutions specific to dryers.

6. EQUIPMENT DESIGN

Table 6.5 Common Failure Scenarios and Design Solutions for Dryers

No.	Event	Consequence	Potential Design Solutions		
			Inherently Safer / Passive	Active	Procedural
Pressure					
	Generally Applicable – High Pressure *(Applicable to all high pressure scenarios)*		Dryer designed to contain overpressure	Deflagration venting Deflagration suppression system Use of inert atmosphere	Operator response to high pressure alarm
	Generally Applicable – Low Pressure *(Applicable to all low pressure scenarios)*		Dryer designed for vacuum conditions		Operator response to low pressure alarm
Flow					
	Generally Applicable – More Flow *(Applicable to all more flow scenarios)*		Alternate type of dryer	Automatic feed trip on loss of ventilation or high concentration of flammable vapor Automatic isolation via quick closing valves of manifold duct system on detection of fire / flammable atmosphere in duct system Automatic shutdown of conveyor on high speed indication Automatic sprinkler system / CO_2 total flooding system Use of inert atmosphere Ventilation system to keep flammable concentration below lower flammable limit	Operator response to high flow alarm Written procedures and training for manual activation of fire protection / inerting system Written procedures and training for manual bonding and grounding Written procedures and training for manual isolation of feed on loss of ventilation

Table 6.5   Common Failure Scenarios and Design Solutions for Dryers

No.	Event	Consequence	Potential Design Solutions		
			Inherently Safer / Passive	Active	Procedural
	Generally Applicable – No / Less Flow *(Applicable to all no / less flow scenarios)*			Automatic shutdown on detection of low circulating flow	Operator response to low flow alarm
1	Buildup of deposits in dryers and ductwork	Potential ignition of material resulting in fire / explosion	Dryer design which minimizes buildup of deposits (smooth surfaces, elimination of potential points of solids accumulation) Use dryer with short residence time (e.g., flash dryer)		Periodic inspection and cleaning Written procedures and training to process most stable materials first when campaigning multiple products to avoid ignition of unstable materials Written procedures and training for determining maximum tolerable material accumulation
2	Inadequate ventilation due to obstructions or closed dampers	Potential flammable atmosphere with subsequent ignition resulting in fire / explosion	Design dampers so that system will handle the minimum safe ventilation rate at maximum damper throttling Provide damper mechanical position stop to prevent complete closure of damper	Limit switch on damper interlocked to introduce inerting gas	Note: manual isolation using quick closing valves is not practical in this application
3	Increase in conveyor speed	Potential generation of solvent vapors from the feed with subsequent ignition resulting in fire / explosion	Ventilation system designed to handle the maximum solvent evaporation rate	Ventilation system flow rate interlocked with the conveyor speed	Operator response to indication of higher conveyor speed Written procedures and training for manual shutdown of conveyer on high speed indication

6. EQUIPMENT DESIGN

Table 6.5 Common Failure Scenarios and Design Solutions for Dryers

No.	Event	Consequence	Potential Design Solutions		
			Inherently Safer / Passive	Active	Procedural
4	Batch operation resulting in a high peak evaporation rate of flammable solvent	Potential flammable atmosphere with subsequent ignition resulting in fire / explosion	Ventilation system designed to handle the peak solvent evaporation rate. Dryer designs where natural circulation is sufficient to keep solvent concentration at a safe level. Use continuous or semi-continuous dryer design	Detection of flammable conditions and adjustment of diluent	Written procedures and training which allow for the unsteady evaporation rates during batch operations
5	Inadequate circulation in dryers	Potential flammable atmosphere with subsequent ignition resulting in fire / explosion	Dryer designs where natural circulation is sufficient to prevent accumulation of flammables		Written procedures and training for manual dryer shutdown on low circulation
6	Excessive atomization in nozzle (spray dryer)	Potential generation of fine resulting in a dust / hybrid fire / explosion	Inlet temperature of heating medium should be sufficiently below the minimum ignition temperature	Pressure control to regulate the nozzle pressure	Written procedures and training to blow lines with nitrogen
7	Manifolding of ventilation exhaust ducts of several dryers	Spread of fire or deflagration from one location to the next	Use dedicated exhaust ducts. Design dryer and ductwork to contain overpressure where practical	Vent individual dryers through conservation vents to prevent back flow. Install flame arresters in dryer vents	Operator action to isolate various ducts on detection of fire / flammable atmosphere
8	Low feed rate to dryer	Potential increased temperature of material in the dryer, possible fire / explosion	Use of heating medium which automatically limits the temperature to which the feed is exposed	Automatic control of heat input to dryer based on feed flow rate. High temperature alarms and shutdown systems. Automatic control of feed rate	Written procedures and training for manual control of feed rate. Operator response to high temperature indication

Table 6.5   Common Failure Scenarios and Design Solutions for Dryers

No.	Event	Consequence	Potential Design Solutions		
			Inherently Safer / Passive	Active	Procedural
Temperature					
	Generally Applicable – High Temperature *(Applicable to all high temperature scenarios)*		Dryer designed for high temperature Dryer designed to contain overpressure Permanent bonding and grounding Eliminate flammables Eliminate ignition sources within the ductwork	Automatic feed trip on loss of ventilation or high concentration of flammable vapor Automatic isolation of associated equipment via quick-closing valves Automatic isolation via quick-closing valves of manifold duct system on detection of fire / flammable atmosphere in duct system Automatic shutdown of conveyor on high speed indication Automatic sprinkler system / CO_2 total flooding system Use of inert atmosphere Ventilation system to keep flammable concentration below lower flammable limit	Written procedures and training for manual activation of fire protection / inerting system Written procedures and training for manual bonding and grounding for feed or product discharge. Written procedures and training for manual isolation using quick-closing valves normally not practical Online flammable gas detection and manual activation of CO_2 total flooding system
9	Condensing of flammable vapor in ductwork	Potential ignition of material resulting in fire / explosion	Dryer design to prevent condensation in ductwork Provision for drainage of ducts (e.g., sloped, low point drains)		
10	Sudden loss of heating medium with vapor condensation	Potential vacuum	Design dryer and duct work for vacuum		Written procedures and training to limit rate of temperature decrease in dryer

6. EQUIPMENT DESIGN

Table 6.5 Common Failure Scenarios and Design Solutions for Dryers

No.	Event	Consequence	Potential Design Solutions		
			Inherently Safer / Passive	Active	Procedural
11	High surface temperature in dryers and ductwork	Potential ignition of surrounding combustibles (including fugitive emissions from the dryer) resulting in fire / explosion	Insulation of external dryer surfaces to reduce surface temperature to a safe limit Limit temperature of the dryer to below the safe temperature limit of surrounding materials Maintain proper clearances between hot surfaces and combustible materials	Fines removal from exit gas (bag filters)	Written procedures and training for good housekeeping
12	Heat generated from mechanical input (i.e., plugging of rotary feeders, paddle dryers, screw conveyors)	Potential fire / explosion	Use dryer component types which minimize mechanical heat input Use non-flammable / high flash point lubricants	Provide torque limiting devices (i.e., shear pins) for mechanical components	Operator response to high and low torque alarms for mechanical devices Written procedures and training to monitor temperature and take action on high temperature alarm
Composition					
13	Attrition of solids resulting in particle size reduction	Potential fire / explosion	Select alternate dryer design which reduces attrition rate		Written procedures and training to keep particle size out of explosive range
Equipment Failure					
14	Lube oil leakage into dryer	Potential fire / explosion	Double mechanical seals Use dryer with no mechanical seals		Written procedures and training for periodic bearing and seal inspection
15	Electrostatic spark (vessel is non-conductive due to glass lining) (double-cone tumbling dryer-glass-lined)	Potential fire / explosion		Automatic shutdown on high outlet temperature	

Table 6.5   Common Failure Scenarios and Design Solutions for Dryers

No.	Event	Consequence	Potential Design Solutions		
			Inherently Safer / Passive	Active	Procedural
16	Flammable dust / vapors above the bed (fluid bed dryer)	Potential fire / explosion	Design freeboard to minimize dust emission		Written procedures and training for manual grounding and bonding for portable units

6.5.2.1   Buildup of Deposits in Dryers / Ductworks (Scenario 1)

Some dryers and drying systems (including ductwork and associated equipment such as cyclones, dust collectors, etc.) are prone to accumulation of deposits on dryer walls and ductwork. Solids often accumulate on spray devices at the top of dryers where the highest dryer temperature is often experienced. Frequent cleaning and monitoring may be required to ensure that these deposits do not overheat and autoignite. Tests should be conducted to evaluate the hazards of dust deposit ignitability. The characteristics of materials deposited on walls or other surfaces may change over time when the materials are continuously exposed to high temperatures or other process conditions.

6.5.2.2   Electrostatic Hazards (Scenarios 15 and 16)

Electrostatic sparks are a common cause of dust and flammable vapor deflagrations. Dryers and drying systems that can generate electrostatic charges must be properly bonded and grounded to drain off these charges and minimize the possibility of deflagrations. Inerting is often needed to prevent the occurrence of a deflagration.

6.5.2.3   Hybrid Mixtures (Scenario 6)

Many drying operations involve the evaporation of a flammable solvent from a combustible powder. This combination of a flammable vapor and combustible powder fines (dust) is called a hybrid mixture. Hybrid mixtures represent a greater explosion hazard than that presented by the combustible dust alone. This increased hazard is characterized by the following:

- The hybrid mixture may explode more severely than a dust-air mixture alone, i.e., the maximum pressure and maximum rate of pressure rise may be greater, even if the vapor concentration is below its Lower Explosive Limit (LEL).
- The minimum ignition energy of hybrid mixtures is usually lower than that of the dust-air mixture alone.
- The Minimum Explosive Concentration (MEC) of a dust is reduced by the presence of a flammable vapor even if the latter is below its LEL. Measurable effects are observed as low as 20% of the vapor LEL.

6.5.2.4   Decomposition (Scenario 8 and 12)

Many powders are thermally sensitive and may decompose at high temperature, resulting in an overpressure or fire. Some dried materials, such as sodium hydrosulfite, may also exothermically decompose when exposed to water. It is very important to determine if organic powders are thermally unstable and, if so, that they be tested for thermal stability to establish a safe operating temperature for the drying operation. The potential for decomposition will depend on the characteristics of the solid, including depth, composition, temperature, duration of exposure, and dryness.

6.5.3 Design Considerations

The choice between different types of dryers is often guided by the chemicals involved and their physical properties, particularly heat sensitivity. As when selecting other equipment, the designer should first ask if the step is necessary; if so, whether this is the correct or safest process step.

- Does the material being processed have to have all of the liquid removed?
- Can the downstream step or customer use the material in liquid, slurry, or paste form?

Some of the hazards in drying operations are:

- Vaporization of flammable liquids
- Presence of combustible dusts
- Overheating leading to decomposition
- Inerting leading to an asphyxiation hazard

For heat-sensitive material, limiting the temperature of the heating medium and residence time of the material is used to prevent decomposition. Inventories of hazardous materials should be minimized. Preventive measures include adequate ventilation and explosion venting, explosion containment, explosion suppression, inerting, elimination of ignition sources, and vapor recovery. Instrumentation may include oxygen analyzers and sensors for temperature, humidity, etc. Effluent gases should be monitored for flammability limits. Design considerations for equipment handling combustible dusts (Ref. 6-27) are:

- Design equipment to withstand a dust explosion.
- Minimize volume filled by dust suspension.
- Minimize (monitor) mechanical failure and overheating (bearing, rollers, mills, etc.).
- Eliminate static electricity and other sources of ignition.
- Minimize passage of burning dust by isolating equipment.
- Provide explosion prevention (e.g., by inerting) and protection (e.g., suppression, venting, or isolation).
- Provide fire protection.

6.5.4 References

6-26. Drogaris, G. *Major Accident Reporting System: Lessons Learned from Accidents Notified.* Elsevier Science Publishers B. V. Amsterdam.1993.

6-27. CCPS. *Guidelines for Safe Handling of Powders and Bulk Solids.* Center for Chemical Process Safety of the American Institute of Chemical Engineers. New York, New York. 2005.

6.5.4.1 Suggested Additional Reading

Abbot, J. *Prevention of Fires and Explosions in Dryers - A User Guide, Second Edition.* The Institution of Chemical Engineers, London.1991.

Bartknecht, W. *Dust Explosions: Course, Prevention, Protection.* Springer-Verlag. New York.1989.

Chatrathi, K. *How to Safely Handle Explosible Dust - Part I.* Powder and Bulk Engineering, p22-28. January 1991.

Chatrathi, K. *How to Safely Handle Explosible Dust - Part II.* Powder and Bulk Engineering, p12-18. February 1991.

Ebadat, V. *Testing to Assess Your Powder's Fire and Explosion Hazards.* Powder and Bulk Engineering, p19-26. January 1994.

Garcia, H. and Guarici, D. *How to Protect Your Drying Process from Explosions.* Powder and Bulk Engineering, p53-64. April 1995.

Gibson, N., Harper, D. J. and Rogers, R. L. *Evaluation of the Fire and Explosion Risk in Drying Powders.* Plant / Operations Progress, p181-189. 1985.

NFPA 654: *Standard for the Prevention of Fire and Dust Explosions from the Manufacturing, Processing, and Handling of Combustible Particulate Solids.* National Fire Protection Association, Quincy, Massachusetts. 2006.

Palmer, K. N. *Dust Explosions and Fires (Powder Technology).* London: Kluwer Academic Publishers. 1993.

Palmer, K. N. *Dust Explosions: Initiations, Characteristics, and Protection.* Chemical Engineering Progress. p24-32. March 1990.

6.6 FLUID TRANSFER EQUIPMENT

This section presents potential failure mechanisms for fluid transfer systems and suggests design alternatives for reducing the risks associated with such failures. The types of fluid transfer equipment covered in this section include:
- Blowers
- Pumps
- Compressors

This section presents only those failure modes that are unique to fluid transfer systems. Some of the generic failure scenarios pertaining to vessels may also be applicable to fluid transfer systems. Consequently, this section should be used in conjunction with Section 6.1, Vessels. Unless specifically noted, the failure scenarios apply to more than one class of fluid transfer systems.

6.6.1 Past Incidents

This section provides case histories of incidents involving failure of fluid transfer systems to reinforce the need for the safe design practices presented in this section.

6. EQUIPMENT DESIGN

6.6.1.1 Startup of Parallel Centrifugal Pumps (Scenario 1)

Parallel high head centrifugal pumps were used to transfer an organic acid stream approximately 1.5 miles from a distillation facility to another manufacturing unit in the same complex. Because both the distillation unit and the destination manufacturing unit had significant inventory capacity, switching from primary to spare pump was not automated since timing was not critical and short breaks in service were tolerable. After one such changeover, the pump taken offline was not properly isolated and drained. Consequently, when the spare pump was started, the offline pump immediately saw full discharge pressure on its seal which caused the offline pump seal to fail, spilling about 500 gallons of material into a contained area until the pump could be shut off.

Lessons learned include adding a check valve in the discharge line of each pump to reduce the likelihood of the scenario occurring. The equipment design basis should consider a range of conditions including startup, shutdown, and unintended operations.

6.6.1.2 Continuous Sulfonation Explosion

During the startup phase of a continuous system for the sulfonation of an aromatic compound, a thermal explosion occurred in a pump and recirculation line. Although the incident damaged the plant and interrupted production, no personnel were injured.

Investigation revealed that, while recirculation of the reaction mass was starting up, the pump and the line became plugged. This problem was corrected and line recirculation was restarted. Four hours later the explosion occurred, resulting in the blow-out of the pump seal, which was immediately followed by rupture of the recirculation line.

Investigation further revealed that during pipe cleanout some insulation had been removed, leaving a portion of the line exposed and untraced. This condition apparently led to slow solidification of the reaction mass and a deadheaded pump. Calculations based on pump data indicated that a temperature of 140°F (60°C) above the processing temperature could be reached within 5 minutes after deadheading occurred.

Lessons learned include the need to positively monitor pump circulation and use direct temperature or pressure measurement to detect the onset of a runaway reaction.

6.6.2 Failure Scenarios and Design Solutions

Table 6.6 presents information on equipment failure scenarios and associated design solutions specific to fluid transfer equipment.

Table 6.6 Common Failure Scenarios and Design Solutions for Fluid Transfer Equipment

No.	Event	Consequence	Potential Design Solutions		
			Inherently Safer / Passive	Active	Procedural
Pressure					
	Generally Applicable – High Pressure (Applicable to all high pressure scenarios)		Downstream piping specified to withstand deadhead pressure	Emergency relief device High pressure shutdown interlock	Operator response to high pressure alarm
	Generally Applicable – Low Pressure (Applicable to all low pressure scenarios)		NPSH maximized Supply tank elevated for fluids close to boiling point	Low pressure shutdown interlock	Operator response to low pressure alarm
Flow					
	Generally Applicable – More Flow (Applicable to all more flow scenarios)				Operator response to high flow alarm
	Generally Applicable – No / Less Flow (Applicable to all no / less flow scenarios)			Interlock to shutdown pump on detection of low flow / low pressure	Operator response to low flow alarm Written procedures and training for starting spare pump / compressor
1	Discharge control valve closed; Downstream block valve closed; Blind not removed on startup; Plugged outlet	Potential to deadhead pump, potential overpressure and / or excessive temperature, potential seal failure, potential loss of containment	Minimum flow recirculation line to ensure a minimum flow through the machine (flow controlled by orifice)	Automatic startup of spare pump (for some scenarios) Localized fire protection Low flow or power shutdown interlock Minimum flow recirculation line (flow automatically controlled)	Written procedures and training to avoid deadheading pump / compressor
2	Blocked suction (valve closed, strainer plugged) (centrifugal pump)	Reduced flow to the inlet of a centrifugal pump causing cavitation, excessive vibration, possible damage to pump seal	Restrictions in suction system eliminated	High vibration shutdown interlock Localized fire protection Low flow shutdown interlock	Operator response to low flow indication and / or high vibration

6. EQUIPMENT DESIGN

Table 6.6 Common Failure Scenarios and Design Solutions for Fluid Transfer Equipment

No.	Event	Consequence	Potential Design Solutions		
			Inherently Safer / Passive	**Active**	**Procedural**
3	Blocked suction (valve closed, strainer plugged) (centrifugal compressor)	Reduced flow through a centrifugal compressor causing surge leading to high vibrations, possible compressor damage	Compressor design other than centrifugal	Automatic anti-surge (recycle system) High vibration shutdown interlock Low flow shutdown interlock	
4	Pump stops	Potential backflow through pump or recycle line	Positive displacement pump Check valve placed at the discharge side	Automatic isolation valve on discharge activated on machine trip or high pressure	Written procedures and training for isolation of non-operating parallel machine
5	Centrifugal compressor stops	Potential backflow via recycle loop resulting in overpressure of low pressure stages	Low pressure stages designed for higher pressure	Emergency relief valve for protection of low pressure stages sized for maximum backflow Restriction to limit back flow	
6	Speed control system failure (compressor)	Potential for compressor overspeed resulting in equipment damage	Solid versus built-up rotor	High speed alarm and compressor overspeed shutdown system	
7	Liquid carryover to compressor	Potential for compressor damage	Liquid-tolerant design (e.g., liquid ring compressor) Knockout drum designed for proper disengaging of liquid	Heat tracing between the KO drum and the compressor Knockout (KO) drum with automatic liquid removal and high level switch to shut down the compressor Online vibration monitoring with automatic shutdown	Operator response to high level alarm in the KO drum
Temperature					
	Generally Applicable – High Temperature *(Applicable to all high temperature scenarios)*		Choice of materials and design to maximum temperature conditions	High temperature shutdown interlock	Operator response to high temperature alarm

Table 6.6 Common Failure Scenarios and Design Solutions for Fluid Transfer Equipment

No.	Event	Consequence	Potential Design Solutions		
			Inherently Safer / Passive	Active	Procedural
	Generally Applicable – Low Temperature *(Applicable to all low temperature scenarios)*		Choice of materials and design to minimum temperature conditions	Low temperature shutdown interlock	Operator response to low temperature alarm
8	Loss of cooling to interstage (compressor)	Loss of upstream / interstage cooling resulting in high enough inlet temperature in subsequent stages of the compressor to cause compressor damage		Automatic shutdown on low coolant flow	Written procedures and training for manual shutdown on low coolant flow
9	Operation on total recycle without adequate cooling	Potential increased temperature		Cooler in recycle loop	
Composition					
	Generally Applicable *(Applicable to all composition scenarios)*		Design all component for expected pressure	Emergency relief device	
10	Composition change of fluid	Potential for high discharge pressure		Automatic pump / compressor shutdown on high discharge pressure detection	Operator action in response to high pressure indication

6. EQUIPMENT DESIGN

Table 6.6 Common Failure Scenarios and Design Solutions for Fluid Transfer Equipment

No.	Event	Consequence	Potential Design Solutions		
			Inherently Safer / Passive	Active	Procedural
11	Particulate matter in feed	Potential for seal damage	Double or tandem seals Pump design to accommodate solids (e.g., diaphragm) Strainer or filter on suction	Automatic back-flushing of strainer Automatic pump shutdown on detection of loss of seal fluid Localized fire protection	Operator response to seal-leak detection alarm Written procedures and training for manual activation of remotely operated isolation valves Written procedures and training for manual cleaning of strainer / filter Written procedures and training for periodic inspection of shaft seals
Equipment Failure					
Generally Applicable *(Applicable to all equipment failure scenarios)*				Explosion suppression systems Flame arresters Inerting or gas enrichment system	
12	Leakage on suction side of blower / compressor	Potential to pull air into system creating a flammable atmosphere	Positive pressure throughout system	Automatic oxygen monitoring interlocked to blower and / or isolation valves on high oxygen measurement Automatic pressure control which limits rate of oxygen infiltration or negative pressure	Written procedures and training for leak testing the suction system prior to startup Written procedures and training for manual oxygen monitoring interlocked to blower and / or isolation valves on high oxygen measurement Written procedures and training for manual pressure control which limits rate of oxygen infiltration or negative pressure

Table 6.6   Common Failure Scenarios and Design Solutions for Fluid Transfer Equipment

No.	Event	Consequence	Potential Design Solutions		
			Inherently Safer / Passive	Active	Procedural
13	Loss of lube oil to blower / compressor	Potential loss of lubrication resulting in bearing / seal failure, increased temperature		High bearing temperature shutdown interlock Low lubrication pressure / level shutdown interlock	Operator response to high temperature indication / alarm on bearings Operator response to low pressure alarm on the discharge of lube-oil pump
14	Loss of seal flush on pump	Potential loss of containment	Pumps that do not require seal flush	Interlock to shutdown pump on loss of seal flush Localized fire protection	Written procedures and training for manual shutdown of pump on loss of seal flush
15	Loss of oil mist on pump seal	Potential loss of containment	Pump seals that do not require oil mist	Interlock to shutdown pump on loss of oil mist Localized fire protection	Written procedures and training for manual shutdown of pump on loss of oil mist
16	Loss of seal flush on compressor	Potential loss of containment		Interlock to shutdown compressor on loss of seal flush Localized fire protection	Operator response to low flow or low pressure alarm on seal flush

6.6.2.1   Deadheading and Isolation (Scenario 1)

Pump and compressor systems should be designed to minimize the probability of deadheading. Deadheading a pump may result in high temperature, high pressure, or both. This situation is especially dangerous if the fluid being transferred is shock sensitive, or prone to exothermic decomposition. Because deadheading of a positive displacement pump or compressor can lead to a buildup of very high pressures, a means must be provided to protect against overpressure, e.g., a pressure relief valve, discharging back to the pump supply vessel. Pump isolation (closed suction or discharge valves) may also present a very serious pump failure scenario, particularly if the pumps are remote start and have the potential to be run extended periods of time in an isolated condition.

6.6.2.2   Cavitation / Surging (Scenarios 2 and 15)

Cavitation in pumps can cause severe damage to the pump impeller and seals, resulting in loss of containment. Cavitation problems usually can be avoided by designing the pump so that the Net Positive Suction Head (NPSH) requirement is met. Design solutions to prevent cavitation include:
- Adequate sizing of suction piping
- Blanketing source vessel

6. EQUIPMENT DESIGN

- Adequate height above pump
- Providing filter and strainers on pump suction

Compressor surge may lead to excessive vibration, high bearing temperatures, and extensive mechanical damage. This risk can be managed by providing automatic anti-surge systems and vibration monitoring systems.

6.6.2.3 Reverse Flow (Scenario 4 and 5)

There are various pump / compressor configurations that may result in the backflow of fluid through the machine. In a parallel configuration, where two or more machines discharge fluid to a common line, the fluid may backflow through the machine that is not in operation possibly causing impeller integrity problems in centrifugal pump applications. Procedures for isolating standby machines help to prevent this problem. In addition, check valves placed on the discharge will reduce the probability of backflow through idle or tripped machines. Some check valves do not completely shut off, and two check valves, in series, may be required. Additional backflow protection via automatic isolation valves may be warranted in fouling service or where the consequence of backflow is severe.

6.6.2.4 Seal Leaks (Scenarios 11, 13, 14, 15, and 16)

Seal leaks are a major source of concern, especially when handling toxic or flammable materials. Centrifugal pumps with double mechanical seals, diaphragm pumps, and various types of sealless pumps may be used for highly hazardous duty. See Grossel (Ref. 6-28) for more details.

6.6.3 Design Considerations

A wide variety of pumps are available including centrifugal, positive displacement, liquid or gas-driven jet, and gas pressurization or vacuum suction transfer systems. Other important criteria to be considered are materials of construction, instrumentation to detect pump-component failure, methods to contain toxic materials within the pump, and methods to control leaks and emissions (Ref. 6-28).

The pumping system should be designed to operate in a manner that prevents the pump from a deadhead operation for more than a very short period of time. Deadheading a pump can result in excessive temperatures that can lead to high vapor pressure or decomposition reactions that could blow the pump apart. Methods to maintain and detect a minimum flow through the pump or a temperature rise in the pump may be required along with a shutdown interlock for heat-sensitive materials. A number of pump explosions have occurred where the material in the pump overheated. Deadheading the pump can cause pump overheating with bearing burnout and flashing of the liquid in the pump and the rupture of downstream piping if the piping is not specified to meet the pump's deadhead pressure. A minimum flow recycle should be provided on pumps if deadheading can result in a serious problem.

It is important to understand that the majority of incidents, such as pump fires, are caused by catastrophic pump bearing failures, which lead to catastrophic seal failures. Tandem or dual seals (although a viable safeguard) do not mitigate the likelihood of catastrophic seal failure caused by bearing failure. Mitigations should focus on preventing the catastrophic bearing failures, such as:

- Limits on minimum flow
- Lube mist bearing lubrication
- Correcting design flaws, e.g., piping strain, poor pump base, etc.
- Online vibration monitoring
- Ensuring proper net positive suction head

Operating centrifugal pumps at severely reduced flows can cause excessive vibration and damage to drivers, piping, and adjacent equipment; a minimum flow recirculating line should be installed to avoid the instability conditions caused by low flow rates. Minimum flow control is usually required for large centrifugal pumps to prevent cavitation in the pump impeller and subsequent damage to the pump. The minimum flow liquid should not pass directly from the pump discharge to suction without consideration of cooling. Excessive heat buildup defeats the purpose of the minimum flow which is intended to prevent the liquid being pumped from vaporizing and cavitation of the pump which causes mechanical damage to the pump. Normally the minimum flow stream passes from the discharge line back to the suction vessel. A temperature sensor in the pump casing and vibration sensors in the bearings may be interlocked to shut off the pump motor due to excessive temperature or vibration. Close attention to the pump seal design and configuration is important to reduce normal wear and leakage for flammable and toxic service. Proper alignment will minimize the chance of mechanical seal failure.

Positive displacement pumps which can be blocked in on the discharge side require a pressure relief device; use of an external relief device is recommended (Ref. 6-28).

Diaphragm pumps do not have a sealing device that can leak. Air-driven diaphragm pumps can be operated at deadhead with no damage to the pump. For highly toxic fluids, pumps with two chambers should be specified, with the volume between the diaphragms monitored and alarmed.

To prevent loss of process fluids, centrifugal pumps should have a sealing system which consists of either double-inside or tandem mechanical seals with a barrier fluid between the seals plus a seal-failure alarm. If collecting the leaking seal is permissible, a secondary seal with a vent and drain gland fixture outside the primary seal is frequently effective in collecting leaked fluids. This secondary seal also offers a gland for inert gas blanketing as well as providing protection if the primary seal fails.

Failures of mechanical seals do occur, though much improvement has occurred in recent years. Mechanical seal problems account for most of the pump repairs in a chemical plant with bearing failures being a distant second (Ref. 6-29). Certain conditions increase the frequency of seal failure; e.g., heat, pressure, corrosion, cavitation, and product characteristics. Other conditions such as particle debris, shaft deflection, bearing wear, vibration, and poor installation can also affect seal life but can usually be minimized by proper pump selection, careful installation, and proper maintenance. Particle debris, particularly during plant startup, can be minimized by using a strainer in the pump suction piping; a T-type strainer is suggested due to the ease with which it can be removed and cleaned with the pump still online. Shaft alignment can be a major source of failure. The appropriate alignment techniques should be used to check the pumps prior to startup, and the alignment should be rechecked if continuing bearing or mechanical seal problems occur on a pump.

6. EQUIPMENT DESIGN

High temperatures decrease lubricity, resulting in increased friction and heat buildup that can promote abnormal wear of the seal face. Temperatures can be decreased by providing a seal flush system which provides filtered and cooled fluid. The pump operating characteristics should be checked to make sure that the appropriate type of lubrication is being used.

Operating pumps in parallel may cause deadheading of one pump, reverse flow scenarios, or thrust bearing failure. If pumps are operated in parallel, then consideration should be given to flow control valves on the discharge of each pump.

Compatibility of the seal fluid with the process fluid should be established. Depending on the seal system used (tandem or double) leakage can occur into the seal fluid or into the process.

Excessive face pressure, either hydraulic or installation imposed, can reduce face lubrication, increase frictional heat buildup, and cause face distortion. Pressure surges and hydraulic shock created by automatic valving can also reduce seal life; therefore, carefully consider system hydraulics. Acid conditions can form acidic metal salts, which can be abrasive to seal faces. A seal flush system should be provided.

Erosion by abrasive particles in the system can contribute to seal failure, particularly particles under 200 mesh size, such as thermal decomposition products in heat transfer fluids. Pump suction strainers may protect the pump from solids debris in the fluid and are used especially during startup and commissioning. However, suction strainers increase overall pressure drop and can reduce NPSH available at the pump inlet. If not carefully evaluated, this pressure drop can cause cavitation that may damage pump internals or reduce pump capacity. Cavitation can cause pressure variation, shaft deflection, vibration, or mechanical shock that will damage seal components. Cavitation problems usually can be avoided by proper system design, especially Net Positive Suction Head (NPSH), and by avoiding entrained gases.

Sealless pumps, both canned-motor and magnetic-drive designs, avoid the seal problem altogether. These types of pumps are driven by a magnetic coupling between the pump and an external rotating motor. The magnets are attached to the pump shaft and the motor shaft, with a non-magnetic shield between them. Magnetic-drive pumps use permanent magnets; canned pumps use electromagnets. Virtually all pump manufacturers now supply magnetic-drive pumps, both centrifugal and gear.

Canned and magnetic-drive pumps are not without their own safety considerations. Most failures of sealless pumps are caused by running them dry and damaging the bearings. A low boiling liquid may flash and a reverse circulation system or bypass stream may be required (Ref. 6-29). If the temperature of the flush liquid increases, the vapor pressure may rise and liquid may flash and the sleeve bearings can run dry. Solids may abrade the bearings of magnetic- drive pumps or may plug small ports in the can area. High temperature can decrease the strength of the magnets.

Sealless pumps are equipped with a more complex hydraulic system involving sleeve bearings and other parts which must receive some attention if the pump is to be kept in good running condition. The specific heat and the rate of change of vapor pressure are two critical physical factors which must be taken into account when designing the pump.

6.6.3.1 Compressors

Compressors run the gamut from small, oil-less fractional horsepower reciprocal units to massive turbine-driven multi-stage compressors. Typical uses of compressors include: compression of process gas, supply of plant air, and compression of air for furnace or fluidized bed combustion, exhaust, ventilation, and aeration. A comprehensive discussion of reciprocating and centrifugal compressors can be found in *Perry's Chemical Engineers Handbook* (Ref. 6-30).

Compressors share several design problems that involve safety: potential overpressure and overheat of the gas, vibration, seal leakage, and liquid intake into the compression chamber. All of these can cause material failure in the compressors or its ancillary piping, causing a gas release to the atmosphere. For reciprocating compressors overpressure is a special problem. While centrifugal compressors will reach a maximum pressure when the compressor is deadheaded, the reciprocating compressor can continue to increase pressure until either material failure occurs or the motor stalls and overheats. For this reason reciprocating compressors are equipped with pressure relief valves. To prevent these potential problems from occurring, the following design features should be considered:

- Use of knock-out drums, cyclones, or inlet heaters to prevent liquids from entering the compression chamber
- The sizing and installation of the proper seals - for large units, this will include seals with a circulating lube oil system, degassing sealpots and piping of the sealpot gases to recovery or treatment
- Piping design, including the proper materials of construction, vent and drain lines, and the use of vibration isolation joints
- Use of appropriate alarm and shutdown instrumentation including vibration switches, low / high discharge pressure, engine overspeed, high discharge temperature, and low oil pressure
- Use or properly sized and located pressure relief devices

Process variables and parameters that determine safe compressor operation and maintenance include: throughput, suction and discharge pressure, rotary speed, gas molecular weight, heat capacity ratio (C_p/C_v), and suction and discharge temperature. In general, during stable operation with a constant rotary speed, the pressure differential across centrifugal and axial compressors decreases with increase in throughput. For a fixed pressure drop, throughput increases with increasing rotary speed. Likewise, for a constant throughput, pressure differential increases with increasing rotary speed.

Potential hazards of high throughput compressor operation, commonly referred to as the "stonewall region," include throughput limits caused by horsepower / torque constraints and insufficient pressure differential to meet the downstream process requirements. Low throughput operation is known as the "surge region." When the throughput falls below a critical value, known as the surge limit, self-sustained oscillations of pressure and flow are induced leading to flow reversal (or slippage inside the compressor) since the compressor wheel fails to impart sufficient kinetic energy to compress gas continuously. Under severe surge, a compressor can exhibit high frequency vibrations and high thrust bearing temperatures which can lead to permanent mechanical damage. A compressor under regulatory control and operating in close proximity to the surge limit can quickly move into surge.

6. EQUIPMENT DESIGN

Compressor controls typically consist of basic process controls, anti-surge controls, compressor optimization controls, and compressor shutdown systems. The first control group is aimed at controlling discharge and suction pressures. The second application employs fast-acting controls to override regulatory controls as the compressor surge region is approached. Compressor optimization is typically computer-based, high level, supervisory control which minimizes compressor energy utilization with respect to regulatory controls set points subject to process and equipment constraints.

Centrifugal compressors require minimum flow control in order to prevent them from going into a surge condition which might cause mechanical damage or destruction of the compressor. Flow measurement should be in the suction piping because there is a better correlation of suction flow with the surge line on the compressor curve than there is with discharge flow. Care should be taken that sufficient straight pipe run is available for the meter run. The use of low permanent head loss differential producing devices, such as venturi and low loss flow tubes, flow nozzles, or averaging pitot tubes should be considered for this application to minimize energy consumption.

It is common practice to manipulate the throughput and pressure differential across the compressor in order to eliminate surge. Venting part of the compressor discharge upon the onset of surge will control surging. However, this is not practical if the discharge is valuable or a pollutant. Alternatively, a portion of the compressor discharge may be recycled back to the compressor suction in order to maintain a minimum compressor flow. Surge encountered during normal operation is most effectively controlled with feed-forward (predictive) and override (detector) controls. Surge feed-forward control uses a predictive model to anticipate the onset of surge and take corrective action by manipulating compressor recycle flow. Surge override control possesses a surge detector acting on time rate of change of pressure differential and throughput. Upon detecting surge, it must be equipped to open the compressor recycle valve quickly.

6.6.3.2 Vacuum Equipment Considerations

Vacuum equipment such as liquid ring pumps, mechanical pumps, and ejectors are used in many chemical process applications. Many of the design considerations used for pressurized equipment also apply to vacuum equipment, but certain specific design safety considerations need to be addressed:

- The system may need to be sealed against the infiltration of air into the vacuum system, which could create a potential flammable or reactive mixture.
- The equipment needs to be designed not only for vacuum but for the highest pressure that the equipment can experience when the vacuum pump fails. If the material in the system is toxic, this may require that the equipment and piping be specified for high pressure as well as vacuum; if less hazardous material is being processed, safety valves, rupture disks, or blowout panels may be used.
- The system should be designed to prevent equipment upstream of the vacuum section from experiencing vacuum if upstream pressure units fail or the upstream units should also be designed for vacuum.
- The exhaust of the vacuum system may require treatment to recover or destroy toxic or flammable vapors from the system prior to final release to the atmosphere.

- The liquid used in liquid ring vacuum pumps may also require treatment prior to release to atmosphere (for example, if it absorbs flammable process liquids).
- Instrumentation should be provided to control and monitor pressure (vacuum).
- Backup of motive steam could cause overpressure in ejectors.
- Loss of intercondenser cooling medium could lead to overpressure of the system.

Ryans and Roper (Ref. 6-31) present a thorough discussion of the design and operation of vacuum systems and equipment.

Dry vacuum pumps are compact and energy efficient compared to other mechanical vacuum pumps because they do not require a working fluid to produce vacuum, so nothing contacts the vapors being pumped. They have been successfully used for pumping corrosive and flammable vapors. Dry vacuum pumps are available as rotary-lobe Roots blowers, claw compressors, and screw compressors. These three all have certain things in common. Thigh clearances result in these pumps running hot and the potential for overheating is inherent in their design. Dissipating the heat of compression is necessary, and temperature control is required. Generally, temperature control is accomplished by using a water jacket or injecting cooled process gas or nitrogen into the working volume of the pump. Occasionally, both methods are used together.

Safety is an issue when pumping flammable vapors and gases because of the potential for an explosion, initiated for example by a spark caused by contact between the rotors and casing. Dry vacuum pump manufacturers address safety in part by designing pumps that will contain an internal explosion. Flame propagation can be minimized by inerting with nitrogen or other inert gas prior to startup.

Autoignition is also a consideration. Dry vacuum pumps run hot, with discharge temperatures for screw compressors sometimes reaching 662-752°F (350-400°C). To cope with this, the latest generation of dry vacuum pumps is designed to run at lower temperature and has precise temperature control.

6.6.4 References

6-28. Grossel, S.S. *Highly Toxic Liquids - Moving Them Around the Plant. Part 1.* Chemical Engineering. 1990.

6-29. Reynolds, J.A. Canned *Motor and Magnetic Drive Pumps.* Chemical Processing, No. 12. 1989.

6-30. Green, D W. and Perry, *R.H. Perry's Chemical Engineers' Handbook*, Eighth Edition, McGraw-Hill. New York. 2008.

6-31. Ryans, J.L. and Roper, D.L. *Process Vacuum System Design and Operation.* McGraw-Hill. New York. 1986.

6.6.4.1 Suggested Additional Reading

Bloch, H.P., Cameron, J.A., James, Jr., R., Swearinger, J.S., and Weightman, M.E. *Compressors and Expanders.* Marcel Dekker, Inc. New York.1982.

Bloch, H.P., and Budris, A. *Pump Users Handbook: Life Extension, Third Edition.* Fairmont Press. 2010.

6. EQUIPMENT DESIGN

Bloch, H. P. Pump Wisdom: *Problem Solving for Operators and Specialists*. John Wiley & Sons, Hoboken, New Jersey. 2011.

Eierman, R.B. *Improving Inherent Safety with Sealless Pumps.* Proceedings of the 29th Annual Loss Prevention Symposium, July 31-August 2, 1995, Boston, Massachusetts. 1995.

Karassik, I. J et al. Pump Handbook, 4th Edition. McGraw-Hill, New York. 2008.

Kletz, T. A. *Lessons from Disaster.* Gulf Publishing Company, Houston, Texas. 1993.

Kletz, T. A. *Learning from Accidents.* Butterworth-Heinemann Ltd., Oxford. 1994.

Ryans, J. and Bays, J. *Run Clean with Dry Vacuum Pumps*. Chemical Engineering Progress, pp. 32-41. October 2001.

Tunna, C. *Pumping Potentially Explosive Atmospheres.* The Chemical Engineer (IChemE), pp. 30-31. May 2005.

6.7 SOLID-FLUID SEPARATORS

This section presents potential failure mechanisms for solid-fluid separators and suggests design alternatives for reducing the risks associated with such failures. The types of equipment covered in this section include:

- Centrifuges
- Filters
- Dust collectors
- Cyclones
- Electrostatic precipitators

This section presents only those failure modes that are unique to solid-fluid separators. Some of the generic failure scenarios pertaining to vessels may also be applicable to solid-fluid separators. Consequently, this section should be used in conjunction with Section 6.1, Vessels. Solid-fluid separation equipment is also often associated with dryers and solids handling and processing equipment. Refer to Sections 6.5 and 6.8 for information on these types of equipment. Unless specifically noted, the failure scenarios apply to more than one type of solid-fluid separator.

6.7.1 Past Incidents

This section presents several case histories involving fires and explosions (deflagrations) to reinforce the need for safe design and operating practices for solid-fluid separators.

6.7.1.1 Batch Centrifuge Explosion (Scenario 9)

A crystalline finished product was spinning in a batch centrifuge when an explosion occurred. The product had been cooled to 19°F (-7°C) before it was separated from a methanol / isopropanol mixture in the centrifuge. It was subsequently washed with isopropanol precooled to 16°F (-9°C). The mixture was spinning for about 5 minutes when the explosion occurred in the centrifuge. The lid of the centrifuge was blown off by the force of the explosion. The overpressure shattered nearby glass pipelines and windows inside the process area (up to 20 meters away), but nearby plants were not damaged. No nitrogen inerting was used and enough time had elapsed to allow sufficient air to be drawn into the centrifuge to create a flammable atmosphere. Sufficient heat could also have been generated by friction to raise the temperature of the precooled solvent medium above its flash point. Because the Teflon® coating on the centrifuge basket had been worn away, ignition of the flammable mixture could also have been due to metal-to-metal contact between the basket and the bottom outlet chute of the centrifuge, leading to a friction spark. A static discharge might also have been responsible for the ignition. Since the incident, the company has required use of nitrogen inerting when centrifuging flammable liquids at all temperatures (Ref. 6-32).

Lessons learned include monitoring the oxygen concentration in conjunction with inerting and sealing the bottom outlet to minimize air entry. Because the ignition source was uncertain (static discharge, frictional heat), this incident illustrates why it often is prudent to assume an ignition source when designing for flammable materials.

6.7.1.2 Dust Collector Explosion

An explosion occurred in a dust collector used to collect a pharmaceutical product from a hammer mill / flash drying operation. The impact hammer mill had been operating for approximately 10 minutes when the operator heard unusual grinding sounds coming from inside the mill. He immediately shut down the mill just as an explosion occurred within the dust collector, located inside the building on the second floor.

The pressure wave caused the explosion vent (a hinged panel) of the dust collector to open, and the explosion products and unburned powder were directed outside the building via a vent duct. However, a screen had been securely fastened at the end of the duct to prevent birds from entering, and as the vent panel swung upward and outward, it struck the screen and opened no farther. It is estimated that the screen prevented the explosion vent panel from opening to no more than 50% of the capacity. With the vent partially obstructed, the access door to the collector failed under pressure and released a dust cloud into the building. The flame front followed the dust cloud through the vent and through the access door, resulting in a fireball at both locations. Also, on the first floor, a fireball was seen exiting the vicinity of the rotary valve outlet at the bottom of a dust collector, which feeds a sifter. There was no secondary explosion on the first or second floor. However, windows were blown out on both floors. The ensuing fire in the dust collector engulfed the wool filter bags (which were burned up) and the remaining powder in the collector hopper, but the fire was quickly extinguished by the automatic sprinkler system inside the dust collector.

A subsequent investigation of the incident revealed that a carbon steel bolt from the inside of the feeder (which feeds wet powder to the hammer mill / flash dryer) fell into the hammer mill. The bolt became trapped inside the 3600-RPM mill, where it heated to above the ignition temperature of the powder. The hot metal ignited some of the powder in the mill which was pneumatically conveyed into the dust collector. In the collector, a dust cloud created by the blow ring (pulse jet) was ignited by the hot powder conveyed in from the hammer mill. An inspection of the feeder revealed that six 3/8-inch carbon steel bolts and nuts were missing.

This incident illustrates that equipment design should assume that an ignition source is available for dusty environments, similar to flammable liquids and gases, and provides relief protection. It is also good design practice to indicate explosion vents on

6. EQUIPMENT DESIGN

installation drawings with explanatory notes describing the clear space needed for vent actuation and for fire ball attenuation. Although it was unfortunate that the dust collector was damaged, much additional damage was probably avoided because good housekeeping minimized the dust available for a secondary explosion. Nuts and bolts located inside of rotating equipment have the potential to cause significant damage if they come loose. Consider the use of tack-welded wire ties or other means to prevent them from disengaging during operation.

6.7.2 Failure Scenarios and Design Solutions

Table 6.7 presents information on equipment failure scenarios and associated design solutions specific to solid-fluid separators.

Table 6.7   Common Failure Scenarios and Design Solutions for Solid Fluid Separators

No.	Event	Consequence	Potential Design Solutions		
			Inherently Safer / Passive	Active	Procedural
Pressure					
	Generally Applicable – High Pressure *(Applicable to all high pressure scenarios)*		Filter design accommodating maximum expected pressure	Emergency relief device Rupture disk upstream of relief valve with appropriate rupture disk leak detection	Operator response to high pressure alarm
	Generally Applicable – Low Pressure *(Applicable to all low pressure scenarios)*		Filter design accommodating minimum expected pressure	Vacuum relief device	Operator response to low pressure alarm
1	Loss of vacuum (vacuum belt filter, vacuum pan filter, rotary vacuum filter)	Potential release of toxic or flammable vapors to atmosphere	Totally enclosed, vapor-tight filter Grounding and bonding	Automatic shutdown operation in response to vapor detection alarm Local exhaust ventilation connected to a control system (vent condenser, adsorber, scrubber, or incinerator)	Written procedures and training for manual shutdown in response to vapor detection alarm
2	Relief device plugged on filter	Potential increase pressure	Flow sweep fitting at inlet to relief device	Automatic sweep of inlet to relief device with purge fluid Heat trace and insulate relief device	Written procedures and training for manual periodic flush of inlet to relief device with purge fluid

Table 6.7 Common Failure Scenarios and Design Solutions for Solid Fluid Separators

No.	Event	Consequence	Potential Design Solutions		
			Inherently Safer / Passive	Active	Procedural
3	High pressure differential across tube sheet	Potential tube sheet buckling, potential loss of containment	Tube sheet designed for maximum possible differential pressure	Relief valve on inlet side of filter High inlet or differential pressure alarm and / or interlock	
Flow					
Generally Applicable – More Flow (Applicable to all more flow scenarios)				Automatic inerting system	
Generally Applicable – No / Less Flow (Applicable to all no / less flow scenarios)				Automatic inerting system	
4	Deposits on walls (tarry or sticky dust) (cyclones, dust collectors, and electrostatic precipitators)	Potential fire	Different type of separator (e.g., wet-type precipitator or scrubber) Fire-retardant filter bags or ceramic cartridges	Fire / explosion suppression	Written procedures and training for periodic cleaning of accumulated flammable dust deposits
5	Loss of feed (clarifier and separator centrifuges, i.e., disc bowl, nozzle bowl, chamber bowl, desludger, opening bowl)	Potential equipment damage caused by vibration	Design that is tolerant to loss of feed (e.g., pusher type centrifuge)	Adequate supply of wash liquid or water automatically as feed is reduced under emergency shutdown conditions	Written procedures and training to provide adequate supply of wash liquid or water manually as feed is reduced under emergency shutdown conditions
Temperature					
Generally Applicable – High Temperature (Applicable to all high temperature scenarios)				Automatic fire suppression system activated by high temperature sensor External automatic fire suppression system	Operator activation of fire suppression system in response to high temperature indication Written procedures and training for manual activation of external fire suppression system

6. EQUIPMENT DESIGN

Table 6.7 Common Failure Scenarios and Design Solutions for Solid Fluid Separators

No.	Event	Consequence	Potential Design Solutions		
			Inherently Safer / Passive	Active	Procedural
Composition					
6	Pyrophoric material used in filter (batch filters)	Potential fire when cake exposed to air when filter is opened	Filter with cake removal by spinning plates and / or sluicing with liquid (filter does not have to be opened up)	Automatic fixed water spray Inerting	Written procedures and training to ensure that filter cake is sufficiently flushed with water before filter is opened Written procedures and training for manual activation of fixed water spray
Equipment Failure					
	Generally Applicable *(Applicable to all equipment failure scenarios)*		Centrifuge design accommodating maximum expected pressure Elimination of flammable solvent Equipment design accommodating maximum expected pressure Permanent grounding and bonding	Automatic external fire suppression system Automatic inerting Automatic isolation of associated equipment via quick- closing valves or chemical barrier (flame suppression) Deflagration venting Internal automatic fire / explosion suppression system	Written procedures and training for pre-inerting prior to restart of a batch centrifuge Written procedures and training for manual activation of external fire suppression system
7	Static electricity (centrifuges)	Potential ignition of flammable vapors resulting in fire / explosion	Avoid non-conductive lined centrifuge Electrically conductive wash liquid Less volatile / flammable wash liquid Non-flammable or high flash point solvent	Automatic shutdown on low pressure or low flow sensor on nitrogen supply line with interlocks to shut down filter or centrifuge	Written procedures and training for manual shutdown of batch centrifuge on detection of low inert gas pressure or flow Written procedures and training for manual bonding and grounding for portable units

Table 6.7 Common Failure Scenarios and Design Solutions for Solid Fluid Separators

No.	Event	Consequence	Potential Design Solutions		
			Inherently Safer / Passive	Active	Procedural
8	Mechanical friction, e.g., out-of-balance basket rubbing against housing or bottom chute overpressure (centrifuges)	Potential ignition of flammable vapors resulting in fire / explosion		Automatic shutdown on proximity / vibration sensor interlocked to shut down centrifuge Automatic shutdown on low pressure or low flow sensor on inert gas supply with interlock to shut down centrifuge	Written procedures and training for manual shutdown of centrifuge on detection of excessive vibration Written procedures and training for manual shutdown on low pressure or low flow sensor on inert gas supply with interlock to shut down centrifuge
9	Electrostatic spark discharge or glowing particles from upstream equipment (cyclones, dust collectors, and electrostatic precipitators)	Potential increased pressure resulting in fire / explosion	Different type of separator (e.g., wet-type precipitator or scrubber) Nitrogen used as conveying gas	Automatic introduction of inert gas via online oxygen analyzer Inerting	Written procedures and training for manual introduction of inert gas on detection of high oxygen via online oxygen analyzer
10	Bearing failure (centrifuges)	Potential equipment damage and possible loss of containment		Automatic centrifuge shutdown on detection of excessive vibration Automatic centrifuge shutdown on detection of lubricating oil low pressure Interlock bearing temperature sensor to shut down the centrifuge at high temperature	Written procedures and training for manual shutdown of centrifuge on detection of high bearing temperature, vibration or lubricating oil low pressure
11	Basket imbalance (batch centrifuges)	Potential equipment damage caused by vibration	Alternate solid / fluid separator designs Continuous centrifuge design Flexible connections to reduce vibration	Control system to admit feed at proper flow rate and appropriate time in acceleration period Vibration sensor interlocked to shut down centrifuge	Written procedures and training for control of feed rate to avoid imbalance of basket and vibration Written procedures and training for shutdown of centrifuge on detection of excessive vibration

6. EQUIPMENT DESIGN

Table 6.7 Common Failure Scenarios and Design Solutions for Solid Fluid Separators

No.	Event	Consequence	Potential Design Solutions		
			Inherently Safer / Passive	Active	Procedural
12	Loss of speed control (centrifuges)	Potential equipment damage caused by vibration	Alternate solid / fluid separator designs	Speed detector interlocked to shut down the centrifuge at overspeed point	Written procedures and training for shutdown of centrifuge on detection of high speed
13	Gasket leak (filter presses)	Potential loss of containment of flammable or toxic material	Different type of filter or centrifuge with fewer gaskets Filter enclosed in splash shield housing Filter located in leak containment trough or in containment vessel Higher integrity gaskets		Written procedures and training to pretest filter for leaks with water before feeding process slurry Written procedures and training for testing compatibility of gasket material with process fluid

6.7.2.1 Dust Deflagrations

Dust deflagrations can occur in cyclones and dust collectors because explosive dust clouds are readily formed inside these types of separators due to turbulence. Dust clouds are created continuously when dust collector bags are shaken or pulsed. Use of nitrogen, rather than air, as the pulsing gas when a combustible dust is being collected may be considered and is used by some companies. Because electrostatic charges are often produced by powders that are pneumatically conveyed to solid-fluid separators, the separators must be adequately grounded and bonded to prevent static sparks. Glowing particles from a previous operation can act as an ignition source when they are transferred into a separator. Because of the great propensity for dust cloud formation in cyclones and dust collectors, they are usually protected by either deflagration venting or suppression systems (Ref. 6-33 and Ref. 6-34).

If flammable dust clouds can also be formed in the electrostatic precipitators by beating the plates and electrodes to remove particles, then deflagration vents should be provided. Using electrostatic precipitators is not recommended when dry combustible dust concentrations in air may exceed the lower explosive limit due to the possibility of ignition by arcing in the precipitator.

6.7.3 Design Considerations

6.7.3.1 Centrifuges

Since centrifuges are subject to the hazards inherent in all rotating equipment, the designer should first consider whether other, safer methods of separation (such as decanters or static filters) can be used. If it is determined that a centrifuge must be used, the design should be reviewed to ensure that it is as safe and reliable as possible.

Potential problems associated with centrifuges include mechanical friction from bearing; vibration, leaking seal; static electricity; and overspeed. Vibration is both a cause of problems and an effect from other sources. The potential destructive force of an out-of-balance load has led to setting lower shutdown limits on the magnitude of vibration as compared to other rotating equipment, such as pumps. Flexible connections for process and utility lines become a must so these vibration problems are not transmitted to connected equipment. Flexible hoses with liners having concentric convolutions (bellows type) avoid the sharp points inherent with spiral metallic liners. By avoiding the sharp point the liner is less likely to cut the exterior covering.

Grounding of all equipment components, including internal rotating parts, must be ensured initially and verified periodically. Grounding via some type of brush or other direct contact is preferred to grounding via the bearing system through the lubricating medium (unless conductive greases are used). Use of non-conductive solvents complicates the elimination of static electricity concerns; use of conductive solvents or anti-static additives should be considered where feasible.

For flammable and / or toxic materials all of the precautions for a pressurized system should be considered. For example, when a centrifuge is pressurized, overpressure protection is required, even if the pressurization is an inert gas. Relieving of the pressure to a closed system or safe location must be considered.

When flammable solvents are used, centrifuges are inerted to prevent fires and explosions. Grossel (Ref. 6-35) discusses various methods of inerting centrifuges.

Many companies install fusible link valves in the feed and wash liquid lines to a centrifuge handling flammable liquids. If a fire should occur in the centrifuge or in the area around it, the heat from the flames melts the fusible link (usually lead) and the valves shut.

6.7.3.2 Filters

One of the primary concerns for filters is the loss of containment of flammable and toxic materials and operator safety during the frequent opening and closing of the equipment (e.g., for changing filter elements or unloading filters). Inherently safer process alternatives should be considered to eliminate or lessen the need for filtration. Self-cleaning, automatic backwashing, or sluicing filters should be considered for pyrophoric or toxic materials as they do not have to be opened or disassembled to remove the filter cake. Solid-liquid filters can be either pressure or vacuum filters. Filters for liquid service should be provided with fire relief valves, as appropriate, and safe operating procedures for out-of-service conditions. Solid-liquid filters that handle mixtures that are either toxic or have other health-hazardous properties should use gas-tight, totally enclosed units. Several types of filters are available in this design. Filters handling mixtures containing flammable liquids may require inerting.

For filters that require frequent cleaning or changing, consideration should be given to provide a parallel filter or bypass line. The design should include capability to take filter offline, have proper isolation for lockout / tagout, depressuring, and draining to safe locations.

Bag house filters are normally low pressure units. They can vary in operating conditions from hot and chemically aggressive to cool and inert. Hot feed may lead to exceeding the temperature rating of the filters and could even result in a bag house fire. As with all filters, not exceeding the design differential pressure is important to both the process stability and safety. As the solid is removed from the gas stream and is subsequently handled for recovery or disposal, all of the conventions and concerns for handling dust, powders and other solids apply. The system should be protected from the potential of dust deflagration by the use of pressure relief or suppression devices. A discussion of safety considerations for these types of systems is found in *Dust Explosion Prevention and Protection* (Ref. 6-36).

6.7.4 References

6-32. Drogaris, G. *Major Accident Reporting System: Lessons Learned from Accidents Notified.*: Elsevier Science Publishers, B.V, Amsterdam. 1993.

6-33. NFPA 68. *Standard on Explosion Protection by Deflagration Venting*, 2007 Edition. National Fire Protection Association. Quincy, Massachusetts. 2007.

6-34. NFPA 69. *Standard of Explosion Prevention Systems*, National Fire Protection Association. Quincy, Massachusetts. 2008.

6-35. Grossel, S.S. *Inerting of Centrifuges for Safe Operation*. Process Safety Progress. R4: Issue 4, pp. 273-278. 2003.

6-36. Barton, J. *Dust Explosion Prevention and Protection – A Practical Guide*. Gulf Publishing, Woburn, Massachusetts. 2002.

6.7.4.1 Suggested Additional Reading

CCPS. *Guidelines for Safe Handling of Powders and Bulk Solids,* Center for Chemical Process Safety of the American Institute of Chemical Engineers. New York, New York. 2005.

ASTM 1986. *Industrial Dust Explosions*. Symposium on Industrial Dust Explosions. Pittsburgh, Pennsylvania. June 10-13, 1986.

IChemE. *Dust and Fume Control: A User Guide*. Second Edition, Institution of Chemical Engineers. London. 1992.

6.8 SOLIDS HANDLING AND PROCESSING EQUIPMENT

This section presents potential failure mechanisms for solids handling and processing equipment and suggests design alternatives for reducing the risks associated with such failures. The types of equipment covered in this section include:
- Mechanical conveyors
- Pneumatic conveying systems
- Size reduction equipment (mills, grinders, crushers)
- Sieving (screening) equipment
- Powder blenders (mixers)
- Solids feeders (rotary valves, screw feeders, etc.)

This section presents only those failure modes that are unique to solids handling and processing equipment. Some of the generic failure scenarios pertaining to vessels and solid-fluid separators may also be applicable to solids handling and processing equipment. Consequently, this section should be used in conjunction with Section 6.1, Vessels, and Section 6.7, Solid-Fluid Separators. Unless specifically noted, the failure scenarios apply to more than one type of solids handling and processing equipment.

The drilling process consists of spraying droplets of liquids into the top of a tower and allowing these to fall against a countercurrent stream of air or inert gas. During their fall, the droplets are solidified primarily by cooling and partly by drying (heat and / or mass transfer), into spherical particles or prills (0.2 - 4.0 mm diameter). Traditionally, ammonium nitrate, urea, and other materials of low viscosity and melting point and high surface tension are treated this way.

Dust explosions have occurred in drilling towers in the past. These explosions are attributed to smaller particles commonly known as mini-drills.

Design considerations include the following:

- Install an extraction, capture, and filter system for ventilation air from areas with dust-generating product handling.
- Use of bag house filters and cyclones to prevent emission of dust-laden air from transfer points, screens, bagging machines, etc.
- Removal of dust emissions by droplet separation techniques for, e.g., knitted wire mesh demister pads, wave plate separators, and fiber pad separators.
- Scrubbing of off-gases with process condensate prior to discharge to atmosphere by using scrubbing devices like packed columns, venturi scrubbers, and irrigated sieve plates. Use of neutralization techniques in wet scrubbers.
- Installation of drilling towers with natural draft cooling instead of towers with forced / induced draft cooling.
- Reduction of dust emissions by adopting an enclosed granulation process instead of drilling process.

6.8.1 Past Incidents

Several case histories involving failures in solids handling and processing equipment are presented to reinforce the need for safe design and operating practices presented in this section.

6.8.1.1 Dust Explosion and Fire

A series of sugar dust explosions at a sugar manufacturing facility resulted in 14 worker fatalities. Thirty six workers were treated for serious burns and injuries - some caused permanent, life altering conditions. The explosions and subsequent fires destroyed the sugar packing buildings, palletizer room, and silos and severely damaged the bulk train car loading area and parts of the sugar refining process areas.

The manufacturing facility housed a refinery that converted raw cane sugar into granulated sugar. A system of screw and belt conveyors and bucket elevators transported granulated sugar from the refinery to three 105-foot-tall sugar storage silos. It was then transported through conveyors and bucket elevators to specialty sugar processing areas and granulated sugar packaging machines. Sugar products were packaged in four-story packing buildings that surrounded the silos, or loaded into railcars and tanker trucks in the bulk sugar loading area.

The first dust explosion initiated in the enclosed steel belt conveyor located below the sugar silos. The recently installed steel cover panels on the belt conveyor allowed explosive concentrations of sugar dust to accumulate inside the enclosure. An unknown source ignited the sugar dust, causing a violent explosion. The explosion lifted sugar dust that had accumulated on the floors and elevated horizontal surfaces, propagating more dust explosions through the buildings. Secondary dust explosions occurred throughout the packing buildings, parts of the refinery, and the bulk sugar loading buildings. The pressure waves from the explosions heaved thick concrete floors and collapsed brick walls, blocking stairwell and other exit routes. The resulting fires destroyed the packing buildings, silos, palletizer building, and heavily damaged parts of the refinery and bulk sugar loading area.

Lessons learned include:
- Sugar and cornstarch conveying equipment was not designed or maintained to minimize the release of sugar and sugar dust into the work area.
- Inadequate housekeeping practices resulted in significant accumulations of combustible granulated and powdered sugar and combustible sugar dust on the floors and elevated surfaces throughout the packing buildings.
- Airborne combustible sugar dust accumulated above the minimum explosible concentration inside enclosed steel belt assembly.

6.8.1.2 Silicon Grinder Fire and Explosion (Scenario 6)

A chemical plant which processed silicon-based chemicals experienced a fire and explosion in a grinder. Raw silicon was received in 1- or 2-inch lumps which had to be ground to a 200-mesh powder before being used in chemical processes. The air-conveyed silicon powder discharged from the grinder passed through a cyclone and then through a bag filter. An explosion and subsequent fire occurred in the system. The fire was extinguished within 15 minutes by a water hose stream. The system had explosion relief, but no sprinklers.

Investigation showed that this incident was caused by hot spot ignition resulting from grinder parts scraping against the inside of the unit. This mechanism was supported by observation of high current draw on the grinder motor before the incident.

Lessons learned include monitoring current draw and possibly interlocking current draw with the motor or activation of a deluge system.

6.8.1.3 Blowing Agent Blender Operation Explosion Incident (Scenario 3)

An explosion occurred in a 3.7-m^3 conical orbiting screw mixer during the blending of Azodicarbonamide (AC) with an aqueous solution of salts to produce an AC formulation. During the batch blending cycle, hot water [176°F (80°C)] is circulated through the blender jacket for several hours, and the vacuum in the blender is released by purging with nitrogen.

The explosion caused the mixer vessel to rupture and two large sections of the top were torn out completely and struck the floor above. The cone section was thrust downwards into the hopper below. There was extensive damage to the building, windows were broken up to 90 meters away by the pressure wave, and missiles were projected up to 120 meters away.

Subsequent experimental testing indicated that the explosion was caused by a decomposition which reached high rates due to a critical degree of confinement. The initiating source of the decomposition was not positively identified, but it was assumed that the heat was generated by mechanical friction due, for example, to the screw rubbing on the vessel wall. Another possibility is that a small metal item found its way into the vessel and became trapped between the screw and the wall (Ref. 6-37).

Lessons learned include the need for good understanding of material reactivity during the design phase. A deflagration suppression system might have prevented the explosion; however this requires knowledge of the decomposition rate and decomposition products.

6.8.1.4 Screw Conveyor Explosion (Scenarios 5, 11, 13)

A deformation occurred in the screw conveyor housing, causing parts of the screw flights to grind against the housing. The grinding produced sufficient frictional heat and sparks to ignite the dust-air cloud in the free space of the conveyor. The primary explosion burst the screw conveyor housing, dispersing a significant amount of additional dust into the air from the freshly filled feed hopper. A secondary explosion was then ignited by the flames of the primary explosion (Ref. 6-38). Three employees were killed, two seriously injured, and a factory building was completely destroyed in an explosion involving skimmed milk powder. The milk powder was fed into a screw conveyor from a feed hopper and was then carried to a blender.

Lessons learned include the need for good understanding of material reactivity during the design phase. A deflagration suppression system might have prevented the explosion; however this requires knowledge of the decomposition rate and decomposition products.

6.8.2 Failure Scenarios and Design Solutions

Table 6.8 presents information on equipment failure scenarios and associated design solutions specific to solids handling and processing equipment.

6. EQUIPMENT DESIGN

Table 6.8 Common Failure Scenarios and Design Solutions for Solids Handling and Processing Equipment

No.	Event	Consequence	Potential Design Solutions		
			Inherently Safer / Passive	Active	Procedural
Pressure					
	Generally Applicable – High Pressure *(Applicable to all high pressure scenarios)*			Pressure relief device	Operator response to high pressure alarm
	Generally Applicable – Low Pressure *(Applicable to all low pressure scenarios)*			Vacuum relief device	Operator response to low pressure alarm
Flow					
1	Blockage of die (extruder)	Potential increased pressure in upstream equipment		Automatic shutdown of motor on overload Automatic shutdown on high pressure at die	Written procedures and training for manual shutdown on motor overload Written procedures and training for manual shutdown on detection of high pressure at die
Temperature					
	Generally Applicable – High Temperature *(Applicable to all high temperature scenarios)*				Operator response to high temperature alarm with manual motor shutdown and quench activation
2	Jamming and frictional heating (rotary valves)	Potential fire	Dust collector bag cages and filters designed to be properly secured to avoid falling into rotary valve Robust bar screen at rotary valve inlet Outboard bearings to prevent failure due to solids contamination	Overload shutdown on the motor driving the rotary valve	Written procedures and training to secure dust collector bags and cages
3	Frictional heating (screw conveyors)	Potential fire	Different type of conveyor (e.g., vibratory conveyor) Gravity and layout	Overload shutdown on the motor driving the screw	
4	Frictional heating (extruders)	Potential fire		Overload shutdown on the motor driving the extruder screw	

Table 6.8 Common Failure Scenarios and Design Solutions for Solids Handling and Processing Equipment

No.	Event	Consequence	Potential Design Solutions		
			Inherently Safer / Passive	Active	Procedural
Equipment Failure					
	Generally Applicable *(Applicable to all equipment failure scenarios)*		Eliminate use of flammable solvents (e.g., aqueous solvents) Equipment design accommodating maximum expected pressure Heavy wall piping and flanges in lieu of tubing and couplings so that system can withstand maximum expected deflagration pressure Permanent grounding and bonding	Automatic fire suppression interlocked to shutdown the belt drive on sprinkler water flow initiation Chokes Deflagration barriers (quick-closing isolation valve or suppressant) in the path from granulator or coater to downstream equipment (dust collector, scrubber) Deflagration suppression Deflagration venting to safe location	Written procedures and training for good housekeeping to reduce dust Written procedures and training for manual activation of fire suppression system Manual bonding and grounding Written procedures and training for periodic inspection and cleaning of combustible materials on walls (housekeeping)
5	Electrostatic spark discharge in end-of-line equipment (silo, cyclone, dust collector) (Pneumatic conveying system)	Potential dust deflagration and loss of containment	Nitrogen in lieu of air for conveying gas (closed-loop system) Dense-phase conveying instead of dilute phase Convey solids as pellets instead of granules or powder. However, avoid transport of pellets containing easily ignitable fines fraction. Additives with high ignition energy Conductive rubber sleeves (boots and socks) when flexible connections are required		Written procedures and training for manual bonding across potential breaks in continuity such as non-conductive rubber socks

6. EQUIPMENT DESIGN

Table 6.8 Common Failure Scenarios and Design Solutions for Solids Handling and Processing Equipment

No.	Event	Consequence	Potential Design Solutions		
			Inherently Safer / Passive	**Active**	**Procedural**
6	Mechanical energy or electrostatic spark (mills, grinders, and other size reduction equipment)	Potential dust deflagration	Fluid energy mill with inert gas instead of air Screens to remove tramp metals and other foreign materials	Magnets to automatically and continuously remove tramp metals and other foreign materials	Written procedures and training for manual removal of tramp metals and other foreign materials
7	Rupture of flexible sleeves (gyratory screener)	Potential dust deflagration	Non-gyratory (rotary) type of screener Outboard bearings to avoid potential source of ignition	Gyratory screener in a separate room with blow-out walls (deflagration vents) Operate under vacuum to avoid escape of dusts into building	Written procedures and training for frequent routine inspection and scheduled replacement of sleeves
8	Frictional heating from slipping belts or chains (bucket elevators and en-masse conveyors)	Potential dust deflagration	Convey solids as pellets instead of granules or powder Increase particle size	Negative pressure for bucket elevators installed inside buildings to minimize dust leakage Hot material detection and automatic quench system	Written procedures and training for frequent routine inspection and scheduled replacement of belts and chains
9	Electrostatic spark discharge or frictional heating (orbiting screw or ribbon rubbing against vessel wall) Overpressure (orbiting screw powder blender, fluid bed blender, or ribbon blender)	Potential dust deflagration	Increase particle size	Inerting Overload shut down on the motor driving the orbiting screw	Written procedures and training to verify adequate purging of bottom bearing
10	Flammable or combustible solvents used (spray granulators and coaters)	Potential dust deflagration or fire	High flash point solvents	Internal deluge water sprays	Written procedures and training to process most stable materials first when campaigning multiple products to avoid ignition of unstable materials

Table 6.8 Common Failure Scenarios and Design Solutions for Solids Handling and Processing Equipment

No.	Event	Consequence	Potential Design Solutions		
			Inherently Safer / Passive	Active	Procedural
11	Fire (screw conveyors or extruders)	Potential equipment damage	Different type of conveyor (e.g., vibratory conveyor) Screens to remove tramp materials	Overload shutdown on the motor driving the screw Temperature sensor in the conveyor trough / barrel automatically tripping the motor and / or activating a water deluge system or snuffing steam	Operator response to high temperature alarm in the conveyor trough / barrel and activation of deluge system or deluge steam Written procedures and training for manual removal of tramp ferrous metals
12	Jammed idler roller, or if the belt jams, as a result of drive rollers continuing to run (belt conveyors)	Potential fire	Fire retardant belts Different type of conveyor (e.g., vibratory type) Sealed roller bearings to minimize ingress of solids	Belt velocity detection interlocked to shut down on low speed	Written procedures and training for manual shut down on detection of low speed
13	Electrostatic sparks igniting (belt conveyors)	Potential fire on the belt	Belts of anti-static material Minimum ignition energy increased Passive static elimination device (e.g., tinsel bar)	Ionizing blower to eliminate static charge	
14	Loss of containment (bucket elevators, screw conveyors)	Potential emission of combustible and / or toxic dusts to the atmosphere or building	"Dust-tight" design Different type of conveyor (e.g., en-masse conveyor)	Negative pressure ventilation to contain and capture any emissions	Written procedures and training for periodic contamination testing of area

6.8.2.1 Pneumatic Conveying Systems (Scenario 5)

Dust deflagrations often occur in end-of-line equipment (e.g., silos, dust collectors, cyclones) of pneumatic conveying systems due to electrostatic sparks. The rubbing of particles against particles and the walls of the pneumatic conveying line generate electrostatic charges on the powder, which are then discharged in the end-of-line equipment, where a dust cloud is often formed, and a dust explosion occurs. A number of preventive and protective measures are commonly used such as using nitrogen in lieu of air as the conveying gas, using dense-phase conveying in lieu of dilute-phase conveying to minimize attrition of the powder, providing deflagration venting or suppression systems for the end-of-line equipment, and good grounding and bonding of the pipeline and equipment. Other measures that can be taken involve modification of the solids being conveyed, such as increasing the particle size (making pellets) or formulating the solids so that they are less friable. Also, it is important to isolate the pneumatic conveying line from end-of-line equipment by a quick-closing valve or suppressant barrier so that the flame front developed in the end-of-line equipment does not propagate backwards into the equipment upstream of the conveying system.

Static ignition mechanisms in recovery bins, silos and related equipment are discussed by Eckhoff (Ref. 6-39). Recommended preventive and protective practices are described in British Standards Institute BS-5958 (Ref. 6-40).

6.8.2.2   Grinders and Other Size Reduction Equipment (Scenario 6)

Size reduction equipment, such as mills and grinders, create turbulent dust clouds due to their operation, which can result in a dust explosion (deflagration). This hazard can be minimized by using fluid energy mills in place of high-impact mills such as hammer mills. Fluid energy mills use a gas, such as air or nitrogen (an inherently safer fluid), to reduce the size of solids. Some types of mills are designed to contain a deflagration; these should be used whenever possible. Care must be taken to prevent the entry of tramp metal and other foreign materials into size reduction equipment. This can be accomplished by installing screens or magnetic separators upstream of size reduction equipment.

6.8.2.3   Gyratory Screeners (Scenario 7)

Dust explosions (deflagrations) have occurred in gyratory screeners (sieves) because dust clouds are readily formed due to the nature of the operation. Because of their vibratory motion, gyratory screeners are connected to process equipment by flexible sleeves (e.g., rubber socks or boots) as they vibrate. If a deflagration occurs, the flexible sleeves could rupture ejecting a burning dust cloud into the room or building, which then can cause a secondary explosion. To minimize this hazard, several things can be done:

- Install the gyratory screener in a room with an outside wall equipped with blow-out vent panels.
- Use a rotary screener, which does not vibrate, in lieu of a gyratory screener.
- Use nitrogen inerting where feasible.

All metal components, including the screening surfaces, should be bonded and grounded because of the vigorous motion of the powder in the screeners and the possible generation of static electricity. Consideration should be given to the use of conductive or anti-static flexible sleeves. Also, for dusts of low MIE, provision of anti-static footwear for operators is recommended (Ref. 6-40).

Leaky flexible sleeves can result in fugitive emissions from gyratory screeners. Leaks can be minimized, or even eliminated, by operating under a slight vacuum, with the screener connected to a dust collector.

6.8.2.4   Bucket Elevators and En-masse Conveyors (Scenario 8)

Bucket elevators and en-masse conveyors contain belts or chains which can loosen and rub against the housing and cause impact sparks or frictional heating, which in turn may cause a dust explosion. Tramp metal that gets into en-masse conveyors can also cause frictional heating which can act as an energy source for an explosion. Sensors for hot material can be installed and interlocked with a water quench system to extinguish the hot solids. Also, it is very important to prevent the propagation of a dust explosion flame into the upstream and downstream equipment connected to conveying equipment. This can be accomplished by installing material chokes such as rotary valves or screw feeders at the inlet and outlet sides of conveyors. It has been found that material chokes (plugs of powder) quench the flame (Refs. 6-38 and 6-39). Quick-closing valves and suppressant barriers can also be used to isolate upstream and downstream equipment from conveyors.

6.8.2.5 Belt Conveyor

Powders being conveyed on a belt conveyor can be ignited by an electrostatic spark if the powder has a low MIE. The electrostatic spark can often be generated by the belt itself, and the use of belts of anti-static (conductive) materials can minimize this problem. Electrostatic charges can also be reduced by use of ionized air or inductive neutralizes, such as static combs and tinsel bars (Ref. 6-41).

6.8.3   Design Considerations

There are various solids handling unit operations crushing, grinding, mixing, classifying and conveying; many of these operations generate combustible dust. All mechanical size reducing or conveying methods carry the risk of overheating due to mechanical failure. Many of these methods also generate static electricity.

The two major hazards of combustible dusts are fire and explosion. Combustible dusts are often easy to ignite and may be difficult to extinguish. Fires and explosion may be prevented by minimizing the accumulation of combustible dusts by collecting and removing them to below the Minimum Explosible Concentration (MEC); control ignition sources, and provide an inert gas atmosphere.

Fires and explosion hazards can be minimized by the use of appropriate preventive measures, such as the following:
- Increasing the particle size of the powder. Larger particle size raises the Minimum Ignition Energy (MIE) and reduces the rate of pressure rise of a dust explosion.
- Using dense-phase pneumatic conveying in lieu of dilute-phase conveying reduces the attrition of the solids conveyed. Dense-phase conveying reduces the static generation per unit mass, and may result in non-flammable mixtures in the transfer line.
- Using low speed mills to reduce dust cloud formation and reduce the potential for high energy metal-to metal-contact.
- Using fluid energy mills in lieu of high-impact mills (e.g., hammer mills); nitrogen can be used as the milling gas rather than air, which in most cases will make the operation inherently safer.
- Using an ionizing spray to dissipate electrostatic charges where possible.
- Designing tightly closed systems that minimize leakage of powders and solids into the surrounding area where they can accumulate.

6. EQUIPMENT DESIGN

Many chemicals are handled as a powder or dust; explosions of dust suspensions and fires of dust suspension or layers of dust are not uncommon. The designer may be able to change the process to avoid generating combustible dust, for example, by using a wet process. The shock sensitivity of the material should be established by testing before selecting size reduction equipment.

CCPS (Ref. 6-42) and NFPA 654 (Ref. 6-43) discuss safety considerations in handling bulk solids and powers.

Some general principles that may apply to equipment handling combustible dusts are:

- Design equipment to withstand a dust explosion.
- Minimize space filled by dust suspension.
- Minimize (monitor) mechanical failure and overheating (bearings, rollers, mills).
- Minimize static electricity.
- Minimize passage of burning dust.
- Provide explosion prevention (e.g., by inerting) and protection (e.g., suppression, isolation). Prevention is preferred over protection.
- Provide fire protection to suppress or extinguish fires.
- Maintain design operating conditions.
- Eliminate sources of ignition.

6.8.3.1 Storage

Storage vessels also include bins and silos used for the storage of solid materials such as pellets, granules, or dusts. The primary danger in the bins comes from dust in the vapor space above the material creating an explosive or ignitable condition. Suspensions of combustible dusts in the vessel vapor space above the material can be ignited leading to fires and explosions. Since dust production typically cannot be prevented, other means of explosion prevention must be applied. Ignition sources should be minimized, and explosion venting of vessels (including bin vent filters or bag-houses) should be considered. Care should be taken during the design of a bin to reduce horizontal surfaces inside the bin where material can remain and create a hazard when the bin is opened for maintenance; the air above such areas has been known to explode while work inside the bins was being performed during normal repairs.

Additionally, the vessels can be inerted in a manner similar to that used for atmospheric storage tanks (Section 6.1.3.4). The pneumatic transfer of solids can also be performed using an inert or a reduced oxygen concentration gas with a closed-loop return to the sending tank. Deflagration suppression can also be provided for bins and silos to prevent a deflagration. Among the principal reasons for providing inerting on reactors and vessels is the desirability of eliminating flammable vapor-air mixtures that can be caused by addition of solids through the manhole or materials having low minimum spark ignition energies, or autoignition temperatures. Also, the pneumatically conveyed stream can first be routed to a cyclone at the top of the silo and then admitted to the silo slowly via a rotary airlock feeder. This minimizes the potential for a dust cloud in the silo.

6.8.3.2   Milling Equipment

Milling equipment may be used in systems where it is necessary to reduce particle size or product agglomeration. One hazard associated with milling equipment is the temperature increase that can be imparted to the material during the milling operation, particularly when product flow through the mill is significantly reduced or interrupted (similar concerns exist for other solids handling operations such as blending and, to a lesser degree, particle size separations such as screening or sieving). This can lead to ignition or decomposition of combustible or unstable materials that could lead to fires or explosions in the milling equipment. Additionally, fires or explosions can result from the presence of combustible dusts typically present in the milling equipment, should other ignition sources be present. Other concerns include the potential for exposure of operating personnel to chemical hazards. A number of design considerations when milling materials that are combustible or are temperature sensitive are:

- Monitoring of milling temperature.
- Shaft speed sensors to detect pluggage in the mill.
- Instrumentation or inspections to ensure product flow, thus limiting material temperature rise to a safe level.
- Static electricity concerns, including proper bonding and grounding.
- Proper area electrical classification.
- Proper selection, location, and maintenance of bearings.
- Removal of tramp materials from the feed to the milling equipment.
- Milling of impact-sensitive materials should generally be avoided.

6.8.4   References

6-37.   Whitmore, M.W., Gladwell, J.P. and Rutledge, P.V. *Journal of Loss Prevention in the Process Industries.*p169-175. 1993.

6-38.   Field, P. *Dust Explosions.* Elsevier Scientific Publishing Company. New York 1982.

6-39.   Eckhoff, R.K. Third Edition. *Dust Explosions in the Process Industries.* Butterworth-Heinemann. Boston. 2003.

6-40.   BS-5958. *Code of Practice for Control of Undesirable Static Electricity: Part 1, General Considerations*, and Part 2, *Recommendations for Particular Industrial Situations.* British Standards Institute. London 1992.

6-41.   NFPA 77. *Recommended Practice on Static Electricity,* 2007 Edition. National Fire Protection Association, Quincy, Massachusetts. 2007.

6-42.   CCPS. *Guidelines for Safe Handling of Powders and Bulk Solids,* Center for Chemical Process Safety of the American Institute of Chemical Engineers. New York, New York. 2005.

6-43.   NFPA 654. *Standard for the Prevention of Fire and Dust Explosions from the Manufacturing, Processing, and Handling of Combustible Particulate Solids.* National Fire Protection Association, Quincy, Massachusetts. 2006.

6.8.4.1 Suggested Additional Reading

ESCIS. *Milling of Combustible Solids: Safety, Evaluation of Feed Materials, Protective Measures with Mills.* Booklet No. 5 (English Edition). Expert Commission for Safety in the Swiss Chemical Industry, Baske, Switzerland. 1994.

FM Global. Property Loss Prevention Datasheet 7-11. *Belt Conveyors.* Factory Mutual Glabal, Norwood, Massachusetts. 2009.

6.9 FIRED EQUIPMENT

This section presents potential failure mechanisms for fired equipment and suggests design alternatives for reducing the risks associated with such failures. The types of fired equipment covered in this section include:
- Process furnaces
- Boilers
- Thermal incinerators (oxidizers)
- Catalytic incinerators

This section presents only those failure modes that are unique to fired equipment. Some of the generic failure scenarios pertaining to vessels and heat transfer equipment may also be applicable to fired equipment. Consequently, this section should be used in conjunction with Section 6.1, Vessels, and Section 6.4, Heat Transfer Equipment. Unless specifically noted, the failure scenarios apply to more than one class of fired equipment.

6.9.1 Past Incidents

This section describes several case histories of incidents involving failure of fired equipment to reinforce the need for the safe design practices presented in this section.

6.9.1.1 Light-Off Error

A safety shut-off valve on the gas supply to a burner remained open after the unit was shut down. There was no indicator to show that the valve was open or closed. On startup, the operator opened the main valve on the gas supply to the burner before lighting the pilot burner. When he tried to light the burner, an explosion occurred.

Lessons learned include the need for positive isolation and confirmation of valve position to safely light burners in equipment.

6.9.1.2 Ethylene Cracking Furnace Overfiring

During operation of an ethylene unit, various light by-products off gases were being collected and recycled to the fuel system. For startup and any other condition during which plant-produced fuel gases could not meet demand for fuel in the cracking furnaces, LPG was available for admission to the fuel system to satisfy demand.

Normally, the firing control system on the cracking furnaces utilized a Wobbe Index analyzer to adjust fuel rate based on heating value. However, the plant operators had disabled the Wobbe Index analyzer and had also disabled the coil outlet temperature cascade to the fuel gas control valve pressure controller.

While operating the cracking heaters on light by-product off gases with a low calorific value, a plant upset resulted in the trip of the cracked gas compressor. The heaters were maintained online with cracked gas routed to a flare. Subsequently, without forward flow of cracked gas to the downstream separation facilities, the production of plant-produced off-gas diminished and LPG was automatically added to the fuel gas system. With the addition of LPG the heating value of the fuel gas increased significantly; this resulted in the overfiring (adding too much heat) of the heaters and major damage to the coil and associated supports.

Lessons learned include the provision of a heater emergency shutdown based on a measurement of coil outlet temperature independent from process controls would have been advantageous.

6.9.1.3 Furnace Tube Failure

A furnace was protected by a relief valve located downstream of the low flow alarm and furnace trip. A blockage in the line exiting the furnace caused the relief valve to lift, which in turn caused the flow through the furnace tubes to drop sharply. Because flow appeared normal at the low flow alarm / trip point, the furnace continued to operate and eventually the tubes overheated and burst (Ref. 6-44).

This incident illustrates the need to carefully locate safety instrumentation during the design phase, and especially to consider downstream and upstream conditions that could cause false or inaccurate measurements.

6.9.2 Failure Scenarios and Design Solutions

Table 6.9 presents information on equipment failure scenarios and associated design solutions specific to fired equipment.

Table 6.9 Common Failure Scenarios and Design Solutions for Fired Equipment

No.	Event	Consequence	Potential Design Solutions		
			Inherently Safer / Passive	Active	Procedural
Pressure					
	Generally Applicable – High Pressure (Applicable to all high pressure scenarios)		Design for maximum pressure	Pressure relief device Deflagration or detonation arresters	Operator response to high burner pressure alarm Operator response to high firebox pressure alarm Written procedures and training for manual heater shut down on indication of high firebox pressure
	Generally Applicable – Low Pressure (Applicable to all low pressure scenarios)		Design for vacuum	Vacuum relief device	

6. EQUIPMENT DESIGN

Table 6.9 Common Failure Scenarios and Design Solutions for Fired Equipment

No.	Event	Consequence	Potential Design Solutions		
			Inherently Safer / Passive	Active	Procedural
1	High fuel gas pressure	Potential flame lift off resulting in fire box explosion if gas flow is reintroduced	Burners with wider turndown ratio. Pilot burners designed with a separate fuel source. Pilot gas supply from the upstream side of the main shutoff valve for all burners	Automatic heater shut down on high firebox pressure or high stack temperature. Automatic heater shut down on high fuel gas pressure	Operator response to high pressure alarm
2	Low fuel gas pressure	Potential flameout resulting in firebox explosion if gas flow is reintroduced	Pilot burners designed with a separate fuel source. Pilot gas supply from the upstream side of the main shutoff valve for all burners	Flame surveillance system to shut down heater on loss of flame. Automatic heater shut down on low fuel gas pressure	Operator response to low fuel gas pressure alarm
Flow					
Generally Applicable – More Flow (Applicable to all more flow scenarios)			Flow restriction orifice	Automatic heater shutdown on high fuel gas flow	Written procedures and training to prevent excessive firing rates
Generally Applicable – No / Less Flow (Applicable to all no / less flow scenarios)				Automatic heater shut down on low process flow (total or individual passes)	Written procedures and training for manual shutdown of heater on low process flow
3	Rapid readmission of air to correct insufficient air situation	Potential fire box explosion		Interlock fuel supply and air supply so that loss of, or significant reduction in air will isolate the fuel supply. "Lead-lag" firing control system to avoid firing without sufficient air	Written procedures and training to limit fuel firing to air availability. Written procedures and training to control rate of air readmission in response to insufficient air flow

Table 6.9   Common Failure Scenarios and Design Solutions for Fired Equipment

No.	Event	Consequence	Potential Design Solutions		
			Inherently Safer / Passive	Active	Procedural
4	Waste gas supply manifold to incinerator	Potential flashback into supply line	Alternative waste gas disposal method (e.g., adsorption)	Automatic control of waste gas concentration Automatic temporary diversion of waste gas to alternative disposal	Written procedures and training for manual control of waste gas concentration Written procedures and training for manual temporary diversion of waste gas to alternative disposal
5	Closure of flue gas damper or trip of induced draft fan	Potential fire box explosion	Firebox designed for shutoff pressure of forced draft fan Mechanical position stop to prevent complete closure of damper Natural draft design to eliminate induced draft fan and / or damper	Automatic heater shutdown on high firebox pressure or high stack temperature Automatic heater shutdown on low oxygen concentration	Operator response to high pressure alarm Operator response to low oxygen concentration alarm
6	Insufficient oxygen (Incinerator)	Potential for incomplete destruction of hazardous materials	Alternate means of disposal of hazardous material Increased stack height to reduce ground level concentration of hazardous materials	Automatic heater shutdown on low oxygen or carbon monoxide concentration Permissive systems that won't allow main burner lighting until pilot confirmation	Operator response to low oxygen or carbon monoxide concentration alarm Written procedures and training for manual sampling of incinerator offgas for concentration of hazardous materials
7	Low or no fuel gas flow (Incinerator)	Potential for incomplete destruction of hazardous materials	Alternate means of disposal of hazardous material Increased stack height to reduce ground level concentration of hazardous materials	Introduction of alternate fuel supply	Written procedures and training for manual sampling of incinerator offgas for concentration of hazardous materials

6. EQUIPMENT DESIGN

Table 6.9 Common Failure Scenarios and Design Solutions for Fired Equipment

No.	Event	Consequence	Potential Design Solutions		
			Inherently Safer / Passive	**Active**	**Procedural**
8	Maldistribution through individual heater passes (process side)	Potential tube rupture resulting in fire outside the firebox	Enhanced tube metallurgy Heavier wall thickness Orifices or venturis to balance parallel tube passes	Automatic control of flow to individual heater passes	
9	Makeup boiler water stops (boiler drum)	Potential tube rupture	Tubes in the convection section designed to operate "dry"	Interlock to shut down firing on low boiler feed water flow	Operator response to flow alarm
10	No pilot flames before opening main fuel supply	Potential fire box explosion	Pilot burners with a separate fuel source Pilot gas supply from the upstream side of the main shutoff valve for all burners		Written procedures and training on lighting heater
Temperature					
Generally Applicable – High Temperature *(Applicable to all high temperature scenarios)*				Automatic heater shut down on high process outlet temperature or high firebox temperature Automatic heater shut down on high stack outlet temperature Automatic heater shut down on high flue temperature Automatic control of firing or outlet temperature	Operator response to high stack temperature or high firebox temperature alarm Written procedures and training for manual shut down of heater on high firebox temperature or high process outlet temperature Written procedures and training for burner adjustment to eliminate flame impingement Written procedures and training for manual shut down on high flue gas temperature Written procedures and training to prevent excessive firing rates

Table 6.9   Common Failure Scenarios and Design Solutions for Fired Equipment

No.	Event	Consequence	Potential Design Solutions		
			Inherently Safer / Passive	Active	Procedural
	Generally Applicable – Low Temperature *(Applicable to all low temperature scenarios)*			Automatic control of firing or outlet temperature	Operator response to low temperature of alarm and manual shutdown of flow to incinerator Written procedures and training for manual shut down of incinerator on low combustion temperature
11	Flame impingement on tubes	Potential tube rupture resulting in fire outside the firebox	Enhanced tube metallurgy Heavier wall thickness Indirect firing		Operator response to tube skin temperature alarm Written procedures and training for visual observation of coils for hot spots
12	Overfiring	Potential tube rupture resulting in fire outside the firebox	Enhanced tube metallurgy Heavier wall thickness Indirect firing	High stack temperature interlock Oxygen analyzer on heater with low oxygen alarm	Operator response to high temperature alarm Operator response to tube skin temperature alarm Written procedures and training for visual observation of tubes for hot spots
13	Firing with insufficient air	Potential afterburning in convection section and flue gas system resulting in heater damage		"Lead-lag" firing control system to avoid firing without sufficient air	Written procedures and training to take corrective action or shutdown heater on indication of high flue gas temperature or low stack oxygen concentration
Composition					
	Generally Applicable *(Applicable to all composition scenarios)*			Automatic shutdown of incinerator on high stack temperature	Written procedures and training for manual shut down of incinerator on high offgas temperature
14	Rapid increase in fuel gas heating value	Potential tube rupture resulting in fire outside the firebox	Dedicated constant heating value fuel gas	Automatic adjustment of firing on process outlet temperature and fuel heating value (on-line Btu analyzer)	Written procedures and training for operation of heater

6. EQUIPMENT DESIGN

Table 6.9 Common Failure Scenarios and Design Solutions for Fired Equipment

No.	Event	Consequence	Potential Design Solutions		
			Inherently Safer / Passive	Active	Procedural
15	Liquid in feed to Catalytic Incinerator	Potential for hot catalyst bed resulting in high temperature or fire	Alternative incinerator design	Feed preheating to vaporize any entrained liquid. Heat tracing of feed system. Liquid knock-out drum with automatic liquid removal	Written procedures and training for manual liquid removal from knock-out (KO) drum
16	Liquid carry over with fuel gas	Potential loss of flame and possible explosion on reignition of gas	Pilot burners with a separate fuel line. Pilot gas supply from the upstream side of the main shutoff valve for all burners	Flame surveillance system to trip heater on loss of flame. Heat tracing of fuel gas system. Liquid knock-out drum with automatic liquid removal	Operator response to high level alarm on liquid knock-out (KO) drum and manual liquid removal
17	Fuel / oxidizer ratio malfunction with multiple equipment to common vents, stacks, heat recovery systems, etc.	Potential explosive mixtures resulting in explosion and fire	Separate vent / exhaust systems for each fired equipment (includes separate flares, scrubbers, absorbers, stacks, heat recovery, etc.)	Measurement systems and shutdown interlocks to detect ratio errors prior to mixing (located far enough upstream to prevent explosive mixtures in common vents)	
Equipment Failure					
18	Delayed ignition on light-off, fuel leakage into the firebox, or insufficient firebox purging	Potential firebox explosion	Continuous pilots for all burners (monitored and alarmed)	Permissive to ensure that fuel and combustion air controls are in proper lighting off positions, before the ignition sequence can proceed. Reliable fuel gas isolation (e.g., double block and bleed). Timed purge prior to light off with interlocks to ensure that all fuel supply valves are closed	Individual burner cocks so that only one burner may be lighted at a time to minimize potential accumulation of fuel prior to light-off. Written procedures and training to ensure that fuel and combustion air controls are in proper position before the ignition sequence can proceed

Table 6.9   Common Failure Scenarios and Design Solutions for Fired Equipment

No.	Event	Consequence	Potential Design Solutions		
			Inherently Safer / Passive	Active	Procedural
19	Corrosion / erosion	Potential tube rupture resulting in fire outside the firebox	Elimination of liquid to burner by using non-condensing gas Enhanced tube metallurgy Heavier wall thickness Sulfur-free fuel	Dewpoint measurement of fuel gas	Written procedures and training to prevent acid dewpoint corrosion Operator response to low stack temperature alarm
20	Forced draft fan stops	Potential firebox explosion	Firebox designed for minimum pressure produced by induced draft fan Alternative design without induced draft fan	Automatic transfer to natural draft operation	

6.9.2.1   Readmission of Air to Firebox (Scenario 3)

Adequate delivery of combustion air to fired heaters at all heat load conditions is essential for safe furnace operation. Firing without sufficient air will result in unburned fuel in the firebox with the potential for subsequent uncontrolled combustion. Firing controls should be configured so that air "leads" fuel on a firing demand increase and "lags" fuel on a firing demand decrease. However, even with a "lead-lag" system, rapid reduction in air availability due to the trip of a fan, for example, may result in insufficient air delivery. An oxygen analyzer with high and low concentration alarms is an important safeguard.

To avoid the accumulation of unburned fuel and a possible positive pressure pulse in the firebox during rapid readmission of air, interlock shutdown via detection of a low air-to-fuel ratio may be warranted. If an automatic air restoration response strategy is used, such as auto-start of a spare fan, suitable system dynamic response analysis should be employed to ensure that sudden loss of air can be effectively managed.

For additional information on fired equipment combustion controls refer to *Instrument Engineers Handbook: Process Control* (Ref. 6-45) and API RP 556 *Instrumentation and Control Systems for Fired Heaters and Steam Generators* (Ref. 6-46).

6.9.2.2   Tube Rupture (Scenarios 8, 9, 11, 12, 14, and 19)

Tube rupture is the second most common failure mode in fired equipment. Overheating tubes drastically reduces their useful life. A pressure vessel may be able to withstand

6. EQUIPMENT DESIGN

several times its design pressure, but a furnace tube may only withstand a few percent increases in its absolute temperature (Ref. 6-44).

6.9.2.3   Closure of Stack Damper (Scenario 5)

Closure of the stack damper during operation or the loss of the induced draft fan can lead to buildup of pressure inside the firebox. This may result in fire / gases coming out of the furnace and risk of personnel exposure and equipment damage. To prevent such a situation it is desirable to maintain an open flue-gas path by putting a minimum position stop on the damper. It may also be necessary to provide a spare induced draft fan or design the furnace to transfer to natural draft operation. If these alternatives are not available, the system should be shut down on detection of high firebox pressure.

6.9.3   Design Considerations

The two main problems with process fired incinerators (thermal and catalytic oxidizers), furnaces, and fired boilers are explosion in the firebox or rupture of process tubes (Ref. 6-46). Tube rupture may be detected by monitoring flow or monitoring the temperature as the tubes overheat. In boilers and boilers used for waste heat recovery, loss of the boiler water level supply could be catastrophic. Reliable level monitoring and control are paramount. Reliable level and control, include the design of a continuous supply of boiler feed water.

6.9.3.1   Corrosion

Corrosion is a major source of tube rupture problems in fired heaters. External corrosion of furnace tubes and other equipment in fireboxes may be caused by:
- Temperature
- Corrosive deposits on tubes
- Flue gas composition
- Physical conditions existing beneath and in any overlying deposit of ash

Oxygen and contaminants in the fuel gas and oils, rather than the fuel itself, cause most of the corrosion in fireboxes. The harmful contaminants are alkali metals (Na, K), sulfur, and vanadium. Although heater tubes usually operate at much lower metal temperatures, consideration must be given to the corrosivity of the process fluid, typical metal temperature, and the fuel used in firing the heater when tube materials are selected.

Corrosion occurs in the convection section when the temperature is lower than the dew point of the flue gases. Proper operation / shutdown procedures are the most effective methods to avoid convection section corrosion.

6.9.3.2   Process Control Instrumentation

Direct-fired heaters are widely used in the process industries. Typical furnace applications include distillation-fractionator pre-heaters and reboilers; steam generators; reactor pre-heaters; and pyrolysis reactors. A comprehensive discussion of direct-fired process furnaces can be found in *Perry's Chemical Engineers' Handbook* (Ref. 6-47). Frequently the process fluids which are being heated in a direct-fired process heater are flammable. Furnace tube failure in the radiant or convective section of the heater could result in serious fire and / or explosion hazard and damage to heater internals.

Incomplete combustion of fuel in the firebox will cause a buildup of combustible gases (unburned fuel or carbon monoxide) which may ignite when sufficient oxygen is present resulting in an explosion within the fire box.

Process variables and parameters that determine safe furnace operation are Coil Outlet Temperature (COT), pass outlet temperature (POT), excess oxygen in the flue gas, combustible gases in the flue gas, flue gas opacity, firebox pressure, firing rate (furnace tube heat flux), coking, stack and bridge wall temperatures, and combustion efficiency. A sound control scheme must supply sufficient air to promote complete combustion, ensure safe operation and maintenance, maintain COT at specified target, balance burner firing, maintain equal POTs, constrain the furnace firing rate to avoid maximum allowable stack temperature, furnace tube temperature, or convection section temperature, and monitor indications of coking over long-term operation.

In the design of safe control systems, constraints imposed on process variables are intended to ensure plant safety and efficient operation. Excessive temperatures lower the strength of carbon steel and alloy materials used in the furnace and may lead to premature failure. Thermocouples can be located in critical areas of the furnace to indicate when temperatures are above safe operating conditions. Constraint controls should be used to override furnace duty or COT controls and maintain the furnace within metallurgical constraints.

In process plants, fired equipment such as furnaces and boilers are a vital necessity. The combustion process must be controlled to maintain the desired rate of heat transfer, to maintain efficient fuel combustion, and to maintain safe conditions in all phases of operation. These combustion controls are normally a part of the basic process control system and typically consist of some or all of the following control functions:

- Firing Rate Demand Control
- Combustion Air Flow Control
- Fuel Flow Control
- Fuel / Air Ratio (Excess Air) Control
- Draft Control
- Feed Water Flow Control (Steam Boilers Only)
- Steam Temperature Control (Steam Boilers Only)

For further details on the implementation of fired equipment controls can be found in API RP 556 (Ref. 6-46).

6.9.3.3 Tube Rupture

A "hot spot" (localized excessive metal temperature) is one major cause of process heater tube failure. Hot spots are generally caused by flame impingement due to incorrect burner adjustment, excessive heater firing rates and / or excessive coking or scaling on the internal tube surfaces, or loss of (or minimal) flow of process fluid in the tubes. Heater instrumentation should provide for detection of failure and automatic shutdown to minimize secondary damage. Such items as stack temperature increase, heater tube pressure, and / or flow loss and loss of outlet temperature can be used to detect a tube failure.

6. EQUIPMENT DESIGN

6.9.3.4   Design Considerations

Furnace and heater design considerations include:
- Providing steam or nitrogen snuffing for control of possible tube rupture events
- Providing pilot burners with a separate fuel system in case of failure of main fuel supply
- Providing flashback protection for burners, including all potential ranges of temperature, pressure, gas composition
- Providing means to prevent liquid slugs from entering burners, e.g., providing enough condensation drums; providing coalescers for liquid droplet removal; providing means to heat trace and insulate the line from the knockout drum; adhering to proper startup, operation, and shutdown procedures
- Preventing flame impingement on tubes, supports, or refractory
- Providing safe firebox purging sequences
- Providing fuel shutoff and startup checking sequence
- Designing the system to transfer to natural draft in an emergency and operate on loss of air-preheater or fan.
- Locating outside

For additional information on Heaters see API RP 560 (Ref. 6-48).

6.9.4   References

6-44.   Kletz, T.A. *Lessons from Disaster.* Gulf Publishing Company. Houston, Texas 1993.

6-45.   Liptak, B.G. *Instrument Engineers Handbook. Process Control, Fourth Edition.* Radnor, Pennsylvania: 2005.

6-46.   API RP 556. *Instrumentation and Control Systems for Fired Heaters and Steam Generators.* American Petroleum Institute. Washington D.C. 1997.

6-47.   Green, Don W. and Perry, R.H. *Perry's Chemical Engineers' Handbook, Eighth Edition*, McGraw-Hill. New York, New York. 2008.

6-48.   API RP 560. *Fired Heaters for General Refinery Service*, Fourth Edition. American Petroleum Institute. Washington D.C. 2007.

6.9.4.1   Suggested Additional Reading

Ghosh, H. *Improve Your Fired Heaters. Chemical Engineering.* 1992.

NFPA.   *Standard for Ovens and Furnaces.*   NFPA 86.   National Fire Protection Association. Quincy, Massachusetts. 2011.

6.10 PIPING AND PIPING COMPONENTS

"Loss of containment from a pressure system generally occurs not from pressure vessels but from piping and associated fittings. It is important, therefore, to pay at least as much attention to the piping as to the vessels" (Ref. 6-49). The purpose of this section is to provide information on safe engineering practices in the areas of detailed piping and

valve specifications, piping flexibility analysis, piping supports, special piping materials of construction, and maintenance in accordance with the proper ASME B31 code (Ref. 6-50). The section focuses on process lines carrying hazardous materials.

Codes of practice and standards address the solutions to common problems but establish only minimum design, fabrication, testing, and examination requirements for average service. Many circumstances relating to service, operation, materials and fabrication, inspection, or unusual design deserve special consideration if the resulting piping systems are to operate safely and be reasonably free from frequent maintenance.

Standards and codes of practice related to the safe design of piping are the following codes issued by the American Society of Mechanical Engineers (ASME) (Ref. 6-50); those also approved by the American National Standards Institute (ANSI) are indicated with an asterisk:

- B31.1*   Power Piping
- B31.2   Fuel Gas Piping
- B31.3*   Chemical Plant and Petroleum Refinery Piping
- B31.4*   Liquid Transportation Systems for Hydrocarbons, Liquid Petroleum Gas, Anhydrous Ammonia, and Alcohols
- B31.5*   Refrigeration Piping
- B31.8*   Gas Transmission and Distribution Piping Systems
- B31.9*   Building Service Piping
- B31.11* Slurry Transportation Piping Systems
- API Specification 5L, Specification for Line Pipe

These various sections provide different margins of safety for pressure piping systems based on service considerations and industry experience.

6.10.1 Past Incidents

This section describes several case histories of incidents involving failure of piping and piping components to reinforce the need for the safe design practices presented in this section.

6.10.1.1 Chemical Storage

A series of explosions was initiated following flame transmission through a complex tank vent collection header system; the first explosion occurred in a tank containing acrylonitrile. It is believed that the pallet in the PV (Pressure-Vacuum) vent had been removed and not replaced during maintenance. The interconnected ducts caused the rapid spread of the fire.

Lessons learned include where it is required to reduce atmospheric emissions via PV vents while retaining the in-breathing capability of the devices; additional vents opening at a slightly lower positive tank pressure can be connected to a collection system. These vent lines can safely be equipped with detonation arresters since if the arrester becomes blocked the tank will not be sucked in while the PV vent remains in service.

6. EQUIPMENT DESIGN

6.10.1.2 Line Pluggage

A line that had been used to blow down wet hydrocarbon formed an ice-hydrate plug, blocking the 18-inch blow down line. As a result of external steaming, the plug loosened and the pressure above it caused it to move with such force that it ruptured the line at a tee.

Lessons learned include the need to consider potential hydrate formation in lines and methods for unplugging. Sloping lines to reduce the potential of plugging and addition of chemicals to prevent hydrate formation are design solutions to be considered.

6.10.1.3 External Corrosion

A valve in a 10-inch liquefied butane line was located in a pit. The pit accumulated rainwater contaminated by sulfuric acid from a leaking line nearby. The bolts on the valve bonnet corroded and gave way, resulting in a massive butane release. The ensuing explosion killed seven people and caused extensive damage.

Lessons learned include the need to consider potential sources of external corrosion during design, both from ambient conditions as well as from adjacent equipment failures.

6.10.2 Failure Scenarios and Design Solutions

Table 6.10 presents information on equipment failure scenarios and associated design solutions specific to piping and piping components.

Table 6.10 Common Failure Scenarios and Design Solutions for Piping and Piping Components

No.	Event	Consequence	Potential Design Solutions		
			Inherently Safer / Passive	Active	Procedural
Pressure					
	Generally Applicable – High Pressure *(Applicable to all high pressure scenarios)*		All piping and equipment designed for maximum expected pressure	Pressure relief device Automatic isolation based on detection of high pressure	Operator response to high pressure alarm
	Generally Applicable – Low Pressure *(Applicable to all low pressure scenarios)*			Automatic isolation based on detection of low pressure	Operator response to low pressure alarm
1	Thermal expansion of liquid in blocked-in line	Potential overpressure of line leading to loss of containment	Elimination of potential for blocking in by removing valves and other closures (e.g., blinds)	Expansion tank	Written procedures and training for draining of all blocked-in lines during shutdown Written procedures and training to leave one end of line open

Table 6.10 Common Failure Scenarios and Design Solutions for Piping and Piping Components

No.	Event	Consequence	Potential Design Solutions		
			Inherently Safer / Passive	Active	Procedural
2	Deflagration and detonation in piping	Potential loss of containment	Dedicated vent lines used where incompatible material mixing may occur Elbows and fittings avoided or minimized, which can cause turbulence and flame acceleration Temperature, pressure or pipe diameter limited to prevent DDT from occurring (e.g., acetylene)	Detonation or suitable deflagration arresters between protected equipment and potential ignition sources Gas flame detection and actuatation of fast closing valve or suppression system Liquid seal drum isolating ignition source (e.g., flare) Multiple rupture disks / explosion vents located at appropriate points on piping Operate outside flammable range, e.g., O_2 analyzer or hydrocarbon analyzer control inert purge or enrichment gas addition	Written procedures and training for inert purging prior to startup
3	Blockage of piping, valves (manual)	Potential increased pressure in systems resulting in eventual loss of containment	Valve car-sealed open	Permissive systems to prevent line-up / blocked flow scenarios	Written procedures and training for proper valve alignment.
4	Solid collection in flame arresters	Potential increased pressure in systems resulting in eventual loss of containment	Parallel switchable flame arresters Piping system sized to maintain minimum required velocity to avoid deposit of material	Removal of solids from process stream (KO pot, filter, etc) with automatic blowdown of solids Tracing of piping to minimize solid deposition Differential pressure measurement across the flame arrester and high differential pressure alarm	Written procedures and training for removal of solids from process stream (KO pot, filter, etc) with manual blowdown of solids Written procedures and training for periodic manual system cleaning Written procedures and training for periodic cleaning via flushing, blowdown, internal line cleaning devices (e.g., "pigs")

6. EQUIPMENT DESIGN

Table 6.10 Common Failure Scenarios and Design Solutions for Piping and Piping Components

No.	Event	Consequence	Potential Design Solutions		
			Inherently Safer / Passive	Active	Procedural
5	Valve in line rapidly closed	Potential liquid hammer and pipe rupture, loss of containment	Slow-closing manual valves (i.e., gate instead of quarter turn)	Closing rate limited for motor-operated valves via appropriate gear ratio Closing rate limited for pneumatic-operated valves via restriction orifice in air line Surge arrester	Written procedures and training to close valves slowly
6	Automatic control valve opens	Potential high pressure in downstream piping and equipment	Limit stop utilized to prevent control valve from opening fully, or a restriction orifice		*Note: scenario occurs too fast for operator action*
7	Block on inlet or outlet of relief device closed	Potential loss of relief capability	Eliminate all block valves in relief path Second relief valve provided with three-way block valve at inlet		Written procedures and training to car-seal open or lock open all block valves upstream and downstream of relief valves per applicable codes and provide administrative procedures to regulate opening and closing of such valves
8	High pressure supply deadheaded at low pressure piping / tank	Failure of low pressure piping, nozzles, etc.	High pressure valving and flanges installed at low pressure isolation (example: class 600 flange and isolation valve on open-top atmospheric tank) Valving after high pressure isolation eliminated (e.g., discharge into top of tank)	Check valve to prevent back flow thru pump to low pressure inlet piping Relief valve on low pressure piping Backflow preventors or auto-starts on pumps to lower the frequency of backflow events caused by loss of pump	
9	Discharge line pluggage involving slurries or polymer positive displacement pumps	Overpressure failure of piping systems		High / low motor amperage shutdown interlock (relief valves ineffective due to pluggage potential) Low / no flow shutdown interlock	

Table 6.10   Common Failure Scenarios and Design Solutions for Piping and Piping Components

No.	Event	Consequence	Potential Design Solutions		
			Inherently Safer / Passive	Active	Procedural
Flow					
	Generally Applicable – More Flow (Applicable to all more flow scenarios)			Automatic isolation based on detection of high flow	Operator response to high flow alarm
	Generally Applicable – No / Less Flow (Applicable to all no / less flow scenarios)				Operator response to low flow alarm
10	Blockage of relief device by solids accumulation (polymerization, solidification)	Potential loss of relief capability	Flow sweep fitting at inlet of relief device. Trace and insulate relief device	Automatic flush of relief device inlet with purge fluid. Rupture disks alone or in combination with safety valves with appropriate rupture disk leak detection	Written procedures and training for manual periodic or continuous flush of relief device inlet with purge fluid
11	High fluid velocity	Potential erosion especially if two phase flow or abrasive solids are present leading to loss of containment	Fittings minimized where erosion can occur. Heavier walls at tees, elbows, and other high abrasion points. Material selection to resist erosion. Sizing of pipe to limit velocities. Tees or long radius elbows used instead of 90° elbow in abrasive solid service. Conductive line		Written procedures and training to limit flow velocity. Written procedures and training for periodic inspection of high wear points
		Potential static buildup resulting in fire / explosion if conditions are appropriate	Appropriately sized lines for maximum expected velocities. Bonding and grounding of lines and equipment (may not be applicable for non-conducting materials)		Written procedures and training on limiting flow velocity

6. EQUIPMENT DESIGN

Table 6.10   Common Failure Scenarios and Design Solutions for Piping and Piping Components

No.	Event	Consequence	Potential Design Solutions		
			Inherently Safer / Passive	Active	Procedural
12	Reverse Flow	Differential pressure on joining lines, drains or temporary connections causing back flow of product resulting in undesirable reaction, overfilling, etc.	Incompatible fittings to prevent unwanted connections Separate lines to final destination	Automatic isolation on detection of low differential pressure Check valve on lower pressure line to prevent reverse flow	Written procedures and training for proper isolation of interconnected lines Written procedures and training for manual isolation on detection of low differential pressure
13	Inadvertent flow (1/4 turn valve opened)	Potential loss of containment	Latching handle Oval / circular handle		
14	Failure to close valves on sample connection, drain and other fittings	Potential loss of containment	"Deadman" (self-closing) valve Latching handle design on valves to prevent inadvertent opening	Automatic closed-loop sampling system	Written procedures and training for double block and bleed valves, valve plugs, caps, blinds, etc. Written procedures and training to immediately reinstall caps and flanges
15	Breakage of sight glasses or other glass components	Potential loss of containment	Eliminate the use of glass components Flow restriction orifice in glass connection Physical protection against damage (i.e., armored sight glass) Sight glasses with pressure design rating exceeding maximum expected pressure	Excess flow check valves to limit discharge due to glass failure	Written procedures and training to normally isolate sight glass when not in use
Temperature					
Generally Applicable – High Temperature *(Applicable to all high temperature scenarios)*				Automatic action in response to high temperature alarm	Operator response to high temperature alarm
Generally Applicable – Low Temperature *(Applicable to all low temperature scenarios)*				Automatic action in response to low temperature alarm	Operator response to low temperature alarm

Table 6.10 Common Failure Scenarios and Design Solutions for Piping and Piping Components

No.	Event	Consequence	Potential Design Solutions		
			Inherently Safer / Passive	Active	Procedural
16	Faulty tracing	Potential increased temperature leading to hot spots resulting in exothermic reaction and loss of containment	Insulating material between tracer and pipe (sandwich tracer)	Electrical tracing with temperature limitation controls Ground Fault Indication (GFI) protection	
17	Temperature control on jacketed piping failure (manual or auto)	Potential increased temperature leading to hot spots resulting in exothermic reaction and loss of containment	Heat transfer media with maximum temperature limited to a safe level (jacketed pipe)		
18	High pressure drop across control valve	Potential flashing / vibration leading to loss of containment	Multiple intermediate pressure letdown devices (valve or orifices) Piping securely anchored Valve located as close to the vessel inlet as possible Valve type suitable for high pressure drop and flashing service		
19	External fire	Potential undesired process reaction (e.g., acetylene decomposition)	Continuous welded pipe Fireproof insulation with stainless steel sheathing and banding	Fire detection system with automatic water spray Automatic closure of isolation valve on fire detection	Operator response to fire detection system and activation of manual water spray

6. EQUIPMENT DESIGN

Table 6.10 Common Failure Scenarios and Design Solutions for Piping and Piping Components

No.	Event	Consequence	Potential Design Solutions		
			Inherently Safer / Passive	Active	Procedural
20	Low ambient temperature	Potential freezing of accumulated water or solidification of product in line or deadlegs	Blowdown lines sloped to avoid accumulation Elimination of collection points or deadlegs Insulation of process lines	Automatic drainage of potential collection points Automatic injection of chemical to reduce freezing Heat tracing of lines	Written procedures and training to maintain a minimum flow through line Written procedures and training for manual draining of potential collection points Written procedures and training for manual injection of chemical to reduce freezing
21	Condensation of steam in cold weather	Potential accumulation of water resulting in steam hammer and line rupture	Securely anchor piping	Heat tracing of lines Install condensate / steam traps to control condensate in steam header	Written procedures and training to slowly warm up downstream piping
22	Excessive thermal stress	Potential loss of containment	Additional support to prevent sagging Expansion loops and joints Insulation of pipe expansion joints		
Equipment Failure					
23	Gasket leak	Potential loss of containment	Double-walled pipe Maximize use of all-welded pipe Minimize use of unnecessary fittings	Ensure proper gasket material is specified and used	Written procedures and training for periodic inspection for leaks

Table 6.10   Common Failure Scenarios and Design Solutions for Piping and Piping Components

No.	Event	Consequence	Potential Design Solutions		
			Inherently Safer / Passive	Active	Procedural
24	Flange leak	Potential loss of containment	Avoid use of underground piping Double-walled pipe Maximize use of all-welded pipe Minimize use of unnecessary fittings Physical collision barriers Proper design and location of piping supports Shielding at flanges to prevent operator exposure	Automated leak detection with shutoff	Procedural restrictions to avoid damage (crane restrictions, climbing restrictions) Written procedures and training for periodic inspection for leaks
25	Valve leak	Potential loss of containment	Proper design and selection of valves	Fusible link valves for automatic closure under fire conditions	Procedural restrictions to avoid damage (crane restrictions, climbing restrictions) Written procedures and training for periodic inspection for leaks
26	Transfer hose leak	Potential loss of containment	Eliminate hose connections (hard piped) Higher integrity hose (e.g., metallic braided) Hose with higher pressure rating	Excess flow check valve upstream and check valve downstream of hose Emergency Isolation Valves (EIVs) installed on both ends of hose	Written procedures and training to pressure test transfer hose before use Written procedures and training for periodic replacement of hoses, gaskets, and o-rings
27	Breakdown of pipe / hose lining	Potential loss of containment	Pipe metallurgy which does not require lining Semi-conductive liner to reduce degradation due to static buildup Thicker liner material Flow limited to avoid static pinholing		Written procedures and training for periodic thickness testing of metal pipe wall Written procedures and training for periodic process stream analysis for metals content

6. EQUIPMENT DESIGN

Table 6.10 Common Failure Scenarios and Design Solutions for Piping and Piping Components

No.	Event	Consequence	Potential Design Solutions		
			Inherently Safer / Passive	**Active**	**Procedural**
28	Corrosion under insulation or external corrosion	Potential loss of containment	Coating and insulation designed to minimize corrosion under insulation. Piping materials upgraded to address potential external corrosion issues		Written procedures and training for periodic thickness testing of metal pipe wall
29	Deadleg line	Potential loss of containment	Deadlegs removed	Heat trace deadleg	Written procedures and training for periodic thickness testing of metal pipe wall. Written procedures and training for identification of deadlegs
30	Loss of cathodic protection for buried lines	Potential increased corrosion resulting in loss of containment	Lines located above ground		Written procedures and training for periodic thickness testing of metal pipe wall. Monitoring of cathodic protection and alarm on detected fault in system
31	Mix point	Potential increased corrosion resulting in loss of containment	Mix points designed to avoid turbulence. Metallurgy upgraded		Written procedures and training for operator inspection of mix points
32	Injection point	Potential increased corrosion resulting in loss of containment	Design injection point to minimize stress / fatigue		Written procedures and training for operator inspection of injection points

6.10.2.1 Blockage of the Relief Path (Scenarios 7 and 10)

Process systems that can be overpressured must never be isolated from adequate overpressure protection. The inherently safer design alternative to providing individual isolation valves at the inlet / outlet points of safety relief devices is to provide a parallel relief path. A parallel relief path uses redundant safety relief devices and a three-way valve, thus ensuring that one relief path is always open. Note that flame arresters located in the relief path may also be a source of blockage, particularly if the process fluid is fouling, or can solidify or polymerize.

6.10.2.2 Deflagration to Detonation Transition (Scenario 2)

Pipelines containing flammable mixtures either normally or under upset conditions may need to be equipped with devices to limit the consequences of an ignition. Where pipelines connect large items of process or storage equipment together it is most important to prevent flame spread via the connecting pipe. The deflagration flame initially produced by an ignition source generally increases in speed as it travels through a pipeline; flame acceleration is enhanced by turbulence promoters such as tees, elbows, and other flow restrictions. After some distance of travel, Deflagration to Detonation Transition (DDT) may occur. This is marked by a sudden increase in flame speed and pressure. As flame speed increases it becomes more difficult to arrest flames; for fast flames and detonations, special flame arresting devices are required. The overall mitigation strategy is highly dependent on the circumstances and should be considered at the earliest possible design stage.

Avoidance of flammable mixtures by design and control is an inherently safer option, often used in conjunction with flame arresting devices. Flammable mixture control is usually achieved by operating below the limiting oxygen concentration (LOC) or the Lower Flammable Limit (LFL) as described in NFPA 69 (Ref. 6-51). Operation above the Upper Flammable Limit (UFL) using an enrichment gas such as methane can offer advantages in some situations such as vapor control systems. Operation below the LFL might be the safest of these strategies where air could leak into a system (for example, at a blower intake), increasing the oxygen concentration. It is important to consider the effects of startup, shutdown and credible upset conditions during which flammable mixtures are produced. If flammable operation cannot be discounted, flame arresting devices should be incorporated.

Devices for gas systems include liquid seals, deflagration and detonation arresters, suppression systems, and fast-acting valves. The first three are the most common. Deflagration flame arresters can only be used under specific circumstances such as at the end of an atmospheric vent line, where DDT on the unprotected side cannot occur. Flame arresters situated in-line must generally be detonation arrester types certified for the actual conditions of use. These devices have pros and cons in terms of installation cost, effectiveness (e.g., risk of failure under upset conditions), and operability (e.g., back pressure, instrumentation, and maintenance needs) which should be considered before the process design is finalized (Ref. 6-52).

6.10.2.3 Loss of Containment (Scenarios 23-30)

Piping and piping components are the most common single sources of flammable and toxic materials release. The Institution of Chemical Engineers reports that 40% of losses are due to pipework failure. Several codes have been established for the design of piping and piping components (Ref. 6-50). To reduce the probability of releases, minimize the use of fittings on lines and glass rotameters and eliminate gauges when practical. For hazardous service, minimize flanges by welding pipes together and do not use threaded fittings. Where flanges are required for maintenance and inspection, proper selection of flanges and gaskets can reduce the risk of leaks.

6. EQUIPMENT DESIGN 277

6.10.2.4 Thermal Stresses (Scenario 22)

Careful attention must be paid to pipe support and flexibility to account for thermal expansion. Designs must address expansion or contraction due to thermal stresses, and also take into account requirements for steam purging, hydro-testing, startup, shutdown, cyclic conditions, etc. Piping flexibility must be provided by the proper design of anchors, supports, and expansion joints. Expansion joints themselves are prone to erosion and cracking.

6.10.3 Design Considerations

6.10.3.1 Piping Specification

Most companies and engineering design firms have detailed piping specifications for use on projects. These piping specifications include:

- Process fluids / materials (influence materials of construction, gaskets, joint design, sealing materials, etc.)
- Ranges of temperatures and pressures (influence line flange class, pipe wall thickness, materials of construction, gaskets, sealing material, piping flexibility, etc.)
- Flow conditions or criteria such as two-phase flow, high pressure drop valves (for noise and vibration considerations), corrosive or erosive fluid properties, or high velocity situations
- Special valving needs (such as plug or vee-ball and VOC emission control valves)
- Fittings, gaskets, fasteners (bolts and nuts)

For a new project, an experienced process engineer should review the process flow diagram with the piping and material specialist in order to address as many considerations as possible prior to development of the detailed design. The piping material specialist can then specify piping details within the piping specifications through detailed commodity codes. This allows such things as special gaskets, seal or trim materials, special pipe bends, or branch connections to be defined.

The yield strength of certain metals used in piping can decrease as the temperature is increased. Many engineers do not have a good understanding of the vulnerability of un-insulated piping to exposure to fire conditions. Failure of piping impacted by flames can occur in about 20 minutes.

A line list is typically developed that contains all system operating conditions and combinations of conditions, such as normal, startup, shutdown, standby, abnormal / upset, emergency, and test must be taken into account. Some systems may have several different modes of operation and could be exposed to different conditions, depending upon system configuration and the phase of plant operation. Operating transients, such as pressure surges or thermal stresses, may be created during startup, shutdown, or reconfiguration.

6.10.3.2 Velocity Criteria

Process and utilities piping are usually sized on the basis of economic criteria (optimum velocity and pressure drop). However, quite often, velocity limitations have to be imposed in order to avoid hazards which could occur because of the following conditions:

- Corrosion
- Erosion
- Vibration
- Noise
- Hydraulic hammer
- Static electricity

6.10.3.3 Valves

The code requirements for valves include ANSI / ASME B16.34 (Ref. 6-53), B16.5 (Ref. 6-54), and MSS standards (Ref. 6-55).

The key to safe valve selection and installation lies in the generic specifications written for the plant, with specific requirements created only for well-defined purposes. The factors that need to be addressed in creating these specifications are discussed below.

- The service that the valve will perform (on / off, throttling, back-flow prevention, etc.), including the pressure drop and the amount of permissible leakage though the valve, will determine the type of valve (gate, ball, diaphragm, etc.) that can be used.
- The need to be able to visually determine the operating position (open / closed) of the valve is often a factor.
- The process fluid conditions the valve must accommodate [chemicals, material phases (including solids), temperature, pressure, and flow rate] will determine the pressure and temperature class, end connection type, and the materials of construction for the valve body, internals, seat, trim, and seals / gaskets. Consideration of corrosion / erosion and temperature stress will be part of the determination.

Regulatory limits on vapor leakage from valves will determine the stem packing requirements. For materials with little or no vapor pressure the standard compressible rope packings can be used. Vapor leakage may be addressed by providing a stuffing box and stem or flexible graphite packing. Backseating the valves will relieve the load on the packing. When complete elimination of packing is required, bellows seal-type valves may be specified.

Valves for normal and emergency operations should have access from grade, particularly if the valve is needed for emergency isolation. Emergency isolation valves should not be located in pipe racks. See Section 7.4.1 for more information.

Check valves are used to prevent reverse flow, such as flow into a plant from storage vessels, reverse flow through a pump, and reverse flow from a reactor. Check valves are selected with consideration of service. Options include ball, piston, spring-loaded wafer, swing, tilting disc, and intrinsically damped. Check valves have had poor reliability and performance issues. Hazardous services (where backflow can create a hazardous situation) should not depend totally on a check valve. Some positive backflow prevention device would then be required, such as instrumented backflow prevention (e.g., tight shutoff control valves or knife-gate valves).

Control valves may fail in-place, fail open, or fail closed. Failure position should be carefully chosen during the design process to ensure a system is taken to a safe state upon failure.

The term "pipe support" is used generically to encompass a whole range of integral and non-integral pipe attachments, variable and constant spring hangers, sliding supports, rod hangers, shock suppressors, vibration dampeners, anchors, pipe support frames, etc. The purpose of pipe supports is to transmit the loads acting on piping systems to building structures or other structures. The designer should also consider the requirements for flexibility in special conditions:
- Steam purging, which may differ from standard operating conditions
- Hydrotesting
- Startup, when temperature may be higher than the operating temperature
- Startup, when attached equipment is cold
- Shutdown
- Cyclic conditions
- Process excursions
- Steam tracing
- Reactive force (recoil) of discharge on vessels
- Reactive forces of relief devices

6.10.3.4 Thermal Expansion

Equipment or pipelines which are full of liquid under no-flow conditions are subject to hydraulic expansion due to increase in temperature and, therefore, require overpressure protection. Sources of heat that cause this thermal expansion are solar radiation, heat tracing, heating coils, heat transfer from the atmosphere, or other equipment. Another cause of overpressure is a heat exchanger blocked in on the cold side while the flow continues on the hot side. Cryogenic systems are particularly vulnerable to such failures.

6.10.3.5 Flanges

Flanges are used to join sections of pipe or connect valves to piping. There are many types and design of flange connections. Of particular concern is the use of long bolts (bolts longer than 3 inches). Long bolts can receive direct flame impingement and expand when exposed to heat. This allows the flanges to leak and feed the fire. Welded pipe joints are preferred; however, standard flange joints should be used before long bolt flanges.

6.10.3.6 Expansion Joints

Flexibility may be provided by including in the piping system mechanical devices specifically designed to absorb expansion-induced piping movements through deformation of their components.

Use of expansion loops is common. The design of an expansion joint can be affected by changes in temperature and in pressure. Expansion and flexible joints are designed for a finite number of cycles, after which fatigue failure becomes probable. Flexible expansion joints are difficult to test and inspect.

It should be noted that the expansion joints should only be considered as the last resort, when all attempts to attain adequate piping flexibility through layout modifications have failed. In such cases, close monitoring of the conditions of the joints must be performed. The concerns with regard to expansion joints are:

- Expansion joints tend to develop cracks when used to absorb large lateral deflections.
- They require additional anchors and guides in controlling thermal movements.
- Due to erosion concerns, expansion joints should not be used in streams with high levels of particulates, although liner sleeves can mitigate this problem.
- For expansion joints handling hazardous materials, double-layer expansion joints with interspatial monitoring should be considered.

6.10.3.7 Vibration

Vibration may cause stresses in a component due to displacement resulting in failure of the component. In addition, vibrations can be transmitted to other equipment and structures. Vibration of piping and components can be classified as either steady state or transient. Transient vibration can be caused by water hammer, earthquake, slug flow, or relief valve thrust forces. Steady state vibration can be caused by pressure pulsations from mechanical equipment subject to pulsating flow, such as reciprocating compressors and pumps, valve chattering, or turbulent flow conditions.

In cases where vibration is present, a stress analysis should be performed to evaluate the impact of vibration on system life. Stress analysis is the calculation of the stress in a component and the comparison to a safe limiting value. The limiting value will be related to time or frequency and is dependent upon the properties of the material. One of the more significant methods of indicating a property of a material is the design fatigue endurance curve. Simplified, the endurance curve indicates failure limits (stress values) based on cycles. Higher cycles require lower stress values; in other words, high stress values result in reduced cyclic life.

6.10.3.8 Heat Tracing

When heat tracing systems are used, they are subject to degradation and malfunction. This can lead to localized overheating and insulation fires. Electric tracing should have over-current and over-temperature protection designed into the system.

6.10.3.9 Special Cases

Some chemicals or situations require unique piping systems. Special attention is devoted to minimizing leaks (especially at piping connections and valves) and avoiding ignition. Pipe stress is not generally affected by specific chemicals. Temperature and pressure requirements of specific chemicals, however, may influence choice of materials of construction. This determination is usually made by a metallurgist. Industry organizations such as the Chlorine Institute and the International Institute of Ammonia Refrigeration (IIAR) provide detailed information on considerations for design, construction, inspection, etc., associated with the special chemicals involved. These include:

- Hydrogen fluoride

6. EQUIPMENT DESIGN 281

- Ammonia
- Oxygen and oxygen-enriched atmospheres
- Chlorine
- Phosgene and other toxic chemicals
- Hydrogen

Considerations could include:
- Chemical compatibility (internal corrosion resistance, corrosion rates, and years of remaining life considerations).
- Gasket systems (chemical resistance, performance limitations, and useful life).
- Materials of construction vs. service. For example:
  - Stainless steel is good in some low temperature services, but subject to chloride stress cracking, making it less suitable for chlorine liquefaction.
  - Carbon steel is a good material for caustic solutions such as sodium hydroxide and potassium hydroxide at relatively low temperatures, but subject to caustic stress cracking as low as 140°F (60°C) depending concentration.

Special cases often require careful consideration of operating conditions outside of "normal" that still can be expected to be encountered. For example, evacuation of a pipeline containing a liquefied gas can get much colder than "normal" requiring a material selection that would be different than selected if evacuation is not considered.

6.10.3.10 Thermoplastic, Plastic-Lined, and FRP Piping

Thermoplastic, plastic-lined, and FRP piping are widely used in the chemical process industries to handle corrosive chemicals (acids and alkalis) as well as hydrocarbons and organic chemicals. This section discusses potential problems with these types of piping and some recommendations on how to eliminate or minimize these problems.

6.10.3.10.1 Thermoplastic Piping

Special materials, such as thermoplastics, should be limited in use to situations where temperature and pressure extremes are not encountered. The use of nonmetallic piping requires consideration of:

- Temperature
  - Do not locate in areas of high or low temperature extremes.
  - Techniques for applying adhesive and joint makeup are affected by temperature.
  - For flammable fluid designs, FRP pipe, but not fittings, may be approved.
- Pressure
  - Prevent pressure surges.
  - Provide vacuum and overpressure relief.
  - Do not use for above ground compressed air.
- Other considerations
  - Isolate from vibrating equipment.

- Protect from sunlight (ultraviolet radiation effect).
- Only a limited number of standards have been developed for design and / or examination.
- Piping constructed of non-metallic materials may require more support; this requires input to and from other design groups.
- Installation may also require special preparation and handling to prevent damage.
- Special joints, connectors and adhesives may be required.

Mruk (Ref. 6-56) discusses the design, application and installation of thermoplastic piping. Secondary containment is also available in fiberglass and thermoplastic systems (Ref. 6-57).

6.10.3.10.2 Plastic-Lined Pipe

Use of plastic-lined pipe requires consideration of these issues:
- Vacuum.
- Installation / joining techniques.
- Fire protection.
- Non Destructive Examination (NDE), such as visual, liquid penetrant, and leak testing.

Two potential problems with plastic-lined pipe that could lead to fire and explosions are:
- They may leak badly at flanges and permeation vents.
- Flange gaskets may not survive a fire.

However, an available connection system for plastic-lined pipe may solve these potential hazards. It is a "high-integrity flange" which confines and directs the permeation vent. It also has a fire-safe metal seated backup to the flange gasket.

If the lined pipe is operating under vacuum conditions, the liner may be pulled away from the outer pipe if the lined pipe is not rated for the vacuum level. This could result in the pipe becoming plugged and overpressuring the upstream equipment. This occurs more frequently with large-size piping. Vacuum ratings for lined pipe are available from most manufacturers.

6.10.3.10.3 FRP Piping

Design practices for FRP piping include:
- Keep the pipe away from high traffic areas where damage from vehicles and equipment impact is likely.
- Keep flange joints to a minimum. Flanges are expensive components and sources of leaks.
- Provide vents at each high point to allow air to be removed from the systems prior to testing and system startup.
- Provide drains at each low point or pocket. Drains with blind flanges will allow the line to be drained if repairs are necessary.

6. EQUIPMENT DESIGN

- Ensure that all supports, anchors, and guides are installed prior to hydrostatic testing. This cannot be over-emphasized since the pipe system can be severely damaged without proper pipe support.
- All valves, valve operators, and other components in the system must be independently supported.
- Valves that require high torques to open and close should be anchored so that the high torque does not damage the pipe.
- Riser supports for vertical runs should be guided or laterally restrained to reduce vibration and effects of wind load. Unnecessary loading in vertical runs should be avoided. Support should be provided to vertical runs in compression, where possible.
- Avoid point loading.
- Provide the minimum support width-bearing stress <85 psi.
- Avoid unnecessary bending.

6.10.3.11 Piping for Two-Phase Flow

The following types of two-phase flows occur often in process plants:
- Gas / vapor-liquid mixtures.
- Solids-liquid mixtures.
- Solids-gas mixtures.

The following sections discuss hazards encountered in two-phase flow and recommended practices to eliminate or minimize these hazards.

6.10.3.11.1 Gas / Vapor-Liquid Mixtures

When designing a process with a gas / vapor-liquid mixture concern, the logical approach is to select a flow regime which provides a stable flow and a minimum pressure drop. It is critical to operate in the flow regime that is least hazardous, e.g., to avoid vibration or other flow phenomena that could result in damage to the piping.

Slug flow can occur at every change of flow direction and may result in increased corrosion / erosion. If a slug flow enters a distillation column, the alternating composition and density of the gas and liquid slugs can cause cycling of composition and pressure gradients in the column. The cycling causes problems with product quality and process control, which in turn could cause damage to connecting equipment. To avoid operating in the slug flow regime, flow regime maps are used that show in which regime the flow is occurring.

Quite often, gas / vapor-liquid mixtures are saturated (at or near their boiling points), e.g., the feed to a distillation column. There usually is also a control valve in the feed line to control the flow rate to the column. It is good practice to locate the control valve at the column feed nozzle so that there is no flashing in the feed piping itself until the feed nozzle is reached. This avoids vibration in the feed piping which, otherwise, could cause damage or failure of piping.

6.10.3.11.2 Solids-Liquid Mixtures (Slurries)

Because of the nature of slurries, proper layout, mechanical design, and selection of piping components (elbows, tees, valves, etc.) are critical to prevent operational problems (which could result in safety hazards), such as piping plugging and erosion.

Piping plugging can be avoided or minimized by:
- Selecting a flow velocity greater than the deposition velocity, but not so high as to cause erosion. Operating experience has shown that for in-plant piping (not transmission pipelines), the metal loss due to erosion (abrasion) is insignificant if the flow velocity is less than about 10 ft/s (3 m/s).
- Minimizing dead-ends and using vertical takeoffs.
- Configuring all valves such that their mechanism is not jammed by solids (e.g., diaphragm valves without an internal weir, pinch valves, and full port ball valves with cavity filler).
- Providing flushing connections for valves.
- Sloping slurry lines towards the receiving equipment to prevent plugging.
- Providing cleanout connections and drains for easy maintenance and cleaning of slurry lines.

The following considerations can minimize corrosion:
- Limit flow velocities to less than 10 ft/s in in-plant slurry lines.
- Use gentle pipe bends (called sweeps) instead of elbows (even long-radius ones). A bend radius-to-pipe diameter ratio of 3-5 is recommended (based on industrial experience). Consider the use of a more corrosion- or erosion-resistant material for bends.
- Thicker walled pipe should be used when slurries are known or suspected to be highly erosive. Many companies use Schedule 80 or Schedule 160 carbon steel pipe and Schedule 20 or Schedule 40 stainless steel pipe, depending on the erosivity of the slurry and desired piping life. Piping with special abrasion-resistant coating should also be considered.

6.10.3.11.3 Solids-Gas Mixtures

Pneumatic conveying systems for solids-gas mixtures have a high risk for fires and explosions (they usually occur in downstream equipment) for the following reasons:
- Static electricity is generated by contact between solid particles themselves and between particles and the pipewall.
- Dust concentration within the explosible range can arise at the delivery point where the solids are separated from the conveying gas (e.g., silos, cyclones, baghouses).
- Heated particles that are created during grinding or drying may be carried into the pneumatic conveying system and fanned to a glow by the high gas velocity. These particles can then cause an ignition in the storage or collection system at the end of the pneumatic conveyor. Tramp metal in pneumatic conveying systems may also cause frictional heating or sparks as it passes through the system.

6. EQUIPMENT DESIGN

- Segments of conveying piping or tubing can become electrically isolated and sparking is possible between conveying line segments and nearby conductive pieces at different potentials.
- Charged powder can leak from joints to the atmosphere and electrostatic sparking can occur, resulting in an explosion.

Guidelines for Safe Handling of Powder and Bulk Solids (Ref. 6-58) discusses design and operating recommendations and practices for pneumatic conveying systems that should be considered to avoid or minimize fires and explosions.

6.10.4 References

The editions that were in effect when these *Guidelines* were written are indicated below. Because standards and codes are subject to revision, users are encouraged to apply only the most recent edition.

6-49. Lees, F.P. *Loss Prevention in the Process Industries*, Third Edition. Elsevier, Inc. Oxford, UK. 2005.

6-50. ASME B31. *Standards of Pressure Piping*. American Society of Mechanical Engineers. New York, New York. 2010. www.asme.org

6-51. NFPA 69. *Standard of Explosion Prevention Systems*. National Fire Protection Association. Quincy, Massachusetts. 2008.

6-52. CCPS. *Deflagration and Detonation Flame Arresters*. Center for Chemical Process Safety of the American Institute of Chemical Engineers. New York, New York. 2002.

6-53. ANSI / ASME B16.34. *Valves--Flanged, Threaded, and Welding End*. American National Standards Institute and American Society of Mechanical Engineers. New York, New York. 1996.

6-54. ANSI / ASME B16.5. *Pipe Flanges and Flanged Fittings*. American National Standards Institute and American Society of Mechanical Engineers. New York, New York. 2009.

6-55. MSS SP-6-2007. *Standard Finishes for Contact Faces of Pipe Flanges and Connecting-End Flanges of Valves and Fittings*. Manufacturers Standardization Society of the Valve and Fittings Industry. Vienna, Virginia. 2007.

6-56. Mruk, S.A. *Thermoplastic Piping*. Chapter D1 in *Piping Handbook,* 6th Edition, M.L. Nayyar, editor. McGraw-Hill Book Company. New York, New York. 1992.

6-57. McCallion, J. *Secondary Containment Takes Off*. Chemical Processing. Pp. 33-38. March 1990.

6-58. CCPS. *Guidelines for Safe Handling of Powder and Bulk Solids*. Center for Chemical Process Safety of the American Institute of Chemical Engineers. New York, New York. 2005.

6.10.4.1 Suggested Additional Reading

ANSI / API STD 603. *Corrosion- Resistant, Bolted Bonnet Gate Valves- Flanged and Butt-Welding Ends, Seventh Edition.* American Petroleum Institute. Washington, D.C. 2007.

ANSI / API STD 607. *Fire Test for Soft-Seated Quarter-Turn Valves, 5th Edition, Includes Errata.* American Petroleum Institute, Washington, D.C. 2005.

ANSI / API STD 609. *Butterfly Valves: Double Flanged, Lug- and Wafer-Type, 7th Edition.* American Petroleum Institute. Washington, D.C. 2009.

ANSI / API STD 610. *Centrifugal Pumps for Petroleum, Petrochemical and Natural Gas Industries, 10th Edition.* American Petroleum Institute. Washington, D.C. 2004.

ANSI / API STD 618. *Reciprocating Compressors for Petroleum, Chemical and Gas Industry Services, 5th Edition, Includes Errata 1 and 2 (2009 and 2010).* American Petroleum Institute, Washington, D.C. 2007.

ANSI / ASME B16.10. *Face-to-Face and End-to-End Dimensions of Valves, 2009 Edition.* American National Standards Institute and American Society of Mechanical Engineers. New York, New York. 1986.

ANSI / ASME B16.20. *Metallic Gaskets for Pipe Flanges: Ring-Joint, Spiral Wound and Jacketed.* American Society of Mechanical Engineers. New York, New York. 2007.

ANSI / ASME B16.24. *Bronze Pipe Flanges and Flanged Fittings, Class 150 & 300.* American National Standards Institute and American Society of Mechanical Engineers. New York, New York. 1998.

ANSI / ASME B16.25. *Buttwelding Ends.* American National Standards Institute and American Society of Mechanical Engineers. New York, New York. 1997.

ANSI / ASME B16.36. *Orifice Flanges.* American National Standards Institute and American Society of Mechanical Engineers. New York, New York. 1996.

ANSI / ASME B16.47. *Large Diameter Steel Flanges.* American National Standards Institute and American Society of Mechanical Engineers. New York, New York. 1996.

ANSI / ASME B36.10M. *Welded and Seamless Wrought Steel Pipe.* American National Standards Institute and American Society of Mechanical Engineers. New York, New York. 2010.

ANSI / ASME B36.19M. *Stainless Steel Pipe.* American National Standards Institute and American Society of Mechanical Engineers. New York, New York. 2010.

ANSI / AWWA. C500-09. *Metal-Seated Gate Valves for Water Supply Service.* American National Standards Institute. New York; New York; American Water Works Association. Denver, Colorado. 2009.

ANSI / AWWA. C504-00. *Rubber-Seated Butterfly Valves.* American National Standards Institute. New York, New York; American Water Works Association. Denver, Colorado. 2001.

ANSI / AWWA. C507-05. *Ball Valves, 6 in. through 48 inches.* American National Standards Institute. New York, New York; American Water Works Association. Denver, Colorado. 2005.

API 941. *Steels for Hydrogen Service at Elevated Temperatures and Pressures in Petroleum Refineries and Petrochemical Plants, 5th Edition.* American Petroleum Institute. Washington, D.C. 1997.

API SPEC 6FC. *Specification for Fire Test for Valves with Automatic Backseats, 4th Edition.* American Petroleum Institute. Washington, D.C. 2009.

API STD 594. *Check Valves: Flanged, Lug, Wafer and Butt-welding.* American Petroleum Institute. Washington, D.C. 2004.

API STD 600. *Steel Gate Valves - Flanged or Butt-Welding Ends, Bolted Bonnets, 12th.* American Petroleum Institute. Washington, D.C. 2009.

API STD 602. *Steel Gate, Globe and Check Valves for Sizes DN100 and Smaller for the Petroleum and Natural Gas Industries, 9th Edition.* American Petroleum Institute. Washington, D.C. 2009.

ASME. *Boiler and Pressure Vessel Code, Section 8, Division 1.* American Society of Mechanical Engineers. New York, New York. 2007.

ASTM A105 / A105M-10. *Standard Specification for Carbon Steel Forgings for Piping Applications.* American Society for Testing and Materials. Philadelphia, Pennsylvania. 2010.

ASTM A182 / A182M-10. *Standard Specification for Forged or Rolled Alloy and Stainless Steel Pipe Flanges, Forged Fittings and Valves and Parts for High-Temperature Service.* American Society for Testing and Materials. Philadelphia, Pennsylvania. 2010.

ASTM G-88-05. *Standard Guide for Designing Systems for Oxygen Service.* American Society for Testing and Materials. Philadelphia, Pennsylvania. 2005.

Beard, C. S. *Final Control Elements: Valves and Actuators.* Chilton Co. Philadelphia, Pennsylvania. 1969.

Branan, C. R. *Rules of Thumb for Chemical Engineers*, 4th Edition. Gulf Professional Publishing, Elsevier, Oxford, UK. 2005.

CGA, *Accident Prevention in Oxygen-Rich and Oxygen-Deficient Atmospheres.* P-14. Compressed Gas Association, Inc. Arlington, Virginia.1992.

CGA, *Oxygen Compressor Installation Guide, Third Guide.* Publication G-4.6. Compressed Gas Association, Inc. Arlington, Virginia. 2008.

CGA, *Oxygen, 10th Edition.* Publication G-4. Compressed Gas Association, Inc. Arlington, Virginia. 2008.

CGA, *Safe Handling of Cryogenic Liquids, 4th Edition.* Publication P-12. Compressed Gas Association, Inc. Arlington, Virginia. 2005.

CGA, *Acetylene, 12th Edition.* Publication G-1. Compressed Gas Association, Inc. Arlington, Virginia. 2009.

CGA, *Cleaning Equipment for Oxygen Service, 6th Edition.* Publication G-4.1. Compressed Gas Association, Inc. Arlington, Virginia. 2009.

CGA, *Oxy-Fuel Hose Line Flashback Arrestors, 5th Edition*. Technical Bulletin TB-3. Compressed Gas Association, Inc. Arlington, Virginia. 2008.

CGA, *Standard for Hydrogen Piping at Consumer Locations, 4th Edition*. Publication G-5.4. Compressed Gas Association, Inc. Arlington, Virginia. 2010.

CGA, *Industrial Practices for Gaseous Oxygen Transmission and Distribution Piping Systems*. Publication G-4.4. Compressed Gas Association, Inc. Arlington, Virginia. 2003.

CGA HB, *Handbook of Compressed Gases, 4th Edition*. Compressed Gas Association. Van Nostrand Reinhold. New York, New York. 1999.

Chlorine Institute. *Chlorine Pipelines,* 6th Edition. Chlorine Institute. Washington, D.C. 2007.

Chlorine Institute. *Piping Systems for Dry Chlorine*. Pamphlet No. 6. Chlorine Institute, Washington, D.C. 1989.

Coker, A.K. *Ludwigs Applied Process Design for Chemical and Petrochemical Plants*, Volume 1, 4th Edition. Elsevier, Oxford, UK. 2007.

Danielson, G.L. *Handling Chlorine-Part 1, Tank Car Quantities*. Chemical Engineering Progress. 60(9)86. 1964.

EJMA (Expansion Joint Manufacturers Association, Inc.). *9th Edition EJMA Standards (Included one copy of Practical Guide)*. White Plains, New York. 2008.

Grossel, S.S. *Improved Design for Slurry Piping*. Chemical Engineering Progress. pp. 114-117. April 1998.

Helguero, V. *Piping Stress Handbook*, 2nd Edition. Gulf Publishing Co. Houston, Texas. 1986.

Kannappan, S. *Introduction to Pipe Stress Analysis*. Wiley. New York, New York. 1986.

Lyons, J.L. *Encyclopedia of Valves*. Van Nostrand Reinhold. New York, New York. 1975.

Mallison, J.H. *Corrosion-Resistant Plastic Composites in Chemical Plant Design*. Marcel Dekker, Inc. New York, New York. 1988.

MSS SP-43-2008. *Wrought and Fabricated Butt-Welding Fittings for Low Pressure, Corrosion-Resistant Applications*. Manufacturers Standardization Society of the Valve and Fittings Industry. Vienna, Virginia. 2008.

MSS SP-53-1999. *Quality Standard for Steel Castings and Forgings for Valves, Flanges and Fittings and Other Piping Components Magnetic Particle Exam Method*. Manufacturers Standardization Society of the Valve and Fittings Industry. Vienna, Virginia. 1999.

MSS SP-67-2002a. *Butterfly Valves*. Manufacturers Standardization Society of the Valve and Fittings Industry. Vienna, Virginia. 2004.

MSS SP-70-2006. *Gray Iron Gate Valves, Flanged and Threaded Ends*. Manufacturers Standardization Society of the Valve and Fittings Industry. Vienna, Virginia. 2006.

MSS SP-71-2005. *Gray Iron Swing Check Valves, Flanged and Threaded Ends*. Manufacturers Standardization Society of the Valve and Fittings Industry. Vienna, Virginia. 2005.

MSS SP-72-2010. *Ball Valves with Flanged or Butt-Welding Ends for General Service*. Manufacturers Standardization Society of the Valve and Fittings Industry. Vienna, Virginia. 2010.

MSS SP-85-2002. *Cast Iron Globe & Angle Valves, Flanged and Threaded Ends*. Manufacturers Standardization Society of the Valve and Fittings Industry. Vienna, Virginia. 2002.

MSS SP-88-2010. *Diaphragm Valves*. Manufacturers Standardization Society of the Valve and Fittings Industry. Vienna, Virginia. 2010.

MSS-SP-58-2009. *Pipe Hangers and Supports - Materials, Design, Manufacture, Selection, Application and Installation*. Manufacturers Standardization Society of the Valve and Fittings Industry. Vienna, Virginia. 2009.

Nelson, H.P. Handling Chlorine-Part 2, Barge and Pipeline Safety. *Chemical Engineering Progress,* Vol. 60, No.9, p. 88. 1964.

Smith, P.R., and Van Laan T.J, *Piping and Pipe Support Systems.* McGraw-Hill. New York, New York. 1987.

Sweitzer, R.A. *Handbook of Corrosion Resistant Piping.* Robert E. Krieger Publ. Co., Inc., Malabar, Florida. 1985.

The Crane Company. *Flow of Fluids through Valve, Fittings, and Pipe.* Technical paper No. 410. Crane Company, Chicago, Illinois. 1988.

Zappe, R.W. *Valve Selection Handbook.* Gulf Publishing Co. Houston, Texas. 1981.

6.11 MATERIAL HANDLING AND WAREHOUSING

This section presents potential failure mechanisms for material handling and warehousing equipment and suggests design alternatives for reducing the risks associated with such failures. The types of equipment covered in this section include:
- Loading / Unloading
- Drumming
- Warehouses

This section presents only those failure modes that are unique to material handling and warehousing equipment. Some of the generic failure scenarios pertaining to other equipment may also be applicable to material handling and warehousing.

Material handling is a procedure-based task. The importance of procedural safeguards and human factors design play a significant role in hazard reduction.

6.11.1 Past Incidents

This section describes several case histories of incidents involving failure of loading and unloading equipment to reinforce the need for the safe design practices presented in this section.

6.11.1.1 Tank Truck Catastrophically Fails

The chemical process operator connected a 3-inch unloading hose to the truck and to the plant's unloading pump. Next, the operator connected a 3/4-inch nitrogen hose from the supply station to a manifold that was just forward of the rear wheels on the trailer. Someone modified the truck's nitrogen padding system and constructed a manifold on the truck that allowed the nitrogen hose connection to be made without climbing the ladder on the truck's tank. It was just a manifold of 30 Feet or less of half-inch tubing.

The plant operator opened the nitrogen valves that were both upstream and downstream of the plant's nitrogen pressure regulator to pad the truck. The truck manifold that was in front of the right rear wheels contained a pressure gauge that read the expected 20 psig (1.36 bars). The operator opened the proper liquid delivery valves, started the unloading pump, and left to handle other chores. The driver stayed with the truck, as was the normal procedure.

At this point the unloading activity appeared normal. In less than an hour, the truck driver checked the sight glass on the truck and informed the control room that the truck was empty. The busy operator passed by and observed that the unloading hose was still vibrating, indicating to her that the tank truck was not quite empty.

People in the area reported hearing a loud rumble noise and a number of employees gathered to witness the catastrophic collapse of the tank truck. There were no injuries, no leaks of hazardous materials, and no damage to the plant receiving the feed stock.

Engineers determined that a nitrogen valve was opened to the hose connected to the tank trailer manifold. The nitrogen gas never entered because there was a closed valve atop the tank. It seems the truck driver misunderstood the recent nitrogen supply piping modifications; the open and closed positions on this oval handle ball valve.

Lessons learned include performing a proper Management of Change and providing training to operation and maintenance staff on the changes implemented.

6.11.1.2 Nitric Acid Deliver to the Wrong Tank

About 2,600 gallons of the 4,000-gallon load of nitric acid was off-loaded on top of the 16,500 gallons of sulfuric acid before the chemical reaction was observed by the driver and the plant employees. Commercial grade nitric acid contains 35% water and the sulfuric acid has a strong affinity for the water. The heat of solution caused an exothermic reaction and the once ambient-temperature acid shot up to about 160°F (71°C). Nitrogen dioxide was liberated from the hot nitric acid creating a blue-white cloud drifting 25 Feet in the air. An emergency shelter in place was called after the accident to protect residents in the nearby community.

Lessons learned include proper labeling of unloading connections, providing different types of connectors, or relocating similar connections away from each other.

6.11.1.3 Truck Delivery Incident

A tank truck of caustic soda arrived with the proper delivery papers, correctly labeled, and with the special hose connections. The operators had the mind-set that they were receiving acid and spent considerable time developing an adaptor to enable them to pump the caustic soda into an acid tank.

Lessons learned include proper labeling of unloading connections, providing different types of connectors, or relocating similar connections away from each other.

6.11.1.4 Sulfuric Acid Unloading

The driver hooked up a utility water hose instead of the compressed air hose to pressure his trailer to deliver the acid. An alert operator observed the potentially dangerous situation before the utility station valve was opened. The housekeeping had been poor. The hoses had been left together on the ground. The driver initially thought he had selected the air hose.

The technical employees understood that mixing water and the 98% sulfuric acid could have caused severe trouble, due to the heat of solution. The generation of heat could have caused a pressure buildup that may not have been controllable by a relief device. Even if the vessel did not rupture, sulfuric acid spray could create severe burns to anyone in the vicinity of the unloading operation.

Lessons learned include proper labeling of utility hoses and providing different types of connectors or colored hoses.

6.11.1.5 Siphoning Destroys a Tank

A chemical plant complex used large volumes of brackish water from an adjacent river for once-through cooling. Occasionally, trace emissions of caustic soda were present in the effluent river water, and the company was concerned with pH excursions.

The process engineer designed a system that would regulate enough hydrochloric acid into the effluent to neutralize the caustic traces to form more environmentally acceptable table salt. A lightweight "off-the-shelf" fiberglass tank about 8 Feet in diameter and 8 Feet high was installed along with the associated piping and controls. The purpose of this modification was to receive, store, and meter out acid to control the pH of the effluent. The atmospheric closed top design tank had two top nozzles.

During normal operation the vessel would receive acid via an in-plant pipeline. It was also equipped to receive acid via tank trucks. One of the top nozzles was the fill line and the other was piped to a small vent scrubber to eliminate fumes given off during the filling operation. The intentions to improve the environment were noble, but the simple vent system design possessed an unrecognized flaw that allowed a minor overfill situation to suddenly and completely destroy the vessel.

About a year after the system was put in service, the tank was filled via a tank truck, instead of being supplied by the usual pipeline. As the delivery tank truck was unloaded, the acid level rose in the small storage tank. The company representative wanted to top off the tank. Before the acid truck was unloaded, acid started to overflow and pour through the 6-inch (15-cm) line into the scrubber. The alert truck driver quickly responded. He abruptly shut the delivery valve on his truck. Unexpectedly, the partial vacuum created by the siphoning action of the overflowing liquid exceeded the tank's vacuum rating and the storage tank was totally destroyed.

Lessons learned include performing a proper Management of Change and providing training to operation and maintenance staff on the changes implemented.

6.11.1.6 Warehouse Fire

A warehouse in the UK stored large numbers of metal drums holding bagged pesticides. In order to detect torn bags quickly and easily, holes had been drilled in the bottom of the metal drums. While this helped the housekeeping efforts, it negated the containment function of the drums. The bags melted during a fire and the pesticide contaminated the firewater, creating a considerable environmental problem.

Lessons learned include performing a proper Management of Change and providing training to operation and maintenance staff on the changes implemented.

6.11.2 Failure Scenarios and Design Solutions

Table 6.11 presents information on equipment failure scenarios and associated design solutions specific to loading, unloading, and drumming.

Table 6.11 Common Failure Scenarios and Design Solutions for Loading, Unloading, and Drumming

No.	Event	Consequence	Potential Design Solutions		
			Inherently Safer / Passive	Active	Procedural
Loading and Unloading					
Pressure					
	Generally Applicable – High Pressure (Applicable to all high pressure scenarios)		Piping / hose designed for deadhead pressure of pump Vessel designed for maximum expected supply pressure	PSV on tank truck, railcar, or marine vessel	Operator response to high pressure alarm
	Generally Applicable – Low Pressure (Applicable to all low pressure scenarios)		Receiving tank designed for full vacuum	Vacuum relief device on tank truck, railcar, or marine vessel	Operator response to low pressure alarm

6. EQUIPMENT DESIGN

Table 6.11   Common Failure Scenarios and Design Solutions for Loading, Unloading, and Drumming

No.	Event	Consequence	Potential Design Solutions		
			Inherently Safer / Passive	Active	Procedural
1	Vent line from tank truck, railcar, or marine vessel blocked or not connected	Potential vacuum in tank truck, railcar, or marine vessel, potential loss of containment		Regulated pad pressure maintained to compensate for transfer rate	Written procedures and training to ensure vent line connected properly
2	Use of high pressure inert gas to transfer material from tank truck, railcar, or marine vessel	Potential to overpressure tank truck, railcar, or marine vessel, potential loss of containment	Flow restriction orifice	PSV on inert gas line	Written procedures and training to monitor pressure gauge on tank truck, railcar, or marine vessel
3	Transfer pump deadhead pressure exceeds transfer hose or piping design pressure	Potential loss of containment	Pump deadhead pressure less than design pressure of hose or piping	Pump shutdown on deadhead conditions (e.g., low flow, low amps, low power)	Written procedures and training to ensure proper valve alignment prior to transfer
4	Overfill storage tank or transport vessel	Overpressure resulting in vessel failure or relief valve lifting (especially an issue with liquefied gases)	Containment	Relief valve discharge directed to secondary containment, scrubber, flare, etc., if toxics or flammables involved Reliable level overfill protection system Volume, weight measurement	Written procedures and training identifying maximum fill volumes / weights (include loading temperature and pressure considerations when appropriate)
5	Low final container pressure	Excess flow devices may not operate		Provisions for inert gas padding	Written procedures and training for identifying minimum final pressure for shipment
Flow					
	Generally Applicable – More Flow *(Applicable to all more flow scenarios)*		Orifice restriction	Automatic response to high flow alarm	Operator response to high flow alarm
	Generally Applicable – No / Less Flow *(Applicable to all no / less flow scenarios)*				Operator response to low flow alarm
6	Excessive fill rate	Potential static accumulation, potential fire / explosion	Flow restriction orifice	Inert receiving vessel	

Table 6.11 Common Failure Scenarios and Design Solutions for Loading, Unloading, and Drumming

No.	Event	Consequence	Potential Design Solutions		
			Inherently Safer / Passive	Active	Procedural
7	Aligned to incorrect storage tank / vessel	Potential inadvertent mixing, potential runaway reaction, potential loss of containment		Automated unloading line valves (opening desired valve only when operator gives permission)	Written procedures and training to verify alignment prior to unloading
8	Quantity of material delivered greater than capacity of storage tank / vessel	Potential overfill of storage tank / vessel, potential loss of containment	Storage tank designed larger than delivery vessel	High level alarm with automatic shutoff of transfer	Written procedures and training to verify capacity prior to transfer
9	Manual valve in delivery system closed	Potential to deadhead transfer pump, potential loss of containment	Pump deadhead pressure less than design pressure of hose or piping	Pump shutdown on deadhead conditions (e.g., low flow, low amps, low power)	Written procedures and training to ensure proper valve alignment prior to transfer
Level					
Generally Applicable – High Level (Applicable to all high level scenarios)				High level alarm with automatic shutoff of transfer operation	
Composition					
10	Receipt of wrong material	Potential incompatibility, potential runaway reaction, potential loss of containment		Automatic shutdown of transfer on tank high temperature or high pressure	Written procedures and training to verify bill of lading for proper material prior to transfer. Written procedures and training to sample and test raw materials prior to acceptance
Equipment Failure					

6. EQUIPMENT DESIGN

Table 6.11 Common Failure Scenarios and Design Solutions for Loading, Unloading, and Drumming

No.	Event	Consequence	Potential Design Solutions		
			Inherently Safer / Passive	Active	Procedural
11	Transfer hose leak / rupture	Potential loss of containment		Excess flow valve (upstream of hose) Automatic shutoff at both ends of loading hose / piping Low pressure interlock on supply / filling line	Written procedures and training to ensure inspection and / or hose replacement at proper intervals Procedures to ensure proper visual inspection of hoses prior to use Preloading pressure test / leak check procedures Written procedures and training to ensure proper visual inspection of hoses prior to use
12	Transfer hose / piping failure due to vessel movement	Potential loss of containment		Excess flow valve (upstream of hose) Automatic shutoff at both ends of loading hose / piping Container movement interlocks Low pressure interlock on supply / filling line Breakaway connections with internal isolation devices	Written procedures and training for controlling truck, railcar, vessel positioning, movement, securing, etc.
Drumming (Drums and Bulk Containers)					
Pressure					
Generally Applicable – High Pressure (Applicable to all high pressure scenarios)					Operator response to high pressure alarm
Generally Applicable – Low Pressure (Applicable to all low pressure scenarios)					Operator response to low pressure alarm
13	Design pressure of transfer pump exceeds drum design pressure	Potential to overpressure drum, potential loss of containment	Design pressure of transfer pump does not exceed design pressure of drums or bulk containers	Relief valve set to below drum design pressure	Written procedures and training to ensure vent is open or vapor control system is operating

Table 6.11 Common Failure Scenarios and Design Solutions for Loading, Unloading, and Drumming

No.	Event	Consequence	Potential Design Solutions		
			Inherently Safer / Passive	Active	Procedural
14	Vent bung cap not removed prior to filling	Potential to overpressure drum, potential loss of containment	Drum placed in curbed area or above sump with slotted cover Containment		Written procedures and training to ensure vent bung cap is removed prior to filling
15	Vent system inoperable or clogged	Potential to overpressure drum, potential loss of containment	Containment	Automatic shutdown of transfer if vent permissive is not valid	Written procedures and training to ensure that vent system is operable and not plugged prior to filling
16	Drum not sealed properly	Potential loss of containment	Drum placed in curbed area or above sump with slotted cover Containment		Written procedures and training to ensure drum is sealed properly
Flow					
17	Operator does not stop transfer to drum	Potential to overfill drum, potential loss of containment		Red-eye high level detection automatic shutoff of filling operation	Written procedures and training to monitor drum level while filling
18	Static electric discharge	Potential fire	Electrical grounding and bonding (with permissive for starting fluid transfer) Minimize free-fall liquid (bottom fill or dip tube)	Inerting	
Composition					
19	Contaminants / foreign material in drum	Potential for reaction in drum, potential loss of containment			Written procedures and training to use clean drums and bulk containers

6. EQUIPMENT DESIGN

Table 6.11 Common Failure Scenarios and Design Solutions for Loading, Unloading, and Drumming

No.	Event	Consequence	Potential Design Solutions		
			Inherently Safer / Passive	Active	Procedural
20	Accumulation of explosive / reactive contaminants	Explosion, corrosion, runaway reaction resulting in loss of containment, fire, etc.	Filling, emptying, venting design considerations to prevent accumulation of undesirable contaminants (e.g., inert padding vs. compressor gas padding, etc.) Upstream design considerations to eliminate / reduce trace contaminants, absorption or entrainment of contaminants, etc.	Padding system design (oil-less, dry, inert, etc.) Padding system air intakes located away from potential contaminant releases	Written procedures and training for controlling transfer methods (such as inert padding vs. compressor gas padding) Written procedures and training for measuring undesired contaminates in raw materials, final products, storage vessels

6.11.3 Design Considerations

6.11.3.1 Loading / Unloading

Loading and unloading of raw materials and finished product is a common activity. This can entail gas, liquid, and / or solid handling via open equipment. This may include pumping of liquids from drums or dumping of solids from other containers into an open vessel, shoveling material into a dryer, or making temporary connections such as at hose stations.

Primary concerns include the loss of containment and the potential for exposure of operating personnel to hazardous materials; the potential for other hazards such as fires or explosions; and the ergonomic issues inherent in manipulating large, heavy containers. The first two concerns are of particular significance in batch operations, since operating personnel are often more frequently and more intimately exposed to the batch processes than is typically the case with continuous processes.

Some commonly applied controls include:
- Providing enclosed charging systems, where feasible
- Use of localized ventilation
- Proper selection and use of personal protective equipment
- Use of mechanical assists for handling drums and other containers
- Procedures and training
- Interlocking vessel openings to prevent opening while the vessel is pressurized

Many of the material hazards present are also present during the drumming of materials out of the process. However, there are additional considerations unique to this operation, including the mechanical handling of massive objects, potential for puncture of containers, and loss of liner integrity. Some of the hazards present in the drumming stage have the potential for overpressurization leading to release of chemicals and operator exposure, underpressurization of drums, or uncontrolled reactions occurring after drumming, leading to potential fires or explosions. Special consideration needs to be given to drummed materials that are shock / heat sensitive as well as drummed materials that degrade over time.

Both manual handling and piping system transfers are used for moving hazardous materials onsite. The kind of material handling system used for hazardous materials is dictated by the process and inventory requirements, the type of container in which the hazardous material is received or shipped, and safety considerations.

6.11.3.2 Manual Handling

Due to the hazardous nature of hazardous materials, they are very often handled in smaller quantities and in smaller container sizes than bulk commodity chemicals. For this reason, manual handling of containers ranging from small cartons to drums and tote bins is commonly required. However, manual handling inherently has different incident potentials than piping system transfers. In particular, the close proximity of the operator to the hazardous material and the potential for dropping or collision during manual handling are significantly different considerations than potentials such as dead-head pumping or blocked-in material in piping systems. Therefore, some companies avoid all manual handling of certain highly hazardous materials. Industry practice with respect to manual handling of hazardous materials is discussed in the following paragraphs.

6.11.3.2.1   Pallets with Hand Trucks

Moving pallet-loads of cartons, sacks, or drums is fairly common practice for most types of hazardous materials. For short distances where stacking is not required or in areas in which forklift trucks do not have sufficient room to maneuver, hand trucks or self-propelled but operator-guided pallet movers can be used. They are less expensive and are less hazardous with respect to collision and flammable vapor ignition; generally, they are only practical for single-building use over relatively short distances.

6.11.3.2.2   Pallets with Forklift Trucks

Forklifts are used more commonly than hand trucks, such as for transferring drums or cartons from an unloading dock to warehouse storage, from storage to process area, from a pack-out area to product inventory storage, and / or from storage to a loading dock. It must be ensured that each forklift truck that might be used for handling a given material has the proper safeguards for use with that material, such as its electrical classification for minimizing the likelihood of ignition of leaking or spilled flammable materials.

It is good practice to move only sealed drums or cartons of hazardous materials by forklift. Some companies have a requirement to open and then reclose the bung of each drum (slowly, while using proper personal protective equipment) before moving in order to detect any unusual pressure buildup inside the drum. However, this may be undesirable for some types of hazardous materials such as peroxide formers and for those having significant toxicity and volatility in addition to reactivity.

6. EQUIPMENT DESIGN

6.11.3.2.3 Other Forklift Transfers

Forklift trucks are also used to handle other types of hazardous material shipping or transfer containers. In particular, cylinders, tote bins, super sacks, specially constructed skid-mounted containers, and containers that are moved directly on the forks of a forklift truck are all in use with hazardous materials. One company uses even larger portable tanks with certain pyrophoric materials; they are transferred by attaching the tank lifting lugs to a crane. All transfer containers will, of course, need to meet regulatory requirements for the container design if used for off-site transport of hazardous materials.

6.11.3.2.4 Roller Conveyors

Roller conveyors are used by some facilities for cartons and / or drum transfers over relatively short distances, such as at a drum filling station. Conveyor systems must be carefully designed and the operating procedures thoroughly reviewed to minimize the potential for incidents such as overturning of a full drum or generating sufficient static electricity to create an electrostatic discharge ignition source for flammable vapors or combustible dusts.

6.11.3.2.5 Transfer to Smaller Containers

It may be desirable to transfer a quantity of highly hazardous material into a smaller container for onsite transport when a full container is not needed, in order to reduce the handling of the larger quantity. However, transfer to a smaller container while inside a storage enclosure should be avoided. There is the potential for spillage of the hazardous material, which could lead to an incident involving the total stored quantity.

6.11.3.3 Piping Specifications and Layout

Many more design details are required for an operation involving transfer of hazardous materials with a piping system than for a manual transfer operation. Details on piping specifications and layout details for several specific hazardous materials are given in Section 6.10. The design of piping systems should ensure provisions to isolate, drain, and purge piping segments for routine maintenance.

Several layout details are common to nearly all hazardous chemical liquid piping systems. Both piping lines and headers are nearly always sloped to allow complete drainage of lines, with no dead legs being allowed. Note that drainage back towards a storage tank creates a risk of contaminating the bulk of the stored material. During operation, reverse flow into equipment containing hazardous materials may lead to uncontrolled reactions and severe consequences. Backflow may need to be protected against by installing failsafe controls such as check valves (sometimes two different types in series) with double block and bleed valve systems or duplication of instrumentation. Process layout and piping runs should be configured to minimize the length of piping between units. Piping must be routed to prevent impact by external forces such as trucks or cranes onsite, and barriers should be installed where piping might be vulnerable to such forces. For critical piping installations, a suggested practice is not only to review the piping design as diagrammed schematically on Piping and Instrumentation Diagrams (P&IDs), but also to review, as part of the design process and / or the process hazard analysis, the actual piping layout configuration as shown on a piping arrangement drawing or a three-dimensional model (computer-generated or scale model).

6.11.3.3.1 Fittings and Connections

Piping system failures nearly always occur at connections, flexible hoses, or small-diameter piping. Screwed connections should be avoided whenever possible, along with potentially vulnerable points such as sight glasses. All-welded connections should be used as much as possible, both to reduce the likelihood of leaks and line failures and to provide better electrical continuity in piping systems where static electricity is of concern. Where flange connections are necessary, higher strength flanges can be used to improve the integrity of the piping system (such as going from class 150 flanges to class 300 flanges). A ground-strap clamp and / or an electrical continuity check across each flange may be needed to avoid accumulation of static electricity.

6.11.3.3.2 Splash Protection

Where the hazardous material being handled is a liquefied gas or corrosive liquid under pressure, splash collars may need to be installed at flanges and valve stems to minimize the likelihood of operator injury if a flange leak occurs or a gasket, seal, or packing fails. Splash protection around pump seals may be warranted to reduce the likelihood of operator exposure if a seal failure occurs; aerosol formation will also be reduced, and some condensation of the liquefied gas is possible.

6.11.3.3.3 Unloading Lines and Flexible Hoses

Flexible hoses provide an obvious convenience for unloading connections and other frequently disconnected lines. However, for some safety-critical installations, flexible connectors are avoided and hard-piped flanged connections are used instead. Either hard-piped systems or flexible connections may employ specially designed connectors such as dripless couplings or breakaway seals. For less hazardous materials, features such as quick-disconnect connectors can be used.

Chemical hoses must be designed for a specific service. Hoses are required to convey compressed gases, slurries, petroleum products, solvents, bulk chemicals, such as acids and caustic, and a full array of other chemicals that may exhibit flammable, toxic, corrosive, cryogenic, or other harsh properties. Chemical hoses are used for loading and unloading truck, tank cars, barges, and ships. They may also operate to bypass equipment that has been taken out of service.

Hoses built to United States standards have a 3.5-to-1 pressure design safety factor. This means that the Maximum Allowable Working Pressure (MAWP) must be one-fourth the hose's minimum burst pressure. This rating is generally specified at 72°F (22°C), and many users are unaware that, as the operating temperature increases, the hose assembly's MAWP decreases. Data for this are available from hose vendors.

Where flexible hoses are used, they are generally reinforced with steel or stainless steel braiding. Flexible unloading hoses in hazardous chemical service are typically visually inspected prior to each usage and replaced in their entirety on a regular basis, such as once a quarter or once a year. Ensuring electrical continuity across flexible connections is a significant concern in systems with static electricity problems. Unloading connections also provide opportunity for cross-contamination and misdirecting of materials. Dedicated and unique connections and hoses are being used for loading and unloading in some installations where contamination or incompatibility is of concern; for example, the decomposition of some organic peroxides can be initiated by acids, unsaturated organics, and other chemicals. Clear and distinct labeling of lines and connections is also very important where multiple unloading connections exist.

Hose assemblies are generally the weakest link in a piping system because they are usually made from materials that promote flexibility. It stands to reason that rigid systems are more durable than flexible lines. However, hose assemblies must be flexible to accommodate piping misalignments, increase efficiency in processes having many connected parts, and act as vibration isolators and dampeners.

Hose specifications need to be based upon a number of requirements. Some hose criteria include size, delivery temperature, maximum pressure, chemical properties of the fluid, hose material, and end fittings. The major components of a chemical hose are the end fittings, an inner core, pressure / vacuum reinforcement, and the protective outer cover. The exact selection of hoses is beyond the scope of this text.

Hoses should be carefully inspected before each use. Obvious signs of damage such as bulging, kinked, or broken covering must be addressed immediately. Naturally, the end fittings should be observed for leak-free tightness. If the material being delivered is extremely flammable or highly toxic, the hose should be tested before each use. Hoses should be properly secured to prevent whipping. If possible, consider using piping instead of hoses.

6.11.3.4 Storage and Warehousing

Storage areas in the plant usually contain the largest volumes of hazardous materials. Frequently storage areas contain flammable liquids or liquefied gases. The main concern in the design of storage installations for such liquids is to reduce the hazard of fire by reducing the amount of spillage, controlling the spill, and controlling the ignition sources. It cannot be emphasized enough that reducing the quantities of hazardous materials is the single greatest method for reducing the hazards of fire or explosion. Minimizing storage quantities also reduces the potential for large spills and further damage. Pipeline feeds from a reliable source can eliminate the requirement for large storage areas.

Solid chemicals may be stored in bulk in bins, hoppers, piles, or containers. Liquid chemicals may be stored in tanks, reservoirs, or specified shipping containers. Gases may be stored in low pressure gas holders, in high pressure tanks or cylinders, or in liquid form in tanks or containers under pressure, refrigeration, or both. Pressure and temperature of storage greatly affects dispersion / emission of liquid or vapor in case containment is lost. Important considerations are separation distances and diking arrangements.

The primary additional safety concern when hazardous materials are stored in containers is the large amount of vehicle and employee traffic associated with the use of containers combined with the hazard caused by constant handling. Storage areas should be designed to allow the smooth flow of traffic without the need to constantly maneuver a forklift or truck. The storage area should be separate from process units and arranged to allow personnel access to inspect all containers for leakage or other damage on a regular basis. The storage of compressed gases and flammable and combustible liquids should meet the local building code requirements.

Incompatible materials should be kept separated so that any spills cannot mix. The storage of containers in rack areas may require specialized fire control systems such as individual sprinkler lines to deliver water or foam directly to each rack level. The placement of drums in processing the area for the dispensing of the contents may not need to meet the same stringent storage specifications, but it will still be necessary to meet all pertinent safety requirements. The process drums area may include safety barriers to prevent traffic from hitting the drums, portable drum sumps to contain any spills, a ventilation system to control fumes, and double valving or a valve and plug to minimize drum leakage.

During operations, most materials require one or several steps of warehousing or other storage outside of tanks or vessels. This type of goods storage can occur in warehouses or buildings (roof and walls), open air, under a roof (no walls), in a tent or inflatable enclosure, or simply in the staging area. Large warehouse storage of hazardous materials in particular may present a danger to people, the environment or plant operations. Warehouse fires have resulted in strict requirements in most European jurisdictions and a reappraisal of North American requirements. Fire and firefighting consequences that relate to the storage of large amounts of hazardous materials as in certain warehouses need to be evaluated to determine if firefighting is appropriate.

Storage and receiving are activities that can greatly contribute to a safe and economic operation. It is here that quality control can be achieved at minimal cost. Label verification and other quality assurance measures can increase the confidence level that the correct chemicals have arrived, thereby potentially circumventing the use of wrong chemicals. Wrongly shipped chemicals can be returned to the manufacturer with minimal or no cost to the batch operation owner. As with all processes and activities it is of great importance to apply the principles of inherent safety, in particular the minimization and attenuation principles.

Materials that can react with each other should be stored in segregated areas. Special attention is needed for corrosive materials which upon leakage from their primary containment (e.g., a plastic bag) can corrode their main container as well as other containers holding different chemicals in adjacent areas. Proper material handling procedures need to be developed and followed and correct tools should be used. For example, the use of forklift trucks with rounded forks to avoid puncturing drums / bags could be considered. Hazards associated with stacked pallets loaded with shrink-wrapped bags of free flowing materials that can topple over when bags have been punctured should be recognized. Storage areas should be inspected on a regular basis and damaged bags, drums, and other type of containers should be isolated and properly discarded by staff using appropriate Personnel Protective Equipment (PPE).

Hazardous chemicals are often stored under an inert material or atmosphere, stored in a diluted form, or stabilized by a chemical additive. These situations require special care; for example:
- Vaporization of solvents covering alkali metals during storage can expose the metals to moisture.
- Vaporization of diluting solvents may increase the concentration of hazardous chemicals to unsafe levels.

6. EQUIPMENT DESIGN

- Low temperatures can cause a phase separation in stabilized solutions in which case one phase can become deficient in stabilizer and subject to runaway reactions. Acrylic acid can crystallize out of stabilized solution, and subsequent thawing of these essentially pure acrylic acid crystals can initiate runaway reactions, often with severe consequences. Thawing of crystallized (frozen) materials needs to be accomplished using established procedures in thaw boxes or similar devices. If established procedures are not available, a safety review needs to be conducted and a procedure developed prior to thawing the material.
- Heat-sensitive materials need to be stored away from heat sources such as heaters and windows where they are subject to solar radiation.
- Shelf life of stabilizers or inhibitors may be limited.
- Some stabilizers or inhibitors require a certain oxygen concentration in the tank head space atmosphere in order to function. Where inerting is required, careful control is necessary to maintain this minimum oxygen concentration in inerting gas while still staying below the minimum oxygen concentration required for combustion.
- Phase changes also mean that pressure and/or vacuum relief needs to be considered in order to maintain the mechanical integrity of the container.

Correct storage requirements, procedures (e.g., first in, first out) and conditions such as temperature control issues including insulation, cooling, heating, and ventilation need to be determined and implemented.

Potential ignition sources need to be eliminated or protected against by proper bonding, grounding, and lightning protection (Ref. 6-59). Good housekeeping is another essential ingredient for the prevention of mix-ups and unanticipated adverse consequences, e.g., fire caused by smoldering dirty rags.

6.11.4 References

6-59. NFPA 780. *Standard for the Installation of Lightning Protection Systems.* National Fire Protection Association. Quincy, Massachusetts. 2011.

6.11.4.1 Suggested Additional Reading

A number of codes, standards, guidelines, and recommended practices promulgated by organizations such as NFPA and API are provided in the reference section. Additional guidance applicable to warehousing includes:

CCPS. *Guidelines for Safe Storage and Handling of Reactive Materials.* Center for Chemical Process Safety of the American Institute of Chemical Engineers. New York, New York. 1995.

CCPS. *Guidelines for Safe Storage and Handling of High Toxic Hazard Materials.* Center for Chemical Process Safety of the American Institute of Chemical Engineers. New York, New York. 1988.

6.12 UTILITY SYSTEMS

This section presents the potential failure of utility systems and suggests design alternatives for reducing the risks associated with such failures. The types of systems covered in this section include:

- Electricity
- Emergency power supply
- Steam / condensate
- Cooling water
- Inert gas
- Instrument air
- Fuel
- Heat transfer fluids
- Process vents and drains

6.12.1 Past Incidents

This section describes several case histories of incidents involving failure of fired equipment to reinforce the need for the safe design practices presented in this section.

6.12.1.1 Nitrogen Backup for a Compressed Air Supply

There was an incident involving an individual in a small enclosed operator's shelter who became suspicious about his surroundings when his cigarette would not stay lit. Some months before, an operator created some relief from the sweltering heat by installing a small, pneumatically driven fan on a direct air stream in the shelter. During the shutdown of an air compressor, a nitrogen supply was utilized. The operator was being exposed to significantly reduced levels of oxygen.

Lessons learned include ensuring that cross-connection between nitrogen and air cannot be made (e.g., provide different fittings).

6.12.1.2 Hydrogen Backs into the Nitrogen System

During a polymer plant shutdown, the operator was supposed to remove the plug from and connect a nitrogen purging hose to a fitting immediately downstream from an isolating block valve connected to a depressurized system. Instead, the employee mistakenly connected to a pressurized fitting immediately upstream from the block valve, into a 350-psi hydrogen system. During the next 2-hours, combustible gas analyzers occasionally detected the presence of flammables in unexpected places. Operators and supervisors responded to the alarms but did not detect any hydrocarbons. Two and a half hours after the error, an explosion occurred in the final degasser. An automatic sprinkler system tripped and extinguished the fire.

This company's approach was somewhat different than the approach described in the previous incident. The polymer plant implemented a procedure to require force-loaded check valves for backflow prevention on any temporary connection to piping that contained flammables. The check valve had to be rated for the highest design process pressure of either fluid.

Lessons learned include cross-connection between process lines and nitrogen should not be made. An independent verification of lineup should be made when process connections are required.

6. EQUIPMENT DESIGN

6.12.2 Design Considerations

This section does not have information on equipment failure scenarios and associated design solutions specific to utilities. Design of plant utility systems is covered in standard references. This section will highlight scenarios in which loss or malfunction of a utility service results in a loss of containment.

6.12.2.1 Electricity

Electricity is supplied for various purposes: to drive equipment and machinery, to operate instrumentation and control systems, to provide heating of process operations and as tracing of piping runs, etc.

Loss of motive power on process equipment may be quite hazardous. Other serious hazards would result from failure of cooling fans or heating loops required to control temperature and pressure or loss of ventilation to prevent buildup of flammable gases. Provision of backup electrical power is routinely addressed in plant design. Electrical system hazards derive from their potential to serve as ignition sources. Electrical area classification is a way to separate flammable materials from ignition sources. Electrical area classification is discussed in Chapter 7.

The biggest hazards include:
- A common cause failure, e.g., loss of electrical power, loss of cooling water pumps, loss of the plant utilities, etc.
- Loss of pumps and compressors.
- Loss of key instruments, emergency lighting, computer controls, and lube oil pumps can be catastrophic and should be addressed through use of uninterruptible power supplies and emergency generators.

It is sometimes necessary to have an emergency or standby power system to protect personnel and plant integrity. Such systems need to be designed such that they can be tested for reliability and readiness.

6.12.2.2 Emergency Power Supply

Emergency power supply is required for some process equipment to allow safe shutdown of the unit or plant; however, it can be interrupted. The application and design of emergency power systems are extensively covered in the IEEE Standard 446 (Ref. 6-60). Diesel generators should be on a schedule to be run at least once a week, allowed to come up to full heat, i.e., run for about a half an hour, and alarm if not successful.

The Uninterrupted Power Supply (UPS) is used for controls and other systems that must have a continuous supply of power. The UPS is designed to be the prime source of power to a critical load. A UPS provides not only continuous power to the critical load, but also isolation from the main AC line (by means of the battery charger) and a regulated source of synthesized AC power from the inverter. The UPS differs from a standby power system in that it is truly uninterruptible since it provides "online" continuous power supply. Standby and backup power systems involve transfers with switching intervals ranging from several cycles to several seconds or more. Therefore, UPS's are the recommended power source for critical process and safety shutdown systems in a chemical plant. An alarm should be provided for faults in the UPS system.

An important aspect related to the application of a UPS is the selection of critical loads that need to be supplied from it. Safety and security systems, as well as operating systems, that need to be connected to a UPS supply may include:

- Distributed Control System (DCS) process computers, with associated video display units, printers, etc.
- Process package unit computer systems
- Plant shutdown systems (DCS and SIS)
- Safety-related instrumentation (gas analyzers, chromatographs)
- Critical controls and interlocks
- Fire alarm and detecting systems
- Communication systems
- Large rotating equipment local control panels

If a power failure occurs, electric pumps and compressors will stop, and process operations will begin deviating from normal operating values. At this point it may be necessary to implement an emergency shutdown sequence in the plant. While shutdown is being implemented, a reliable and continuous power supply is required to bring the process to a stable condition where it is safe and does not jeopardize the integrity of equipment. All the instrumentation and control devices that would be called upon to operate under an emergency situation should be identified and supplied power from a UPS in order to perform an orderly plant shutdown or maintain the plant in a safe standby condition.

It is widely justified to power devices that monitor both personnel safety and plant integrity from a UPS bus. Fire alarm and detecting systems also fall within this category. Modern process package units, such as the Pressure Swing Adsorption (PSA) unit for hydrogen recovery, are furnished with a separate process computer to control valve settings, and yield parameters and safety sequences; feeding this load from the UPS is recommended.

Sizing of the UPS and its application should be performed carefully to select the critical process and safety-related systems that must remain operational without overburdening the UPS with less critical loads that could be served equally well from other standby power systems.

Another scenario is the momentary loss and immediate restoration of power. Under such circumstances instrumentation will likely react to the loss of power by opening or closing valves. When the power is restored a significant process upset may be introduced when these controllers attempt to return to normal operation. This is particularly true for pneumatic systems.

6.12.2.3 Steam / Condensate

Steam is frequently used as part of the chemical process, as well as to drive machinery, provide heating to the process or heat tracing, and as a safety measure to control a process reaction by snuffing and purging operations. Loss of heating steam may need to be addressed, e.g., if cooling of reactants would cause condensation or solidification and thus create unsafe conditions. Often, steam is considered a secure backup to electric drivers. If this is not assured through a highly reliable design, plant-wide emergency

6. EQUIPMENT DESIGN

shutdown can occur upon power loss. Consideration for operating steam systems during power outages is a common design philosophy.

Design considerations for steam systems include:

- Thermal expansion loops.
- The potential for vacuum when steam condenses (1600 to 1 reduction in volume).
- The power (and potential forces) of hydraulic water hammer when starting up and shutting down steam systems.
- The need to remove and return condensate from steam heaters.
- The value of isolation block valves on steam and condensate headers at battery limits or other strategic locations.
- Thermally insulating steam lines for personnel protection.
- Steam piping and tracing design should address adequate flexibility and avoid condensate pockets.
- Sparing in case of loss of one boiler.

The problems that should be addressed due to steam loss in the plant will include:

- Loss of heat in endothermic reactors.
- Loss of heat in tanks where steam coils are used to keep material liquid.
- Loss of process motive power because of steam-driven pumps and eductors.
- Freezing of steam traced piping and vents.
- Loss of steam for purging.
- Loss of mixing steam to the flare units.

6.12.2.4 Cooling Water

Loss of cooling water may lead to the development of serious process hazards. Alternate power supplies or pumping arrangements (both steam and motor drivers) are usually provided to allow for only partial loss. Spare pumps should be provided in the event of a pump failure.

Corrosion of cooling water systems is commonly prevented by the addition of corrosion inhibitors or oxygen scavengers or by pH adjustment. Unusual ambient conditions in the plant or process may require special considerations (such as when highly salinated sea water is used for cooling). Since cooling water systems are the primary service for equipment cooling to remove process heat, clean, unfouled conditions are a must to avoid failures of the equipment or the cooling water system itself. The potential should be considered for process leaks into the cooling water causing possible toxic or flammable or corrosive conditions at the cooling tower. Water leaking from a heating coil in a hot storage tank can cause froth-over. Consideration should be given to fire protection for cooling towers. Fires have occurred in cooling towers during operation while water is flowing.

Isolation valves are usually provided for individual pieces of equipment so that the entire system does not need to be taken out-of-service in the event of a leak. The isolation valves also are useful when troubleshooting to identify which exchanger is leaking.

6.12.2.5 Inert Gas

Explosion hazards can be reduced by preventing the formation of flammable mixtures; this is done by replacing air with an inert gas to minimize oxygen concentration. Atmosphere control is used in reactors, storage tanks, flare headers, centrifuges, driers, and pneumatic conveyors.

The inert gas system should be designed so that all potential deviations from design conditions are outside combustible limits. To ensure safe operation of the system, the oxygen content should be monitored and provided with interlocks to shut down the system if the oxygen level rises. It is preferable to use duplicate analyzers of different types to monitor quality of the purge system gases. Maintain reliable operation of the oxygen sensor by filtering out particulates, condensing out vapors and remaining corrosive gases (Ref. 6-61).

A list of safety design concerns for inerting systems (Ref. 6-62) which might be carried forward through the life of the facility include:
- Pressure indication on equipment being inerted
- Check valves to prevent back-flow and contamination
- Flow indication to verify inerting is adequate during pump-out
- Capability to test the system regularly
- Nitrogen vacuum break and block valve in line to vacuum source on systems
- Purge gas or steam used in flare systems to prevent flashback
- Furnace purge timer set for 4 to 6 changes of furnace volume

Gases such as nitrogen and carbon dioxide are frequently used to replace oxygen and allow process pressures to be maintained. Non-reactive atmospheres are frequently required for process reasons, for example, polymerization reactions. In some cases, nitrogen may react and argon is used. A minimum oxygen level may be required to activate a polymerization inhibitor, for example, in reactive monomer systems.

Air is removed prior to startup by replacing it with nitrogen, carbon dioxide, or gas from an inert gas generator. The process of bringing the equipment online is complex and requires integrating many systems, to avoid reintroducing air or water.

Inerting can be used to reduce the possibility of tank fires in fixed roof tanks by preventing the formation of a flammable vapor mixture in the vapor space of a tank. An inert gas, usually nitrogen or carbon dioxide, is used to replace the air in the tank's vapor space, removing the oxygen needed to support combustion. The inert gas is usually fed into the tank by one of two control methods - a pressure demand system or constant flow regulation. Care must be taken to ensure that the inerting system is sized to deliver the maximum flow of gas needed during operations in order to prevent vacuum in the tank. It is also necessary to make a decision on what should happen if the inerting system fails in the off position. A vacuum safety valve can be installed so that air is used as the backup for the inerting gas to prevent tank failure so long as no additional hazards are introduced; some companies, however, prefer to allow the tank to collapse rather than introduce air into a tank containing flammable materials. Another alternative is the use of a bladder tank as backup.

The inerting system should be designed so that there are no low point pockets in the inert gas supply line downstream of the pressure regulator. It must also be noted that

6. EQUIPMENT DESIGN

inerting a tank does not prevent the release of material vapors into the tank's vapor space. Material vapors will diffuse into the inerting gas until equilibrium is reached, just as it would with air. This is important to remember when designing tank purging systems and when estimating the toxic and volatile organic compounds (VOCs) material releases for the plant.

Another purpose of inerting is to control oxygen concentrations where process materials are subject to peroxide formation or oxidation to form unstable compounds (acetylides, etc.) or where materials in the process are degraded by atmospheric oxygen. An inert gas supply of sufficient capacity must be ensured. The supply pressure must be monitored continuously.

The designer should consider the need for additional measures to supply inert gas. Particular attention must be given to the following situation: In the case of locally high nitrogen consumption (e.g., when a large kettle is inerted), the pressure in the main line may drop so far that the mains could be contaminated by gases or vapors from other apparatus connected at the same time. Depending upon the application, the quality of inert gas (e.g., water content, contaminants) can be important to process safety. The required level of inerting must be ensured by technical and administrative measures, for example:

- Control and monitoring of inert gas flow and inert gas pressure
- Continuous or intermittent measurement of oxygen concentration
- Explicit information in the standard operating procedures or in the process computer program for the correct procedure to achieve a sufficient level of inerting
- Control of the health hazards of nitrogen asphyxiation to operators entering tanks and reactors being inerted

6.12.2.6 Instrument Air

High quality instrument air is required for proper operation of instrumentation and control systems. In particular, moisture must be extremely low to avoid corrosion and freeze up problems. Compressed air systems must also be free of oil. Air systems, if not properly dried, can contain moisture that can lead to internal corrosion of piping and instrument failure from rust.

It is often considered cost effective to "back up" the instrument air supply with another compatible fluid. The backup fluid is often nitrogen. From an instrumentation standpoint, this poses no significant problems. However, from a personnel safety standpoint, a little recognized but significant hazard is introduced to control rooms, buildings, or vessels: the possibility of lowering the oxygen content in the enclosure to dangerous and possibly lethal levels. At an oxygen concentration of 16% or less, humans will exhibit symptoms of respiratory distress (Ref. 6-63). Further, two breaths of a pure inert gas can be fatal. Older control rooms are particularly susceptible because pneumatic equipment operates by continuously leaking instrument air.

If instrument air is backed up with an inert gas, then the design should provide the following:

- Label instrument air lines as containing both inert gas and air
- An appropriate alarming whenever the backup is activated

With the trend to electronic instrumentation and distributed control system instruments, the central control room is less vulnerable to this hazard than when instrumentation was pneumatic. The main problem in new plant construction is in small buildings, such as analyzer houses, that are not normally occupied. These buildings, though "climate controlled" for the instrumentation, do not usually have elaborate HVAC systems; they use smaller units not usually designed to "change" as well as condition the internal air. Rather than adding a backup system, more care should be taken in designing an adequate secure instrument air supply.

Sometimes during maintenance, plant air is used to ensure air movement through vessels or other enclosures, either by a blower or eductor. In cold climates where the plant air must be low in moisture, the instrument air and plant air system are the same. A backup system for instrument air consists of air cylinders. If high capacity is required, nitrogen may be used; however, provisions for safe breathing atmosphere must be addressed.

Over the years, a number of schemes have been provided to backup for compressed air systems with a gas from a highly dependable nitrogen system, if the occasion would arise. Depending on the system, various approaches to safely handling this situation have been effectively employed. In modern plants there can be installations in which small local buildings that shelter chromatographs and other instruments that depend on compressed air for a number of reasons, including purging the housing. If these buildings can receive any backup nitrogen, they are provided combinations of procedural checks and with bright eye-catching lights to warn of a temporary use of nitrogen. Any such system must be very carefully engineered, fully reviewed, and periodically audited for compliance to the procedures.

One company that relies upon a compressed nitrogen supply to back up the compressed air system has this long-standing procedure: The operator monitors the failing air supply and sounds the area evacuation siren, just prior to switching gas supply systems. The siren alerts all nonessential personnel to leave and especially anyone working within several of these small specialty buildings which rely on compressed air. The operator quickly travels to each small, enclosed instrument building and assures himself that no one is in the building. Next, the operator padlocks the door until this situation has passed. In addition, each building has an oxygen monitor mounted on the exterior wall adjacent to the entrance door.

6.12.2.7 Fuel

Failure of fuel systems (process gas, natural gas, waste streams, etc.) can affect many processes requiring temperature and pressure control. Affected systems may include:
- Boilers
- Furnaces
- Engine drivers
- Compressors
- Gas turbines
- Fired reboilers

The flammability hazards of these fuels are usually addressed in routine design, but often ignored in temporary or emergency operations. For that reason, multiple interlocks, precise air-freeing operations, and other controls are used to make operating

6. EQUIPMENT DESIGN

fuel systems as fail-safe as possible. Also, combustible gas or oxygen analyzers are commonly used to provide necessary information to plant operators.

6.12.2.8 Heat Transfer Fluid

Heat transfer fluids have wide application in the chemical, petroleum, and solar energy fields. Their operating temperature range makes them suitable for heat transfer for both heating and cooling over a broader range than is easily attainable with steam or water. Heat transfer fluids are used to optimize heat transfer in a number of applications where high temperatures are required, where temperature variations must be precisely controlled, or where it is essential to prevent water (or steam) from contacting process chemicals. Examples of such processes include reaction vessels for organic chemicals, pharmaceuticals, resins, and plastics; reboilers for regenerating amines and glycol; dies and molds for injecting and extruding plastics, and regeneration gas for solid desiccants.

These systems have the potential for great destruction, as they involve the pumping of hot flammable liquids in conjunction with one or more unfavorable factors, such as: the heat transfer fluid is above its flashpoint; systems have large hold-ups and high flow rates; piping and user equipment are located throughout the plant; and piping and user equipment are adjacent to other important equipment or nearby combustibles. Although heat transfer fluids provide better heat transfer control than other fluids, leakage (even minor) can have serious results if released fluids contact exposed hot surfaces on process equipment. Backflow of process fluids into the heat transfer fluid system (or vice versa) should be evaluated during plant design.

Molten salt technology is a catch-all phrase that include some very diverse technologies; electro-chemistry, heat transfer, chemical oxidation / reduction baths, and nuclear reactors. All of these technologies are linked by the general characteristics of molten salts:

- Can function as solvents
- Have good heat transfer characteristics (heat capacity)
- Function like a fluid
- Can attain very high temperatures
- Can conduct electricity
- Some molten salts have chemical catalytic properties

Molten salts, depending on their composition, have a range of temperatures; however, most tend to melt into a liquid at temperatures around 1474°F (801°C). This liquid is stable, has a heat capacity similar to water (by volume), and flows much like water does. The major differences are the obvious higher temperatures attainable in the molten salt state and when the salt solidifies (freezes) it contracts versus expanding like water. Thus, molten salt freezing in a pipe would not burst the pipe as water would.

Both sodium and chlorine are notoriously reactive; sodium is one of the most electropositive substances (wants to lose an electron) and chlorine one of the most electronegative (wants to take an electron). These two opposite substances readily join to form stable sodium chloride via a strong ionic bond.

6.12.2.8.1 <u>System Design Considerations</u>

Design of the heat transfer fluid system is determined by process design, for example, batch versus continuous process; philosophy of process control; etc. Temperature ranges, peak temperature, and peak loads are factors influencing choice of heat transfer fluid. Available steam or other utility temperature should be considered. Prevention of leaks and temperature regulation are critical design criteria. Manufacturer's literature should be consulted for final application of heat transfer fluids.

Instrumentation and controls applicable to heat transfer fluid systems are comparable to conventional process control systems. Most instrument systems are intended to control the heating or cooling mechanism at both the heater or vaporizer and the energy using units. The heater controls are required to regulate the firing in proportion to either the fluid flow or the outlet temperature. In certain situations the controls are simplified to an on / off or high / low mechanism depending upon the degree of accuracy required by the process. However, since the most critical variable in the operation of the heat transfer fluid is temperature, it is generally recommended that units be equipped with modulating temperature controllers. Proper energy delivery is further achieved by installing individual temperature controls at each user. Manufacturer's literature should also be consulted.

As the fluid degrades, generally flash point, fire point, and autoignition temperature of the fluid decrease; this increases the hazard. As discussed by Ballard and Manning (Ref. 6-64) regular analysis of the fluid is important. For other heat transfer fluids, consult the manufacturer for specific analysis. Automatic sampling devices should be considered. To establish a degradation curve for each specific system, testing is conducted more frequently at first. Essentially, fluid change-out must be determined on a case-by-case basis. Continued operation with degraded fluid can be disastrous, for example, irreparable fouling of the heat transfer surfaces. On the other hand, discarding usable fluid is wasteful. Knowing how fast the heater performance is deteriorating and the extent of the fluid degradation is a key factor in deciding when to change the fluid.

No matter how long the system is designed to operate, it will have to be deinventoried (emptied) and hydrocarbon freed (cleaned) occasionally for normal maintenance and inspection. The system should be designed so that the "normal" equipment, especially circulating pumps, can also be used for deinventorying. However, some components of the system, notably filters and the circulating pumps, will have to be deinventoried more frequently, while the balance of the system continues to operate. These pieces of equipment may require special considerations for emptying and cleaning, in addition to those which apply to the entire system.

Non-absorbent insulation should be selected and applied after leakage and pressure tests have been completed. Where leaks are likely to occur, either use no insulation, non-absorbing insulation, a spray shield, or insulation treated to prevent penetration by heat transfer fluids.

6.12.2.8.2 Safety Considerations for Design of Heat Transfer Systems

Some other features that may need to be considered are:
- Volatiles in the system. At startup, heat-up should be slow to allow for volatiles (water, for example) to be vented.
- Compatibility / reactivity with process fluids. Welded tube sheets or other special design may be considered.
- Tracing with heat transfer fluid circuits requires unique application techniques. Manufacturer's literature should be consulted.

6. EQUIPMENT DESIGN

- Heat transfer fluid circuits may fall under ASME Section 1, "Boiler Code," requiring additional pressure relief considerations.
- Ethylene (or propylene) glycol / water systems may have further design criteria because of the potential for corrosion of bundles to result in cross-contamination. Freeze protection may be required.
- Decomposition products may form deposits on metal heat transfer surfaces, causing localized overheating and failure of the metal.
- Consideration should be given to conducting special leakage testing in addition to a hydrostatic test (Ref. 6-65). Consult the manufacturer for detailed testing procedures.

6.12.2.9 Process Vents and Drains

Process vents and drains, including emission control devices, are often overlooked but are important elements in the safety of batch systems. Inadequate attention to these items can result in incompatible chemical mixtures within the system formation of combustible atmospheres, or overloading of emission control equipment. Some items requiring special attention are:

- Elimination of pockets or traps in pipelines
- Identification and consideration of all process fluids or equipment that could simultaneously drain or vent into common pipelines or equipment
- Need to prescrub the stream being vented prior to mixing with other streams
- Proper selection of materials of construction
- Determining the need for a vent seal tank

6.12.3 References

6-60 IEEE 446. *Recommended Practice for Emergency and Standby Power Systems for Industrial and Commercial Applications*. The Institute of Electrical and Electronics Engineers, Inc. New York, New York. 1995.

6-61 Halpern, G.S., Nyce D., and Wrenn.C, *Inerting for Safety*, 20th Annual Loss Prevention Symposium, Paper No. 82C, New Orleans, American Institute of Chemical Engineering. New York, New York, 1986.

6-62 CCPS. *Guidelines for Technical Management of Chemical Process Safety*. Center for Chemical Process Safety of the American Institute of Chemical Engineers. New York, New York. 1992.

6-63 AIGIH. *Industrial Ventilation, A Manual of Recommended Practice for Design*, 27th Edition, ASHRAE / American Conference of Governmental Industrial Hygienists. New York, New York. 2010.

6-64 Ballard, D., and Manning W.P., *Boost Heat Transfer Systems Performance*, Engineering Progress, Vol. 86, No. 11. p 51-59, 1990.

6-65 FM Global. *Heat Transfer by Organic and Synthetic Fluids*. Property Loss Prevention Data Sheet 7-99. Factory Mutual Global. Norwood, Massachusetts. 2009.

6.12.3.1 Suggested Additional Reading

Cadick, J., et al. *Electrical Safety Handbook*. McGraw-Hill. New York, New York. 2005.

IChemE. *Hazards of Electricity and Static Electricity*. BP Process Safety Series. Rugby, U. K. 2006.

7

PROTECTION LAYERS

With all process designs, there is the potential for failures to occur. Process safety incidents of greatest concern typically involve the loss of containment, where the release of a hazardous material could lead to fire, explosion, or toxic release with the potential to harm employees, the public, or the environment.

This chapter describes engineering design protection layers. It is important to remember that human interaction can also provide a layer of protection, which requires procedures and training in order to be effective. However, in many circumstances automated response is necessary due to the quick response time required; in other words operator response cannot occur fast enough.

A preventive safeguard stops the occurrence of a particular loss event after an initiating cause has occurred, i.e., a safeguard that intervenes between an initiating cause and a loss event in an incident sequence. A mitigative system is designed to reduce the consequences of an incident in an effort to maintain a safe and operable plant. Mitigative systems provide a layer of protection after there has been loss of containment or the incident has progressed to a point that the preventive safeguards will not be of value. The layer of protection concept is shown in Figure 7.1.

This chapter deals with protective layers whether preventive or mitigative, such as:

- Safety Instrumented Systems (SISs) to shut down a process based on preset process conditions, i.e., high temperature in a reactor will shut down heating and activate cooling.
- Pressure relief systems to prevent overpressure of equipment or vacuum, i.e., a pressure relief valve will open to a safe location (e.g., flare) in the event a control valve failed closed in the overhead system of a pressure vessel.
- Equipment isolation and blowdown to limit the amount of material that can be released in the event of a leak, i.e., a remotely operated valve on the inlet of a pump can be closed in the event of a seal leak.
- Detection and alarm systems to detect a release of flammable or toxic material and provide an alarm so that action can be taken by operations or emergency response personnel.
- Fireproofing to protect structural components so that in the event of a fire structural supports will not fail.
- Explosion suppression and isolation systems to detect an internal fire or explosion and provide quenching or isolation.
- Fire protection to control and extinguish fires.
- Effluent control to manage and control runoff or vapors and ensure that hazardous wastes can be managed.

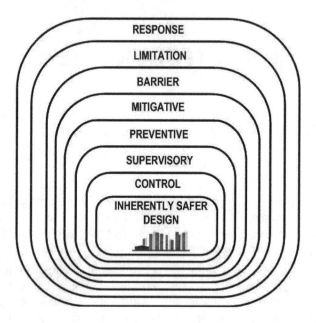

Figure 7.1 Protection Layers

7.1 IGNITION CONTROL

7.1.1 Electrical Area Classification

Flammable liquids, vapors, and gases, combustible dusts, and ignitable fibers are hazardous materials found in many manufacturing facilities. These materials are necessary for, created by, or unavoidable by-products of the manufacturing process.

It has long been recognized that electrical equipment has the potential to act as an ignition source for such materials; primarily the energy for ignition is generated in the form of a spark or due to the temperature generated by the operation of the device. The concept of the electrical classified area and the divisions within the area was created to provide a graded measure of the risk of an ignition event based upon the probability of the existence of a flammable mixture within the area.

A hazardous (classified) location is a space containing any of the following:
- An atmosphere in which an ignitable concentration of flammable gas, vapor, or dust is present or might occasionally be present
- Electrical equipment on which combustible dust might accumulate and interfere with heat dissipation from the equipment
- Surfaces (especially horizontal surfaces) that contain easily ignitable fibers

Most general purpose electrical equipment has an unacceptably high probability of igniting flammable concentrations of vapors, dusts, and fibers. Consequently, a range of electrical equipment has been developed to minimize this probability by means of special design and construction features. This includes limiting the surface temperature of the

7. PROTECTION LAYERS

equipment, minimizing the potential for sparking, and controlling vapor travel into or out of an electrical enclosure, which include upset and cleanout / turnaround conditions.

When selecting heat-generating, electrical equipment for a hazardous (classified) location, its hottest external-surface operating temperature should be compared to the ignition temperature of the surrounding gas, vapor, or dust. Lowering of the ignition temperature for organic dusts that dehydrate or carbonize should be considered. Care should be taken to ensure that these special features of the equipment match the flammability and ignition characteristics of the materials to which it is likely to be exposed.

Appropriate precautions should be taken to maintain hazardous (classified) location equipment in a manner that does not jeopardize the integrity of protection. Guidance offered by the manufacturer, listing agency, or NFPA 70B (Ref. 7-1) should be followed.

Where practical, the possibility of electrical equipment igniting a combustible gas, vapor, or dust should be reduced by:
- Eliminating the use of hazardous materials. Alternate processes or material substitutions may accomplish this.
- Maintaining the mechanical integrity of process equipment so as not to allow material to be released into the workplace.
- Limiting hazardous (classified) areas by using pressurization, ventilation, barriers, enclosures, or other suitable means.
- Locating electrical equipment outside of hazardous (classified) locations or replacing electrical operators with manual or pneumatic operators.

Locations are classified according to the properties of the material being used and its surrounding atmosphere. Elements that affect area classifications may include availability of flammable or explosive material, operating temperature and pressure, flash points, autoignition temperature, vapor density of the material, resistivity of dust or fibers, explosive pressures, dust layer ignition temperature, open or sealed conduit, and ventilation. Definitions for flammable and combustible liquids are given in NFPA 30 (Ref. 7-2).

Each room, section, or area must be considered individually in determining its classification. Normal activities such as draining liquids, disconnecting hose, drumming, and sampling can affect the electrical classification. The overall classification of the area should also be considered. For example, consider the control building within a processing unit. Although the process unit may be electrically classified, the control building could be pressurized, making it non-classified.

7.1.1.1 Traditional Electrical Classification System in the U.S.

Hazardous (classified) locations have traditionally been designated by Class, Division, and Group. Equipment used in areas so designated is selected and systems are designed, based on requirements established for the classification. This approach is defined in Article 500 of the NEC (Ref. 7-3) and API 500 (Ref. 7-4).

Three distinct classes of hazardous (classified) locations have been established:
- *Class I hazardous atmospheres* are characterized as areas containing flammable vapors escaping from a flammable or heated combustible liquid and areas containing flammable gases.

- *Class II areas* contain combustible dust suspended in air or combustible dust accumulations that can interfere with heat dissipation from electrical equipment or can be ignited by that equipment.
- *Class III areas* contain accumulations of fibers.

For a Class I or Class II area, a Division 1 location is likely to contain the hazardous condition during normal operations or frequently because of maintenance and repair. A Division 2 location is likely to contain the hazardous condition only under abnormal circumstances, such as process upset or equipment failure. These two divisions, which are based on the likelihood of an atmosphere being hazardous, control or prescribe the design, construction, and operating features of equipment in that area. Engineering practice tolerates lower levels of protection where there is less likelihood of a hazardous material being present. Thus, Division 1 locations require equipment built to higher standards than equipment built for Division 2 locations.

For Class III areas, the division classification is based on whether the area is used for processing or storage. A manufacturing area may be Division 1 location: a warehouse is likely to be a Division 2 location.

Equipment protective features also depend on the degree or severity of a hazard to which equipment is exposed. For convenience, Class I hazardous materials are typically placed into one of four groups, depending on their physical properties and characteristics. Dusts, which are Class II materials, are similarly grouped by degree of hazard.

7.1.1.2 Equipment

Once a hazardous location has been classified, appropriate electrical equipment should be chosen for that area. In general, equipment must be approved for use in that hazardous classified area. Testing labs, such as Underwriter's Laboratory (UL) test, label, list, or approve equipment suitable for installation in accordance with their legislated code.

Listed equipment for hazardous (classified) areas is marked to show the code-specified environments where it can be safely used. These markings often include the maximum surface temperature of the equipment under normal operating conditions.

The best-known type of hazardous location of electrical equipment is explosion-proof equipment. This equipment is suitable for use in certain Class I, Division 1 locations and in Class I, Zone 1 locations when listed for use in those atmospheres. Explosion-proof equipment is not suitable for use in Class I, Division 1 locations where ignitable concentrations of gases or vapors can exist for long periods of time, or in Class I, Zone 0 locations. Explosion-proof equipment is designed to contain explosions without allowing the escape of enough energy to ignite the hazardous atmosphere in the area.

In recent years, electrical area classification has become more focused on risk of release and distance away from the release point than the risk of flammable vapor / air mixture. Therefore, equipment types are often mixed inside buildings or units, instead of all being explosion proof.

Comparable equipment suitable for use in Class II, Division 1 locations is called dust ignition proof. Dust-tight equipment is designed for use in Class II, Division 2 locations. These terms should not be confused with equipment designated "dustproof." Dustproof equipment is constructed or protected so that dust will not interfere with its successful operation. This term does not imply the equipment is suitable for use in a hazardous (classified) area.

While explosion-proof and dust-ignition-proof enclosures are most frequently used in hazardous areas, there are other National Electrical Manufacturer's Association (NEMA) type enclosures for electrical equipment located in non-hazardous areas.

7.1.2 Purging and Pressurized Enclosures

Another option to allow the use of conventional (not intrinsically safe) electrical arcing equipment in hazardous areas is to create an enclosure that is less hazardous (or nonhazardous) by means of dry air or nitrogen purging and pressurization systems. Arrangements for purging are discussed in NFPA 496 (Ref. 7-5). Positive-pressure ventilation is addressed in NFPA 70, Article 500 (Ref. 7-3). Appropriate safeguards should be provided against ventilation failure.

For example, a local panel for a large process compressor may require a large number of electrical components like relays, switches, and push buttons that won't fit into a cast metal enclosure. A valid alternative is to design a pressurized sheet metal enclosure and create and maintain a non-hazardous atmosphere inside the enclosure. Several types of purged enclosures (known as X, Y, Z types) are described in NFPA 496 (Ref. 7-5) and in Table 7.1.

Table 7.1 Purged and Pressurized Electrical Equipment Enclosures

Type	Description
X	Reduces enclosures classification from Division 1 to non-hazardous
Y	Reduces enclosure classification from Division 1 to Division 2
Z	Reduces enclosure classification from Division 2 to non-hazardous

A purged enclosure requires:
- A source of clean, dry air or an inert gas such as nitrogen
- A compressor (or compressed air cylinder) or mechanical ventilation system to maintain positive pressure inside the enclosure
- Interlocks to prevent the power from being applied before the enclosures have been purged and to de-energize the system if the pressure falls below a safe value
- Provisions to confirm adequate purge pressure (rotameter, instrumentation, etc.)
- Provision to detect and alert operations if normal minimum purge is lost

Purged enclosures are difficult to maintain; therefore, they should be used as a last resort. NFPA 496 (Ref. 7-5) explains the different types of purging systems that can be used depending on the degree a hazardous area is declassified to a less hazardous one or to a non-hazardous area.

Pressurization is mostly used in areas with large volumes such as a control room or a switchgear building. In this case, the fresh air intake is positioned to ensure clean air. A draft fan maintains internal positive pressure.

7.1.3 Low Energy Electrical Equipment for Hazardous Locations

Another kind of electrical equipment suitable for use in hazardous locations is equipment whose maximum possible energy output is insufficient to ignite the hazardous material. The electrical input to this equipment should be controlled by a specially designed electrical barrier. Such electrical equipment should be compatible.

Intrinsically safe equipment and wiring are defined as incapable of releasing sufficient electrical energy at standard temperature and pressure to cause ignition of a specific hazardous substance in its most easily ignited concentration. Intrinsically safe equipment is primarily limited to process control instrumentation with low energy requirements. Several commercial devices in this category are listed by Underwriters Laboratories, Inc. (UL), and Factory Mutual Engineering Corporation (FMEC). It should be noted that intrinsically safe equipment and its associated wiring should be installed so they are purposely separated from non-intrinsically safe circuits by vapor impermeable barriers (Ref. 7-1).

With the age of handheld electronic devices such as cell phones, radios, etc., it is important that these devices are intrinsically safe or designed for the electrical classification in which they are used.

Design tests and evaluation of intrinsically safe systems are provided in ANSI / UL 913 (Ref. 7-6). Notes regarding installation of such devices are summarized in NFPA's National Electric Code (Ref. 7-3). It is permissible to use general purpose enclosures as housing for intrinsically safe wiring and apparatus instead of the more expensive explosion-proof or dust-ignition-proof enclosures that are mandatory for conventional (not intrinsically safe) arcing equipment. This represents an improvement in that a safe system is provided while avoiding the long delivery and high cost of explosion-proof fittings and boxes or dust-ignition-proof enclosures.

Non-incendive equipment and wiring are incapable of releasing sufficient electrical or thermal energy, during *normal* operating conditions, to ignite a specific hazardous atmosphere mixture (Ref. 7-1).

American National Standards Institute (ANSI) / UL 913 (Ref. 7-6) defines low energy intrinsically safe electrical equipment and associated apparatus permitted in Division 1 areas. Non-incendive electrical equipment is permitted in Division 2 locations. Table 7.2 describes intrinsically safe and non-incendive equipment and identifies permitted uses.

7. PROTECTION LAYERS

Table 7.2 Intrinsically Safe and Non-incendive Equipment

Type	Description	Where Used
Intrinsically Safe	Will not ignite the most ignitable concentration of the hazardous material at 1.5 times the highest energy possible under normal conditions, under 1.5 times the energy of the worst single fault, and under the energy of the worst combination of two faults.	Class I, Division 1 Class I, Zone 0 Class II, Division 1 Class III Locations
Non-incendive	Will not ignite the most ignitable concentration of the hazardous material under normal conditions.	Class I, Division 2 Class II, Division 2 Class III Locations

Note that intrinsically safe equipment approved for use in the European community might not pass UL 913 tests for intrinsically safe designation in the U.S. These standards are similar, but not identical. Integration components intended for different codes or systems should be avoided unless approved by an appropriately qualified electrical engineer.

Electrical equipment suitable for classified locations can be expensive and hard to maintain. Alternatives to using this equipment are sometimes available. These options include eliminating hazardous materials, separating the hazardous location from electrical equipment, or moving electrical equipment outside the hazardous location. It is frequently possible to locate much of the equipment in less hazardous or in non-hazardous locations and, thus, to reduce the amount of special equipment required.

Special precautions are required to maintain equipment used in hazardous (classified) locations. Examples are identified in NFPA 70B (Ref. 7-1). If maintenance work voids the listing applicable to a device, the device should not be reenergized in a hazardous (classified) area. A replacement device should be obtained. Special attention should be given to replacement and proper tightening of enclosure bolts, covers, and other fastening devices following maintenance.

7.1.4 Ventilation / Exhaust

All enclosed spaces with the potential for flammable atmospheres should be ventilated, preferably at a rate of not less than six air changes per hour. Low level exhaust ventilation is important to minimize the potential accumulation of vapors. A rule of thumb for design is 1 cfm/ft^2 of floor area with exhaust points no higher than 12-inches (30 cm) off the floor. The ventilation system should be designed in accordance with NFPA 30 (Ref. 7-2).

Where ventilation is installed to effect a reduction in the area electrical classification, the rate should be no less than 12 air changes per hour for an adequately ventilated area in accordance with API RP 500 (Ref. 7-4). The number of air changes required should be supported by appropriate calculations. Areas containing ignition sources, such as control and switchgear buildings, gas turbine acoustical enclosures, and power generators, should be pressurized in accordance with NFPA 496 (Ref. 7-5) if located in an electrically classified area.

7.1.5 Static Electricity

Static electricity is caused by the movement of electrons when two dissimilar substances in contact are separated. While they are in intimate contact, a redistribution of the electrons is likely to occur across the interface, and as equilibrium is achieved, an attractive force is established. When the two substances are separated, work should be done in opposition to these forces. The expended energy reappears as an increase in voltage between the two surfaces. When an electric charge is present on the surface of a non-conductive body from which it cannot escape, that charge is called static electricity. Likewise, if the body is conducting, but the charge cannot escape because it is in contact only with other non-conductors, that charge is also called static electricity. The charge may be either positive or negative, depending on an excess or a deficiency in electrons (Ref. 7-7).

Static electricity discharges and unexpected electrical currents are frequently overlooked as potential sources of ignition that should be controlled. Some of the conditions that may result in sufficiently intense electrical discharges or arcing are:

- Flow of liquids in piping
- Pneumatic conveying of dusts, powders, or particulates
- Splash or free-fall filling of tanks, vessels, or containers
- Mixing and blending of powders
- Use of wet steam
- Moving non-conductive rubber belts, e.g., conveyors or drive belts
- Personnel wearing non-conductive shoes
- Static generated by clothing
- Atmospheric lighting strikes
- Stray electrical currents from faulty equipment, improperly applied electric welding leads, or other sources

Electrostatic charges continually leak away from a charged body. This mechanism, *dissipation,* starts as soon as a charge is generated and can continue after generation has stopped. Electrostatic charges accumulate when they are generated at a higher rate than they dissipate. The ability of a charge to dissipate from a liquid is a function of the following:

- The conductivity of the product being handled
- The conductivity of the container
- The ability of the container to bleed a charge to ground

Conductive fluids such as methanol will discharge significantly faster than toluene. It is important for the design engineer to understand the static electricity properties of the materials in the process.

Static electricity is possible whenever materials are transferred. Some materials have more capability to generate static electricity than others. If the material is readily ignitable, such as flammable vapors or flammable / combustible dust, then steps are required to reduce the risk during design and operation. The primary risk control measure when flammable liquids and combustible dusts are flowing is to prevent accumulation of electrical charges. This is of particular concern in systems that are open to the air during material transfer, where a material / air mixture may exist.

7. PROTECTION LAYERS

Free fall of liquids into vessels or containers during transfer can create static electricity. Two primary methods of minimizing static are to:
- Provide a dip tube that extends into the liquid (generally within 6-inch of the bottom)
- Consider bottom filling the vessel or container

The minimum ignition energy depends on the composition of the mixture and can be as low as 0.2 mJ for many common hydrocarbon fuels and even lower for reactive hydrocarbons like acetylene. This low energy threshold means even a small electrical spark or static discharge can ignite a hydrocarbon vapor cloud.

Loading and unloading operations of ships, barges, tank cars, and tank trucks or, in the case of solid material, hopper cars or trucks are susceptible to static electricity generation. Filling operations should use down-comers or run down the side of the container to avoid splashing that causes static. Transferring from drums to small containers and some processing operations in open-topped vessels can also be at risk.

For more information, refer to API Recommended Practice 2003, *Protection against Ignitions Arising Out of Static, Lightning and Stray Currents* (Ref. 7-8), and *Avoiding Static Ignition Hazards in Chemical Operations* (Ref. 7-9).

7.1.5.1 Grounding and Bonding

A static charge can be removed or allowed to dissipate. The predominant means to prevent accumulation of electrical charges is bonding, grounding, or a combination of both. Other means include prevention of free fall of liquids or solids through the air.

Process structures and the equipment and vessels in them should be effectively grounded to ensure the dissipation of static, stray, and induced charges encountered in normal and abnormal operations and lightning strikes. Process equipment, vessels, and piping should be bonded together as well as grounded. Equipment and vessels in open ground level processing areas should have their own grounding systems. Normally, equipment and vessels in steel multi-level process structures are grounded to the process structure's steelwork tying them to the structure's grounding system (Refs. 7-8 and 7-9).

The key issue is to design grounding and bonding systems so that the long term integrity of the grounding or bonding path is maintained. The issue is more one of integrity of the path to ground over time and in a variety of conditions than the electrical conductive capacity. Solid conductors are used for fixed connections to ground. A stranded or braided conductor is used where the wire should be frequently moved or connected and disconnected. Uninsulated conductors are recommended to facilitate inspection of the integrity of the conductor. Redundant paths to ground should be provided. Trucks either loading or unloading should be grounded.

Insulating flange is a generic term of pipe flange joints with requirements in sealing performance of buried steel pipes and electrical insulation properties of electro-corrosion pipes. It includes a pair of steel flanges, an insulating and sealing article between two flanges with each end welded. The insulating flanges are specially designed and produced for petroleum, chemical, and other industries.

If an electrically charged liquid is poured, pumped, or otherwise transferred into a tank or container, the unit charges of similar sign within the liquid will be repelled from each other toward the outer surfaces of the liquid. The surfaces in contact with the container walls and the top surface adjacent to the air space, if any, will receive the charge. It is this latter charge, often called the *surface charge,* which is of most concern. If the potential difference between any part of the liquid surface and the metal tank shell should become high enough to cause ionization of the air, electrical breakdown may occur and a spark may jump to the shell. This spark across the liquid surface is in an area where flammable vapor-air mixtures are normally present. Bonding or grounding of the tank or container cannot remove this internal surface charge (Ref. 7-7).

7.1.5.2 Fill Rate

API RP 2003 (Ref 7-8) recommends that a slow start (limited flow velocity to 3 f/s) be used until the pipe outlet is covered to the recommended depth. Since this normally requires operator action to control the flow, operation may not be entirely splash free during the initial stages of filling and procedures and training are used to mitigate the risk. However, until this dip pipe is covered by a substantial depth of liquid, splashing may still occur.

In principle, the procedural element could be replaced by an active system controlling flow rate by monitoring liquid depth in the tank. A completely passive system for avoiding splash filling might involve maintaining a minimum liquid level in a tank via appropriate elevation of the product outlet pipe. However, even if a tank is dedicated to one product and minimum liquid level can be maintained, the presence of a stagnant layer in the tank base may make this solution impractical for product quality reasons.

7.1.6 Lightning

Lightning strikes have resulted in fires in processing facilities. They can also be the cause of electrical and computer control system malfunctions and result in process upsets.

Open structural steel process structures normally do not require specific lightning protection since the columns, beams, joists, and stringers are all metal, electrically continuous down through the structure, and bonded to the building or structure's grounding system as required. Buildings of masonry construction or steel frame buildings with non-metal side wall cladding or non-metal roof or top decks usually require lightning protection.

Building structures that are non-conductive can be equipped with air terminal ("lightning rod") conductors and ground terminal systems to safely direct lightning strikes to ground. Buildings in the design stage should utilize conductive building supports or rebar in concrete walls and floors to provide conductive paths. The conductive path to ground for lightning charges should have less than 1 ohm resistance. NFPA 780 (Ref. 7-10) provides additional guidance on lightning protection.

Tanks, vessels, and equipment handling flammable or Class II combustible liquids and constructed of 3/16-inches (4.8-mm) or thicker metal that may be exposed to direct lightning strikes are not normally required to have lightning protection. This presumes that the tank bottom is grounded and the ground conductors are periodically inspected, tested, and replaced if deteriorated. Even so, the highly charged condition of the tank wall can result in an arc if the pathway to ground is interrupted at any point. The resulting arc, if in a flammable vapor space, can readily cause an ignition. For example, vapor leaking from poorly maintained seals on floating roof tanks has been ignited by lightning. Sheet steel less than 3/16-inches (4.8-mm) in thickness might be punctured by a severe lightning strike (Ref. 7-10). Lightning protection is not required on process structures since they are already grounded.

7.2 INSTRUMENTED SAFETY SYSTEMS

The process industry has adopted automation to improve product quality and production rates, to reduce the potential for operator error, and to decrease resource requirements. Process automation includes many different instrumented systems designed to control the process and maintain safe operation. These systems range from simple hardwired systems to complex programmable electronic systems. While the latest in instrumentation and controls often requires less support once implemented, the increased complexity may require a significantly greater level of attention during assessment, design, inspection, testing, and maintenance.

Instrumented Safety Systems (ISSs) implement functions that are identified during a PHA for process safety risks. These functions are referred to as Instrumented Safety Functions (ISFs) in an update to existing standard at ISA. This standard, ISA 84.91.01, *Identification and Mechanical Integrity of Instrumented Safety Functions in the Process Industry* (Ref. 7-11), provides requirements for identifying instrumented safeguards and ensuring that they provide the required risk reduction through adequate mechanical integrity.

Some ISFs may be implemented in the BPCS if an analysis of the BPCS design and management demonstrates that it can achieve the safety requirements. However, Safety Instrumented Functions (SIFs) are a specific type of ISF covered by the international standard IEC 61511 (Ref. 7-12) and the U.S. standard ANSI / ISA 84.00.01-2004 (Refs. 7-13, 7-14, and 7-15). SIFs are implemented in a Safety Instrumented System (SIS) that is designed and managed to achieve a specified safety integrity level. These standards require that the SIS be independent from the BPCS to ensure protection in the event of BPCS failure.

7.2.1 Safety Instrumented Systems

The International Electrotechnical Commission (IEC) issued the first international standard for SIS in 2003. This standard - IEC 61511 - was adopted as a U.S. standard by the International Society of Automation (ISA) and titled ANSI / ISA 84.00.01-2004. The standard follows the process safety management life cycle and provides detailed requirements for achieving and maintaining SIS performance. ISA TR84.00.04, *Guidelines for the Implementation of ANSI / ISA 84.00.01-2004* (Ref. 7-16) provides practical guidance on implementing the standard and specific examples on how to execute particular life cycle activities.

ANSI / ISA 84.00.01 (Refs. 7-13, 7-14, and 7-15) / IEC 61511 (Ref. 7-12) requires a management system for identified SIS. This management system defines how an owner / operator intends to address the SIS performance throughout its life cycle. The essential roles of the various personnel assigned responsibility for the SIS should be defined and procedures developed, as necessary, to support the consistent execution of their responsibilities.

ANSI / ISA 84.00.01 / IEC 61511 establish a numerical benchmark for the SIS performance known as the Safety Integrity Level (SIL) and provide requirements on how to design and manage the SIS to achieve the target SIL. Achieving the SIL requires rigorous analysis, design, operation, maintenance, testing, and management.

7.2.1.1 SIS Life Cycle Overview

The SIS life cycle can be presented in many different ways. Figure 7.2 is an illustration of the ANSI / ISA 84.00.01 / IEC 61511 life cycle process, which covers how an owner / operator plans to assess, design, engineer, verify, install, commission, validate, operate, maintain, and continuously improve their SIS (Ref. 7-12). Due to its broad scope, the standard has many general requirements, which are organized in 19 clauses.

The 10-steps of the life cycle are:

Step 1. Development of a Management System (Clauses 5 through 7)

Step 2. Execution of Hazard and Risk Analysis (Clause 8)

Step 3. Allocation of Safety Functions to Protection Layers (Clause 9)

Step 4. Design and Development of Safety Functions Allocated to Non-SIS Layers (Clauses 8 and 9)

Step 5. Development of a Safety Requirements Specification for the SIS (Clauses 10 and 12)

Step 6. Design and Engineering of the SIS (Clauses 11 and 12)

Step 7. Installation, Commission, and Validation (Clauses 14 and 15)

Step 8. Operation and Maintenance (Clause 16)

Step 9. Modification (Clauses 5 and 17)

Step 10. Decommissioning (Clause 18)

7.2.1.2 Safety Integrity Level

When the hazards analysis identifies the need for an SIS, the SIL also needs to be determined. The SIL is an order of magnitude metric based on the SIS's Probability of Failure on Demand (PFD), e.g., SIL 1 SIS has a PFD between 0.01 and 0.1, SIL 2 SIS has a PFD between 0.001 and 0.01, and SIL 3 SIS has a PFD between 0.0001 and 0.001. The PFD reflects the likelihood that the system will fail to respond appropriately when the demand occurs. The lower the probability of failure on demand, the higher the availability of the SIS to act in the manner it was designed.

7. PROTECTION LAYERS

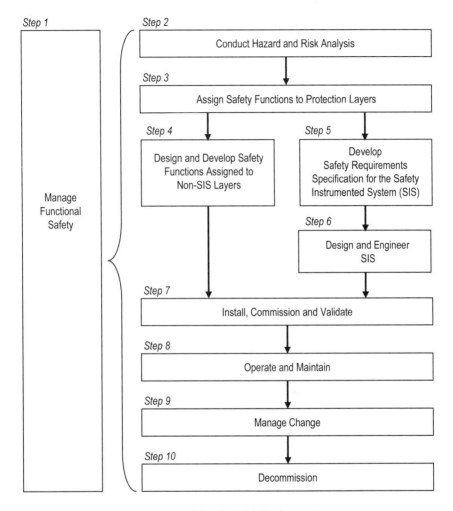

Figure 7.2 ISA 84.01 / IEC 61511 Life Cycle Process

Using reliability analysis techniques, the overall availability of the SIS can be mathematically computed to verify that the assigned SIL is achieved. This method uses all possible states (combinations) of the SIS - from the fully operational, to partially failed, to completely failed. Analysis can be performed using hand calculations, but there are many computer programs available to reduce the time required to complete the calculation and the potential for error. ISA TR84.00.02, *Safety Integrity Level (SIL) Verification of Safety Instrumented Functions* (Ref. 7-17), provides guidance on performing the calculation.

As would be expected, performance of the SIS deteriorates over time without periodic inspection, preventive maintenance, and testing. Performing scheduled testing of the SIS is required. The minimum testing frequency is generally determined during the SIL verification process, but the chosen testing frequency must be sufficient to ensure that the SIS achieves the required risk reduction in the operating environment.

ISA TR84.00.03, *Mechanical Integrity of Safety Instrumented Systems (SIS)* (Ref. 7-18), provides guidance on developing and executing a mechanical integrity program for SIS.

7.2.2 Engineering Aspects of Instrumented Safety Systems

Achieving the required risk reduction and reliability from automated systems, such as Instrumented Safety Systems (ISSs), requires a life cycle approach and a multi-disciplined effort. Only some life cycle activities are performed during the design process. These include:

- Hazard and risk analysis (Section 7.2.2.1)
- Development of the process requirements specification (Section 7.2.2.2)
- Development of the safety requirements specification (Section 7.2.2.3)

7.2.2.1 Hazard and Risk Analysis

Depending on when the study is performed, the information available varies widely. The choice of hazard and risk analysis method is often influenced by the available information. At the research and development stage, the information may be preliminary, such as process flow diagrams along with material and energy balances, while at the operational stage, a complete design package is available, e.g., as-built piping and instrumentation diagrams (P&IDs), design basis for the protective systems, and operating and maintenance procedures.

When projects are in scoping and front-end loading phases, the cost of making changes to the design or to the protection layers is relatively low, and numerous options for risk reduction can be evaluated. As the project proceeds toward full implementation, the cost associated with making changes increases dramatically and the options for change decrease sharply. Consequently, it is important to begin hazard identification early in a project life to identify options that lower the cost of safe operation.

During process design, the risk associated with potential hazardous events should be reduced, when practical, through inherently safer design. The process designer should consider inherently safer design strategies, such as minimize, substitute, moderate, and simplify, to reduce the risk or to eliminate the potential event.

The work process, shown in Figure 7.3, relies on a team to identify hazardous events and to develop and verify the risk reduction strategy. To accomplish this, the team should include personnel that have expertise in the operation of the process unit and its protection layers. Hazardous events may be caused by events internal or external to the process, including those hazardous events that are the result of a failure to maintain the process within safe operating limits. For more detailed guidance on hazard evaluation, refer to *Guidelines for Hazard Evaluation Procedures* (Ref. 7-19).

The successful execution of the remaining life cycle phases is dependent on appropriate identification of IPLs. Potential common cause failure between the initiating causes and the protective functions or between different protective functions should be considered for each hazardous event. Common cause occurs when protection layers share management systems, equipment, controllers, support systems, utilities, interfaces, etc.

7. PROTECTION LAYERS

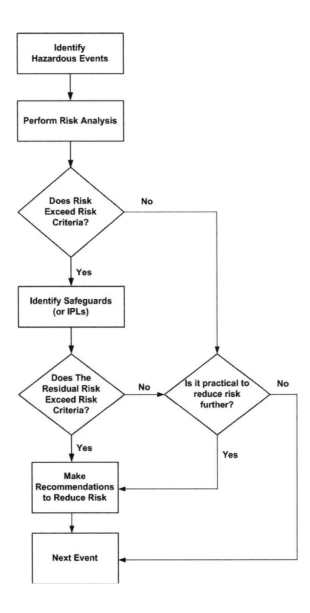

Figure 7.3 Hazard and Risk Analysis Work Process

A Layer of Protection Analysis (LOPA) is one method for evaluating the effectiveness of protection layers in reducing the frequency and / or consequence severity of hazardous events. LOPA is one of the methods suggested by ANSI / ISA 84.00.01 / IEC 61511 for conducting the risk analysis to determine if an SIS is required.

Some protection layers are more effective than others in reducing risk. A protection layer may only partially mitigate the event, resulting in an undesirable though reduced severity event. Some layers are not sufficiently effective in detecting and responding to the particular types of events. Other layers can achieve predictable risk reduction based on proven design and management practices. Protection layers are known as IPLs when they are designed and managed to achieve the following three core attributes:

- *Independence* - The performance of a protection layer is not affected by the initiating cause of a hazardous event or by the failure of other protection layers.
- *Reliability* - The probability that a protection layer will operate as intended under stated conditions for a specified time period.
- *Auditability* - Ability to inspect information, documents, and procedures which demonstrate the adequacy of and adherence to the design, inspection, maintenance, testing, and operation practices used to achieve the other core attributes.

More information on protection layers can be found in *Layer of Protection Analysis: A Simplified Risk Assessment Approach* (Ref. 7-20) and *Guidelines for Independent Protection Layers and Initiating Events* (Ref. 7-21).

7.2.2.2 Development of the Process Requirements Specification

Input from process engineering and operation personnel is essential to the correct specification of ISSs. The process requirements specification is a collection of information covering the operability, functionality, reliability, and maintainability requirements for the process. This work process finalizes the risk reduction strategy and is used to develop the safety requirements specification. The process safety information and hazard and risk analysis findings are used along with input from operations to define the following:

- Operability requirements, including operating modes, process composition, operating conditions, external conditions, and criticality
- Functionality requirements for each process operating mode, including a definition of the process measurements, safe state actions, process actions and process safety time
- Reliability requirements, including sensitivity to spurious trips
- Maintainability requirements, including proof test opportunities and availability of compensating measures

The process requirements evolve throughout the process and ISS life. During the design phase, verification is performed as part of the project execution to ensure that the process requirements specification continues to achieve the hazard and risk analysis requirements. Over the ISS life, Management Of Change (MOC) should be followed when changes to the process requirements specification are proposed.

7. PROTECTION LAYERS

7.2.2.3 Development of the Safety Requirements Specification

The Safety Requirement Specification (SRS) is developed by a control systems or instrument engineer to document the design basis necessary for the ISS to achieve the functional and risk reduction requirements defined in the hazards and risk analysis. The SRS is then given to the designer to implement.

Before starting the SRS, the hazard and risk analysis for the process unit should be reviewed to understand the initiating causes for the hazardous events and the protection layers used to reduce the process risk below the risk criteria. The ISS should be independent of the initiating cause and any other IPL used for risk reduction in the same accident scenario. The SRS should meet the intent of the hazard and risk analysis team, which recommended the ISS, and should fulfill the process requirements specification.

Personnel should review the process requirements with a process representative to ensure that operability, functionality, maintainability, and reliability constraints are understood. The risk reduction allocated to any SIF yields the SIL requirement for the SIF and its equipment. ANSI / ISA 84.00.01 / IEC 61511 provide specific requirements for claiming that an SIF achieves a specific SIL.

The process requirements provide the general functional description that is turned into an ISS architecture comprised of many elements. Each element consists of hardware and software that is configured for specific functional and / or physical characteristics. The elements are physically organized, interconnected, or integrated into an overall system architecture.

Three major subsystems are generally implemented:

1. Field sensors
2. Logic solver
3. Final elements

ISS field sensors detect the process condition. A logic solver processes the available information and executes decisions based on the detected condition. Final elements take action on the process.

Figure 7.4 provides an example of how a typical ISS is organized. An ISS consists of input subsystems (e.g., a single transmitter or voting switches), logic (e.g., discrete, calculation), and output subsystems (e.g., a pump motor control circuit or dual block valves). The action taken by the ISS can be manual (e.g., response to safety alarm) or automatic (e.g., via SIF).

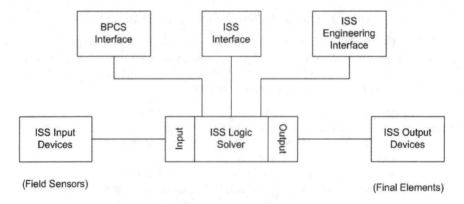

Figure 7.4   ISS Elements

As illustrated, ISSs are complex systems comprised of multiple devices that must operate correctly when required. Redundant subsystems may be needed to achieve the target risk reduction. Redundancy involves the use of two or more pieces of equipment to perform the same function, e.g., multiple sensors to detect the same process condition. To ensure adequate performance, ISSs should be implemented using equipment that has been proven to work in the operating environment. ISFs should also be designed to fail to the safe state on loss of signal, out of range values, loss of communications, power failure, instrument air failure, and loss of other utilities.

When completed, the SRS defines:

- Architecture and configuration requirements
- Mechanical integrity requirements
- Means used to detect ISS failure and expected system and operator response to detected failure
- Conditions required to safely bypass the ISS (includes override or manual operation) and any compensating measures to be in place during bypass
- Conditions required for safe reset of the ISS

7.2.2.4 ISS Documentation

As with any process safety system, documentation needs to be maintained and kept current. A comprehensive design dossier describing all aspects of the ISS should be developed for distribution to operations and maintenance management.

7.3   PRESSURE / VACUUM RELIEF SYSTEMS

Emergency relief devices and systems are designed to prevent catastrophic failure of a vessel caused by equipment overpressure or vacuum. The design goal of all layers of containment is to minimize the actuation of relief devices. This section deals with causes of overpressure, relief devices available, and problems encountered in sizing relief systems.

Pressure safety valves and relief devices are addressed on the premise that the Maximum Allowable Working Pressure (MAWP) and design temperature of the

7. PROTECTION LAYERS

equipment being protected are proper. The relief device design should be consistent with the system's temperature and pressure.

7.3.1 Relief Design Scenarios

The designer of overpressure protection systems should consider all scenarios that constitute a hazard under the prevailing conditions and evaluate them in terms of the pressure generated and / or the rates at which the fluids should be relieved. The scenarios under consideration may cause a release from a single piece of equipment or from multiple equipment items. Overpressure may result from a single failure or multiple failures, and the probability of occurrence of multiple events leading to relief should be considered in the design.

The scenarios leading to overpressure are discussed in Table 7.3 (Ref. 7-22).

Table 7.3   Relief Design Scenarios

Scenario	Description
External fire	The main result of fire exposure is heat input, causing thermal expansion or vaporization or thermally induced decomposition, resulting in pressure rise. An additional result of fire exposure is overheating a vessel wall to high temperature in the vapor space where the wall is not cooled by liquid.
Blocked outlet	Operation or maintenance errors (especially after a plant turnaround) can block the outlet of a liquid or vapor stream from a process equipment item, resulting in an overpressure condition.
Operational failure	Manual valves which are normally closed to separate process equipment and / or streams can be inadvertently opened, causing the release of a high pressure stream or resulting in vacuum conditions.
	Control valves downstream of high pressure vessels containing liquid could fail open resulting in excessive flow of liquid generating a high vapor flow to the downstream vessel.
	Control valve, even with the proper fail-safe design, could be switched from automatic to manual and then overlooked while excessive pressure builds up in the downstream equipment.
Equipment failure (hardware failure such as tube rupture or control system failure)	Heat exchangers and other vessels should be protected with a relieving device of sufficient capacity to avoid overpressure in case of internal failure.
	There are two failures that commonly occur in air coolers, fan failure or louver failure.
	Failure of a control valve in wide open position causing a high pressure fluid to enter a lower pressure system.
Process upset, such as runaway reactions or excessive exothermic reactions	Runaway temperature and pressure in reactor vessels can occur as a result of several factors. Some of these are loss of cooling, feed or quench failure, excessive feed rates or temperatures, runaway polymerization, contaminants, catalyst problems, or instrument and control failures (e.g., agitation failure).
	Design pressure of equipment located downstream of a centrifugal pump is normally set at pump cutoff head combined with maximum suction pressure. However, if the downstream equipment has a low design pressure, or if the pump is a positive displacement type, a relief valve may be set at the design pressure of the equipment and sized to relieve pump capacity.

Table 7.3 Relief Design Scenarios

Scenario	Description
Process upset, such as runaway reactions or excessive exothermic reactions (Continued)	Design pressures for interstage receivers and recycle gas circuits and their associated relieving requirements depend upon the type of compressor used, the compressor performance curves, anti-surge controls, settling-out pressure considerations, and number of stages used. The design pressure of the firebox of a forced-draft furnace should be set to withstand the overpressure generated by the fans with the stack dampers fully closed.
Utility failure	One of the most commonly encountered causes of overpressurization is cooling water failure. Different scenarios should be considered for this event depending on whether the cooling water failure affects a single equipment item (or process unit) or is plant-wide. Power failure will shut down all motor driven rotating equipment such as pumps, compressors, air coolers, and reactor agitators. As with cooling water failure, power failure can have a cascading negative effect on other equipment and systems in the plant. Different scenarios should be considered for this contingency depending upon whether the power interruption is local (to a single equipment item), to a unit substation, or plant-wide. The consequences of instrument air failure should be evaluated in conjunction with the failure mode of the control valve actuator. It should not be assumed that the correct air failure response will occur on these control valves (fail open, closed, or in position). The loss of reflux or recirculation on fractionation towers is typically caused by power failure to the pumps, by a pump trip, or when a control valve fails closed. The relieving rates should be analyzed based upon heat balances around the fractionator to account for the loss of this heat sink. A failure in the inert gas system can lead to overpressure of equipment and should be considered in the relief calculations.
Thermal expansion	Equipment or pipelines which are full of liquid under no-flow conditions are subject to hydraulic expansion due to an increase in temperature and, therefore, require overpressure protection. Sources of heat that cause this thermal expansion are solar radiation, heat tracing, heating coils, heat transfer from the atmosphere or other equipment.
Vacuum	Ejectors, where steam is used as a motive force, may create a vacuum condition primarily during maintenance.
Storage tanks that operate at or near atmospheric	NFPA 30 (Ref. 7-2) and API Standard 2000 (Ref. 7-23) provide guidance for design of this type of overpressure protection. In particular, NFPA 30 focuses on flammability issues, while API Standard 2000 focuses on both pressure and vacuum vent requirements. A common tank failure scenario is insufficient vent capacity (either pressure or vacuum) to allow for all operating cases, plus rapid climatic changes. Adherence to API Standard 2000 is recommended.

7.3.2 Pressure Relief Devices

The most common method of overpressure protection is through the use of safety relief valves and / or rupture disks that discharge into either an open system, i.e., to the atmosphere, to a containment vessel, or to a disposal system such as a flare or scrubber.

7. PROTECTION LAYERS

The following sections provide brief descriptions of pressure relieving devices and guidelines for their use based upon their performance and service characteristics.

7.3.2.1 Safety Relief Valves

Conventional safety relief valves are used in systems where built-up back pressures typically do not exceed 10% of the set pressure. The spring setting of the valve is reduced by the amount of superimposed back pressure expected. Higher built-up back pressures can result in a complete loss of continuous valve capacity. The designer should examine the effects of other relieving devices connected to a common header on the performance of each valve. Some mechanical considerations of conventional relief valves are presented in the ASME code (Ref. 7-24); the manufacturer should be consulted for specific details.

Balanced safety relief valves may be used in systems where built-up and / or superimposed back pressure is high or variable. A balanced valve's capacity is not affected by back pressure until it rises to about 30% of set pressure. Most manufacturers limit back pressure on balanced valves to 45 - 50% of the set pressure. Care should be taken that back pressure developed does not exceed the mechanical limit of the bellows at higher set pressures. This consideration may limit the maximum back pressure permitted for a given service.

7.3.2.2 Pilot-Operated Relief Valves

In a pilot-operated relief valve, the major relieving device (the main valve) is combined with and controlled by a self-actuating pressure relief valve (the pilot control unit). The pilot is a spring-loaded valve that senses the process pressure and opens the main valve by lowering the pressure on the top of an unbalanced piston, diaphragm, or bellows of the main valve. Conversely, once the process pressure is lowered to the blowdown pressure, the pilot closes the main valve by permitting the pressure in the top of the main valve to increase.

Pilot-operated relief valves are commonly used in applications where the built-up or superimposed back pressure is too high (50%) for conventional or balanced relief valves. They are also used where a large relieving area at high set pressure is required or where the operating pressure is close to the set pressure. In fact, pilot-operated valves are frequently chosen when operating pressures are within 5% of set pressures and a close tolerance valve is required. The main disadvantage of these valves is that they are normally temperature limited by the elastomer or plastic piston seal materials and limited to non-corrosive and non-fouling services. The pilot control unit and its tubing can easily be blocked by dirty or fouling process materials.

7.3.2.3 Rupture Disks

A rupture disk is a non-reclosing device actuated by inlet static pressure and is designed to function by the bursting of a pressure retaining disk. A rupture disk assembly consists of a thin, circular membrane, made of metal, plastic, graphite, or a combination of materials, that is firmly clamped in a disk holder. It is designed to withstand differential pressure up to a specified level at which it will burst and release the pressure from the system being protected. It can be installed alone or in combination with other pressure relief devices. A rupture disk can be installed in front of a pressure relief device to protect against plugging. Where a rupture disk is installed before a relief device, a pressure indicator should be installed to alert personnel that the rupture disk has burst.

Rupture disks are available in several types and designs and can be used in pressure or full vacuum. Choice of types is based on safety and operating considerations and vendor alternatives should be closely evaluated.

The burst tolerance of rupture disk devices is typically 5% for set pressures over 40 psig, compared with tolerance of ± 3% for pressure relief valves at set pressures over 70 psig; however, disks can be made to closer tolerances for special applications. In addition, manufacturing tolerances exist which can affect the stamped burst pressure on the rupture disk. Consideration of the operating temperature of the rupture disks needs proper specification due to the reduced metal strength at elevated temperatures. Generally additional care is required in the selection of graphic and knife blade rupture disks due to premature failure concerns.

7.3.2.4 Liquid Seals

Liquid seals are U-tube hydraulic loops used in systems whose design pressure is slightly above atmospheric pressure. Liquid seals are typically used to prevent oxygen ingress into equipment operating close to ambient pressure, overpressure, and vacuum. Seal depth and diameter are sized to pass the design relieving rate at the required design pressure. When designing a liquid seal, the following criteria should be considered:

- Seal loops should be filled with water, oils, or other process compatible fluids and a means to monitor level.
- Continuous seal fluid should be provided to ensure adequate seal, especially after a blowout. Location of discharge for the seal fluid should be able to handle toxicity, flammability, etc., due to contamination by the relieving vapor.
- Winterizing should be provided where necessary.
- Seal depth should exceed the maximum normally expected system pressure by a suitable margin to allow for pressure surges.

A disadvantage to the use of a liquid seal is the inertia of the liquid. A liquid seal should not be used in situations of rapid pressure change.

7.3.2.5 Pressure-Vacuum Relief Valves

For some applications, primarily atmospheric and low pressure storage tanks, Pressure-Vacuum Relief Valves (PVRVs) are used to provide pressure relief. These units combine both a pressure and a vacuum relief valve into one assembly that mounts on a single nozzle on top of the tank. These valves frequently operate under normal tank working conditions at low pressure differences but they should also be sized to handle the maximum possible pressure requirements for the tank. API Std 520 (Ref. 7-25) and API Std 2000 (Ref. 7-23) can be used as references for sizing. For emergency pressure relief situations, an additional safety relief valve can be placed on the tank.

7.3.2.6 Vacuum Relief Devices

Occasionally, a vessel may experience vacuum conditions due to excessive condensation or upset process conditions and, therefore, should be protected from collapse due to vacuum. In this case, the design engineer has the option to design the vessel for either

7. PROTECTION LAYERS

full or partial vacuum, provide a vacuum relief device, or permit ingress of air, nitrogen, or fuel gas to the vessel to prevent a vacuum from developing. If vacuum relief is from a header, it should be assured that the header does not contain condensable vapors. Designing for full vacuum is the inherently safer approach whenever practical. Glinos and Myers (Ref. 7-26) discuss the sizing of vacuum relief valves for atmospheric distillation columns.

7.3.3   Sizing of Pressure Relief Systems

A critical point in design is determining whether the relief system should be designed for single-phase or two-phase flow. Two-phase flow frequently occurs during a runaway reaction, but it may also need to be considered in a less complicated system such as a vessel with a gas sparger or bottom fire on a raised vessel.

7.3.3.1 Vapor Service

API Std 520, Part I (Ref. 7-25) recommends formulas for calculating the discharge area of a relief valve for gas or vapor for low pressure steam exhaust. API Std 520, Part I, should be consulted for unusual situations in which deviations from ideal gas law behavior are significant.

7.3.3.2 Liquid Service

Liquid capacity certification is required for pressure relief valves designed for liquid service. The procedure for obtaining capacity certification includes determining the coefficient of discharge for the design of liquid relief valves at 10% overpressure. Valves that require a capacity in accordance with the ASME Code may be sized using the equation given in API Std 520, Part I.

Where liquid-full equipment can be blocked in and continued heat input cannot be avoided, a pressure relief device (thermal relief) should be provided. This is particularly important for long lines that can be warmed by the sun. The rate of expansion depends primarily on the rate of heat input and the liquid properties. Liquid expansion rates for the sizing of relief devices that protect heat exchangers, condensers, and coolers against thermal expansion of trapped liquids can be approximated using an equation from API Std 520 (Ref. 7-25).

7.3.3.3 Flashing Liquids

Simpson (Ref. 7-27) and Leung (Ref. 7-28) have presented methods for sizing safety devices for two-phase flow, including flashing flow. The Design Institute of Emergency Relief Systems (DIERS) has also developed methods for sizing relief systems for two-phase flow (Ref. 7-29).

7.3.3.4 Relief Valve Inlet and Outlet Sizing

The inlet line of a relief valve should be designed to prevent excessive pressure loss which can lead to unstable valve chatter. Design criteria for inlet lines to relief valves can be found in codes and standards. Typical inlet size should limit friction loss to <3%. Additionally, there should not be any obstruction or reduction on the inlet line. Consult the ASME Code (Ref. 7-24) for basic requirements for pipe sizing and limitations; consult API Std 520, Part II (Ref. 7-25) for additional information.

The outlet line size of a relief valve discharging to atmosphere is generally dictated by back pressure, velocity limitations and environmental considerations. Sizing of relief valves discharging into a closed system, for example, a flare, is impacted primarily by back pressure considerations. Design guidelines for sizing outlet lines should limit the friction loss to 10% and should consider multiple relief devices operating simultaneously so that the back pressure can be adequately calculated. Refer to *Safe Design and Operation of Process Vents and Emission Control Systems* for additional information (Ref. 7-30).

Inlet and outlet piping for a rupture disk should be the same size as the disk.

7.3.4 Sizing of Rupture Disks

The sizing criteria for rupture disks are similar to those of relief valves. When rupture disks are used in conjunction with relief valves, a sizing factor must be used to derate the effective relief capacity of the disk / safety valve assembly. The designer should consult with the disk manufacturer or engineering standards to arrive at this factor.

7.3.5 Other Considerations

7.3.5.1 Location

Normally, relief valves are installed at the top of vessels. Barring any code requirements, it is permissible to mount relief valves on the outlet piping from a vessel. In some towers handling corrosive or dirty fluids, relief valves are best installed below the packing or trays since there is a potential for column plugging. In other cases, it would be advisable to install relief valves at a point in the tower which provides the most advantageous temperature, phase, or density for relief, thereby avoiding possible disposal problems.

The discharge point for relief devices should also be a consideration in designing relief protection. The nature of the discharge and the point of the discharge may present significant hazards.

These designs should include consideration of when it is acceptable to relieve to the atmosphere, when it is necessary to relieve to an effluent disposal system, and when to relieve inside or outside a building, etc. For example:

- Chlorine vaporizer relief protection discharge into a caustic scrubber
- Distillation system relief protection to a safe location, e.g., flare system for flammables
- Fluorocarbon refrigerant relief outside enclosed building
- Avoid steam relief valve discharge at head level into walkways or buildings to avoid possible asphyxiation, etc.

7.3.5.2 Spares

Sparing of relief valves is now an installation approach that allows online maintenance of the valve by switching. Use of spares should be accompanied by certain restrictions, including:

7. PROTECTION LAYERS

- Installation in parallel and isolated by full-port, three-way or transflow valves at the inlet and outlet. Full-port block valves can also be used to isolate relief valves installed in parallel.
- Providing bleed valves between the relief valve and the inlet block valve.
- Key locks, where appropriate, to assure a proper isolation sequence. Locks should be approved by administrative policy and procedures.
- Procedures to ensure that after a valve relieves and the spare is being used, the performance of the valve that relieved is checked.

One relief valve can protect more than one piece of equipment connected by piping in the following cases:

- The relieving path between the equipment pieces and the relief device is free of any potential blockage or block valves, unless car-sealed open.
- The pressure of the relieving path at the time of relief should assure that ASME Code limits on equipment overpressure are not exceeded for any of the protected equipment.

7.3.6 Methods of Overpressure Protection for Two-Phase Flows

One design consideration of particular significance is whether the relief system should be designed for single-phase (vapor or liquid) or two-phase (vapor and liquid) flow. The Design Institute for Emergency Relief Systems (DIERS) has developed methods to predict when two-phase flow might exist and the application of various sizing methods in emergency relief design (Ref. 7-29). The DIERS Project Manual is the best source of detailed information on these methods (Ref. 7-31).

The most significant findings of the DIERS program are the ease with which two-phase vapor-liquid flow can occur during an emergency relief situation and the requirement for a much larger (by 2-10 times) relief system.

Two-phase flow is dependent on the physical properties of the material being vented (surface tension, solids content), heat input rate to the vessel with resulting vapor formation (and bubble rise rate), and liquid level in the vessel. Two-phase flow frequently occurs during a runaway reaction, but it may also need to be considered in a less complicated system such as a vessel with a gas sparger or a unique fire case. Vapor-liquid mixtures can also form in the relief system as venting occurs. In addition, two-phase flow can occur intermittently during a release.

The DIERS methodology is important as a means of addressing situations such as two-phase flow. Because of uncertainties in application of these techniques to sizing relief systems, however, a prudent course for the designer is to use the most conservative calculation. *Guidelines for Pressure Relief and Effluent Handling Systems* (Ref. 7-32) addresses calculation methods for selection of relief device size, prediction of flow, etc.

7.4 EQUIPMENT ISOLATION / BLOWDOWN

7.4.1 Equipment Isolation

Process fires will continue and may escalate until the flow of fuel is stopped, the fuel is fully consumed, or the fire is extinguished. Isolation valves are used to reduce or isolate inventories of flammable gases or liquids.

Isolation valves may be located near the property line, the edge of a process unit, or the liquid outlet of a vessel. Valves should be installed on all hazardous materials lines entering or leaving the facility to ensure the facility can be isolated in the event of a spill or fire. Similarly, valves should be located at or near the battery limit of each unit or outside dike walls for the same reasons and for safety and ease of access.

These battery-limit unit isolation valves should be clearly identified and installed in easily accessible locations that are a safe distance from potential fire sources. Generally, 25 - 50 feet (8 - 15 m) provides an acceptable separation distance. These isolation valves can also be used for turnaround purposes and should be located at or near ground level.

Where emergency isolation valves are provided on the suction of pumps and compressors or on vessels with large flammable liquid holdup, the emergency isolation valve should be provided with remote activation. Generally 50-feet (15 m) provides an acceptable separation distance. The location should also consider the ability to reach the remote activation switch given the radiant heat load. Additionally, fail-safe valves can be used.

Equipment such as pumps, compressors, tanks, and vessels associated with large inventories of flammable gas or liquid (>5,000 gallons) should be provided with equipment emergency isolation valves to stop the flow of material if a leak occurs. For example, the decision to add emergency isolation valves to the inlet and outlet of a compressor is dependent on the flammability of the gas, its pressure, and the quantity of gas in the associated piping and vessels.

Isolation valves can be operated remotely or locally, and automatically or manually but should be placed such that operators and emergency responders can activate them safely.

All isolation valves should be clearly and uniquely identified by sign or color or other means.

7.4.2 Depressurization

When the shell of a vessel is exposed to extreme heat on the outside and the inside is in contact with vapor, metal temperatures may reach levels where tensile strength will be reduced such that rupture may occur even though the pressure does not exceed the set point of the pressure relief valve [for ordinary carbon steel, this temperature is ~900°F (482°C)]. Depressurization provides fire protection for process units by reducing the contained pressure at such a rate that:
- There is a significant reduction in the chance of rupture of any steel pressure vessel exposed to fire.
- The driving force behind pressure jet flames is rapidly decreased.
- The leakage rate of liquid spills decreases, allowing containment of pool fires.

Depressurization may also be used as protection during startup, shutdown, and short-term upsets by reducing the pressure to prevent a pressure relief valve discharge. However, vapor depressuring should not be used as a permanent operating control in lieu of correcting the causes of a process upset. There are several advantages of depressurization, including the following:
- An auto-refrigeration effect may be produced in a pressure vessel, which provides cooling of liquid contained in the vessel.

7. PROTECTION LAYERS

- The need for a liquid blowdown system is eliminated. By retaining liquid in the vessel as a heat sink, any increase in temperature of the wetted shell is minimized.

It should be noted that depressurization may not be practical when the vessel design pressure is less than 100 psig (690 kPa) because valves and piping can become unreasonably large and costly or when the vapor depressuring load governs the size of pressure relief and flare headers. Refer to ANSI / API Std 521, *Pressure-Relieving and Depressuring Systems* (Ref. 7-22), and *Guideline for Protection of Pressurized Systems Exposed to Fire* (Ref. 7-33).

7.4.2.1 Design

Depressurization design considerations include the following:

- Vessels should be depressured to at least 50% of the design pressure within approximately 15 minutes for vessels exposed to pool fires with wall thicknesses of 1-inch, (2.5 cm) or greater. Jet flame impingement requires more rapid rates.
- Vessels with thinner walls usually require a greater depressurization rate.
- The depressurization rate percentage is based on the wall thickness, initial metal temperature, and the rate of heat input from the fire (Ref. 7-22).
- Bare pressure vessels in a process unit protected by vapor depressuring may not require protection by fixed water spray systems.

Some companies limit the application to facilities operating over 250 psig (1724 kPa), while others depress all light hydrocarbon processes.

7.4.2.2 Depressuring Flow Rate

To calculate the vapor flow rate needed to accomplish depressuring, the maximum expected operating pressure of the vessels under consideration should be used as the initial pressure, and 50% of the design pressure or 100 psig, whichever is lower, after a maximum of 15 minutes as the final pressure (Ref. 7-22). When estimating flows for sizing depressuring valves and piping, it is important to consider:

- Effect of the initial depressuring rate on the closed pressure relief system and flare
- Average depressuring rate used to determine depressuring time
- Variations in temperature, pressure, vapor composition, and possible liquid entrainment into the relieving system during the depressuring period

7.5 EFFLUENT DISPOSAL SYSTEMS

A waste gas incinerator at a chemical plant near Houston recently experienced a flashback with a pressure wave in the suction vent gas system, resulting in extensive damage to the flame arrester, fan, valves, and incinerator piping. A well-designed system was overcome through an unforeseen combination of failures which defeated the safeguards which were already in place. The air supply normally contained organic emissions at concentrations designed to be less than the lower flammable limit. Through a combination of automatic and operator responses to a trip of the waste gas feed, a fuel-rich stream was suddenly introduced into the incinerator, creating a "slug" of fuel which allowed flame from the burner to blow back into the windbox and the combustion air header. The flame front generated a pressure wave which then blew apart the flame arrestor, fan, valves, and piping. This incident shows that even well-designed systems may be overcome. Determination of actual failure mode is complicated by the safeguards already in place. More importantly, it demonstrates the need to consider the use of in-line detonation arresters or explosion vents for assurance of passive protection of vapor lines in flare, incinerator, and blowdown systems.

Selection of the disposal system is determined by characteristics of the effluent such as physical state (vapor / liquid / solid), pressure and temperature, and boiling point; quantitative factors such as flow rate, duration of discharge, total quantity of material to be discharged; hazardous properties (toxicity, flammability, buoyancy); nuisance factors (noise, odor); as well as the location of the disposal system (in relation to meteorological conditions, local populations, and local regulations and ordinances). If the effluent is nontoxic, it can be discharged to the atmosphere; however, many non-toxic materials should not be discharged to the atmosphere because of the potential for environmental damage, fire, explosion, odor, or noise. Further treatment may be required in accordance with the Clean Air Act's New Source Performance Standards.

Standards for emergency relief system or Volatile Organic Compound (VOC) emissions control are governed by the Code of Federal Regulations (CFR) under the Clean Air Act's New Source Performance Standards. Flares meeting these conditions are assigned a destruction efficiency equal to 98% of the organic materials by the EPA.

These standards are detailed in 40 CFR 60.18 and include requirements for:
- Minimum Btu values for the flare gas, for non-assisted and steam- or air-assisted flares
- Maximum flare tip speeds, which vary with the Btu value of the flare gas
- Continuous monitoring for the presence of a flame

7.5.1 Flares

Federal, state, and local permits are required to construct and operate flares. The minimum amount of information required for the permitting process includes normal and design maximum flow rates, estimated gas composition and Btu value, normal maximum flare tip velocity, a description of the flame tip monitoring system, and the location and height of the flare. In some cases regulatory authorities may require that the flare emissions be modeled for ambient air effects. Regulatory authorities may also require smokeless (zero visible emissions) operation up to a prescribed percentage of the flare's design maximum emission. Aircraft warning lights may be another regulatory requirement.

Several types of flares are available in the market for application in process plants. Four of the most common flare types in the process industry are (Ref. 7-23):
- Elevated flares
- Ground flares
- Low Pressure flares
- Burn pits

7. PROTECTION LAYERS

The following common design criteria for flare systems need to be considered by the designer:

- Regulatory limits on release of toxic, corrosive, and flammable substances; noise; smoke (federal, state, or local venting permits).
- Location and spacing in relation to process units, storage areas, grade level, and personnel. Criteria are based on radiant heat flux and ground level concentrations of toxic or corrosive components of the flare gas combustion products.
- Ability to remove liquids entrained in the flare gas.
- Prevention of oxygen from entering the system, especially via relief devices. Maintenance of relief valves should be performed using procedures that prevent air from entering the system.
- Flashback protection to prevent internal explosions in case flammable vapor-air mixtures are generated. Air may be present from backflow through the stack or inlet piping after a release of hot process gas (a hot blow).
- Provision for pilot ignition systems and their controls to be located safely.
- Provision for purging the flare header with fuel gas or an inert gas.
- A separate flare system for oxygen-containing streams might be preferable to avoid introduction of streams containing air or oxygen into the main flare header. This practice avoids the potential for explosion if flammable concentrations are possible.
- Exit velocity; excessive exit velocity can cause flame detachment or flameout.
- Materials of construction should be addressed, especially in regard to low temperatures or corrosive or reactive chemicals.

Design of elevated flares is dictated by radiation at grade level and the possibility of falling sparks. Sizing criteria and calculations of elevated flares are detailed in ANSI / API Std 521 (Ref. 7-22).

Although pressure relief valves are sized to accommodate individual peak relieving loads, the relief system design requires that a cumulative relieving load from valves discharging simultaneously be determined. This load is used to determine the back pressure obtained in the relief system, fluid velocity in sections of the relief header and at the flare tip, and the level of thermal radiation and noise at grade. Since back pressure may affect the performance of a pressure relief valve, the relief header system (PSV tailpipe, subheader, and main flare header) is sized to limit the back pressure at the valve outlet and thus maintain the required capacity of the pressure relief valve. The maximum allowable back pressure is a function of the type of pressure relief valve and its set pressure. The actual back pressure obtained at the relief valve outlet is a function of line size and its associated relieving rate in each section of the relief header system. The flow rate in each section of the relief header is different depending on the location, number, and capacity of each pressure relief valve which is expected to discharge into the relief header at the same time.

Typical common mode failures such as fire, cooling water failure, and power failure are generally involved in the simultaneous discharge of several relief devices. Consequently, the controlling loads generated by one of these emergencies should be evaluated for design of the flare headers as well as the equipment items in the system.

These failures should be further analyzed to determine if the effect is plant wide or local, if other standby equipment is available to pick up the service, if automatic startup spares are available, or if standby power supplies are provided. The relief loads for one contingency (e.g., cooling water failure or power failure) may not be additive, and therefore proper transient analysis may reduce the overall controlling load.

Special consideration should be given to situations where relief devices can discharge flashing liquids or where a combination of cold liquid and hot vapor discharge may result in vaporization of the liquids. Such situations may generate additional vapor loads, beyond those corresponding to the relieving loads. Mechanical effects due to uneven thermal stress should be considered.

Instrumented shutdowns of equipment and heat sources can appreciably reduce flare design loads. This not only reduces environmental problems associated with flaring or scrubbing but also reduces the cost of chemicals wasted during plant upsets. When considering the relief loads resulting from instrumented shutdowns, it may be assumed that all trips will function. However, it is recommended design practice to assume that the trip on an equipment piece contributing the largest noncumulative relieving load will not function. It is also suggested that this philosophy be supported with a quantitative assessment of reliability.

7.5.2 Design Considerations for Flares

7.5.2.1 Flare Header Design

The following are general guidelines for flare header design (Ref. 7-34):

- Extensive measures should be taken to avoid pockets in the flare header and associated piping.
- Piping (discharge piping, subheaders, and headers) should be free draining to the knockout drum.
- Consider intermediate knockout drums in or near process units if the flare stack is located in a remote area of the plant.
- Sectionalizing is not a requirement and is avoided in some organizations to avoid maintenance problems with valves and possible misoperation or malfunction. Line blinds sometimes are used where sectionalizing is required.
- Flare headers may collapse if a large volume of liquid is inadvertently discharged into the header, exceeding the capacity of the piping supports. To prevent such events, it is advisable to use criteria such as specifying the pipe as half-full of liquid or otherwise ensure that the header can support the weight of the liquid and absorb the impact of any liquid slugs.
- Pressure relief headers should not be routed from one operating area through another area where operators frequently perform maintenance.
- Flares handling combustible vapors from multiple relief valves should not be used for venting air or steam during startup or at any time loss of flame is likely.
- Avoid freezing or solidification of liquids such as water, high pour point, or high viscosity oils, polymers, or other materials during low ambient temperatures; heat tracing and drains may be required.

7. PROTECTION LAYERS

7.5.2.2 Knockout Drums

The flare knockout drum collects relief loads and separates liquid droplets from vapor releases. Depending on its composition, this liquid may be returned to the process for further recovery or later vaporized and routed to the flare. An overview of methods of sizing knockout drums and various other types of blowdown / knockout drums and vapor-liquid separators used in the chemical industry is given in ANSI / API Std 521(Ref. 7-22) and *Safe Design and Operation of Process Vents and Emission Control Systems*
(Ref. 7-30). Other considerations in the design of a knockout drum are:

- A steam coil, jacket, or other means of heating is sometimes provided in the drum to prevent high viscosity liquids from becoming too viscous to drain or be pumped.
- The drum should be sloped towards the liquid outlet nozzle.
- For cold climate locations, methods for freeze protection are recommended in the event that the knockout drums capture some water.
- Consideration should be given to the reactivity of all chemicals which might be encountered, especially when external heating is applied.
- Consideration should be given to liquid entrainment and carryover to the flare if the velocity is too high or holdup time is short.

7.5.2.3 Flare Stack

The sizing of a flare requires the determination of stack diameter and stack height for the maximum simultaneous load from the source(s). Determination of the maximum simultaneous load is a complex problem requiring an understanding of interactions among loads and an agreement on a philosophy of design. Several factors govern the stack sizing, including velocity, pressure drop available, wind effect, dispersion of flammable and / or toxic gases, and ground level heat flux. A detailed sample calculation for sizing a flare stack is presented in ANSI / API Std 521 (Ref. 7-22).

7.5.2.4 Flare System Safety

Safety concerns in flare design involve the risk of explosion or fire due to improper flare design or operation. Routine scenarios encountered during maintenance and operation should be carefully considered to avoid contamination of relief systems with oxygen or reactive materials that may rapidly polymerize, releasing large amounts of heat or plugging the flare. These scenarios should be carefully documented, and training should be provided for operations and maintenance personnel.

The flare is an open flame and a major source of radiant energy. The flare should be located to minimize the chance that flammable vapors from a storage tank leak or unit rupture will contact the flare. Flare placement and height should minimize the radiation exposure of storage tanks, process units, and personnel working in the area.

The entrainment of air into the flare header can cause the vapor in the header to burn or explode, causing fires or rupture in the process systems. The possibilities of entrainment and the consequent flashback can be minimized by the use of a seal drum, molecular seal, and sweep gas to prevent air from traveling down the line.

In order to prevent the risk of explosion to the flare, protection can be provided by seal drums, header purging, or use of a dry seal such as a molecular seal, especially when the flare gas is lighter than air, e.g., hydrogen. Flame arresting devices may be installed in headers of the flare system to prevent propagation of any flashbacks which might occur.

Some tanks, wastewater treatment facilities, and other units may continuously vent to the flare without the use of relief valves. Use of direct venting of low pressure tanks and pressure relief devices to flare headers is risky due to potential for overpressure or back mixing during emergency events in other equipment. Care should be taken in the design of flare headers to make sure that these units, and units with low pressure relief valves, cannot overpressure and rupture during high volume relief situations. In some cases it may be necessary to increase the pressure rating of the individual unit to above that seen in flaring situations or use a separately dedicated low pressure stack.

Improperly sized knockout drums can lead to the presence of liquids at the flare tip during high levels of flaring. This dangerous situation can cause an explosion at the flare tip, extinguishing the flame or ejecting burning liquids into the air.

A method to monitor the pilot and provide a reliable system to reignite the pilot burners should be provided. The most frequent cause of pilot failure is loss of fuel gas flow; this is often due to a plugged line or filter. Provide a means to ensure that the fuel gas is clean and to verify flow to the pilot. Another cause of loss of flame is blowout on low pressure flares in high winds.

7.5.3 Blowdown Systems

Condensable vapors, contaminated aqueous effluents, and various other liquid streams generated due to plant emergencies require disposal. These "blowdown" systems include plant oily water sewers, chemical sewers, closed drain header systems for flammable liquids or special materials, quench blowdown drums, blowdown drums, effluent disengaging drums, or other facilities capable of handling the additional loads. Systems for routine deinventorying are not in the scope of this section.

The method of disposal is determined by the hazardous properties of these fluids, such as toxicity, and temperature, viscosity, solidification, and miscibility. The objective in design of blowdown systems is to not create a new problem while solving the disposal problem. Commonly used blowdown systems are described below.

The primary safety considerations related to blowdown drums are that they should be designed to handle overpressures that could result from continuing runaway reaction or from an external fire. Design of the vessel to withstand deflagrations is addressed in NFPA 69 (Ref. 7-35). It is good engineering practice to design the blowdown drum for a minimum design pressure of 50 psig (Ref. 7-22).

7.5.3.1 Equipment Drainage Systems

During upset conditions or shutdowns, process equipment items should be drained of their contents to allow personnel safe entrance. Disposal of small inventories of fluids depends on their volatility and toxicity; frequently discharge is to a sewer or to the atmosphere, provided the material is not hazardous or toxic. Compatibility considerations are of utmost importance.

7. PROTECTION LAYERS 347

7.5.3.2 Disengaging Facilities

Tube failure in heat exchangers using cooling water or steam invariably causes contamination of these utility systems with organics or other fluids if the process side pressure is higher than utility pressure. These utilities should be treated before they are recycled for further use. Treatment is performed in separate disengagement drums for each contaminated utility system. Disengagement drums normally operate at atmospheric pressure; therefore light ends flash and should be safely vented (possibly to a flare). Similar treatment may be warranted.

The following are guidelines for design of disengaging drums:
- The inlet liquid line size is based on the maximum liquid rate to the drum.
- The vapor outlet is sized for the vapor load generated by flashing from the maximum quantity of feed to the drum. This vapor may discharge to a flare or to the atmosphere at a safe location.
- The minimum design pressure of the drum should be 50 psig (Ref. 7-22).
- High liquid level alarms should be provided.
- Condensed organics should be skimmed and pumped to a suitable recovery system.
- Drum liquid holding time is determined by liquid / liquid (organic / water) separation requirements.

7.5.3.3 Quench Drums

A quench drum is used to cool and partially condense vapors discharging from relief devices by spraying water or other suitable liquid directly into the gas stream. By condensing organics, this type of drum reduces flare loads and vapor loads to other downstream facilities and reduces the reaction mass carried over. Quench drums are used to reduce the amount of organic emissions to meet federal, local or state regulations. Condensed fluids may be pumped back to the process area for treatment or recovery. The vent vapors (noncondensables) may be discharged to a flare, scrubber, or the atmosphere if appropriate.

A disadvantage of a quench drum is the requirement for a substantial amount of liquid. This will increase the size of the drum and produce large amounts of contaminated quench liquid. Use of this type of drum is limited by the type of organics present in the effluent; that is, it cannot be used for water-miscible organics, liquid low boilers, or fluids below 32°F (0°C).

The following are guidelines for the design of quench drums:
- A single drum may be used for more than one process unit. Consideration should be given to chemical compatibility and continuity of process operation if this drum is out of service.
- Single or multiple headers from various plant locations may enter the drum. Closed liquid headers should be run separately to the drum.
- The quench liquid should not react with the hot relieved fluids.
- The quantity of quench liquid is determined by the heat balance, assuming that the final temperature of the condensed fluid is 10 to 20°F (-6 to -12°C) below

condensing temperature. Continued reaction in the drum should also be considered in the heat balance.
- The materials of construction should be based upon the corrosive properties of the relieved fluid and the quench medium and operating temperature.
- A heating coil may be included in the drum to prevent solidification of condensed material at low temperatures or freezing of water by low boiling vapors.
- Instrumentation for pressure, temperature, and level control should be provided.
- Vapor and liquid loads to the quench drums are determined on the basis that all relieving devices from process units will discharge under one controlling contingency only (e.g., cooling water or power failure).
- Design pressure of the drum should be a minimum of 50 psig.
- Operating pressure of the drum should be based on the hydraulics of the discharge system and the downstream requirements in order to vent the vapors to a flare stack or other destination.
- If the drum is treated as a pressure vessel, it should be provided with means of overpressure protection.
- The liquid holdup volume of the drum should be sized for expansion of the quench liquid, collected condensate, and collected liquid carryover.

7.5.4 Incineration Systems

Incineration is the burning of waste in a closed system under carefully controlled conditions, such as a kiln or furnace. The efficiency of the process is measured by destruction of toxic or hazardous components. Incineration technology has been applied to liquid, sludge, solid, and gaseous wastes. Federal, state, and local regulatory agencies are considering or have already implemented regulations governing safety, design, and limitations on incinerator emissions. These standards are contained in 40 CFR 264 Subpart 0 for incinerators burning hazardous wastes and 40 CFR 761.70 for burning wastes containing Polychlorinated Biphenyls (PCBs) and 40 CFR 266 Subpart H for boilers and process furnaces burning hazardous wastes.

Thermal oxidizers are used to treat vent streams containing organic vapors with a wide range of concentrations. A thermal oxidizer consists of a refractory lined chamber that has one or more gas- or oil-fired burners located at one end. The burners are used to heat the gas stream to the necessary temperature to oxidize the materials, typically 1,300 to 1,800°F (704 to 982°C). The vent gas stream does not usually pass through the burner itself, unless a portion of the gas stream is used to provide the oxygen or fuel value needed to support combustion.

The primary design criterion needed to properly specify the system is a waste heat load and material balance showing the number and types of wastes to be incinerated. The heat balance should be as complete as the material balance for the system; however, it should be recognized that incinerator systems handle varying loads in terms of both quantities and composition. In many industrial applications, where high Btu wastes are common, the limiting factor of the incinerator unit will be the heat load on the system, not the material transfer.

7. PROTECTION LAYERS

The design should include provisions for handling the effluents in case the effluent disposal / treatment facilities are inoperative. Consideration should also be given to the schedule of operations (for example, batch or continuous process) and the procedures for disposal of waste loads from the incineration system.

Incinerator hazards are similar to those involved in combustion processes located near flammable materials. Storage areas for materials, particularly liquids and sludges, should be designed to prevent flammable or detonable material from coming in contact with an ignition source, including the incinerator itself. Fire detection and protection equipment should be the same as that used in the rest of the plant. Additional care should be taken to ensure that incompatible wastes are not mixed in one vessel. This has been the cause of several waste storage fires.

In all cases, consideration should be given to installation of detonation arresters for last-resort, passive protection against deflagrations and detonations in vapor lines.

7.5.5 Vapor Control Systems

Vapor control systems are intended to collect vapor during transfer or storage. Use of these systems is increasing, primarily to meet environmental requirements. Economic factors may determine whether to destroy or recover the collected vapors. Recovery may require additional equipment and piping, e.g., to refrigerate and recover the condensed vapor. If the loading system handles a variety of fluids, a separate recovery system might be required for each vapor, or the recovered vapors might be an unusable mixture of compounds.

In general, a vapor control system design and installation should eliminate potential overfill hazards, overpressures and vacuum hazards, and sources of ignition to the maximum practical extent. Each remaining hazard source which is not eliminated should be specifically addressed in the protective system design and operation requirements. See *Guidelines for Pressure Relief and Effluent Handling Systems* (Ref. 7-32) for additional information.

7.6 EMERGENCY RESPONSE ALARM SYSTEMS

A system for reporting and alerting plant personnel, the plant fire brigade and outside responders is important for any site, both large and small. The size of the facility will affect the complexity of the system.

This system should be as simple as possible to minimize the potential for confusion in emergencies. The preferred design is a multiplex system that alarms in the control room or some other 24-hour constantly attended location and activates visual devices, such as strobes or beacons, and communication and audible notification devices.

A number of factors should be considered in the design of a reliable detection and alarm systems, including:
- Data on the nature and arrangement of power sources
- Coverage provided by the system
- Alarm function on loss of system operability
- Suitability of detection devices for the risk involved
- Testing and maintenance procedures to ensure a reliable system

- Systems that should be activated or shut down on detection
- Consequence associated with false alarms

Where a detection system is part of an automatic, fixed fire extinguishing system, complete compatibility between the systems is essential.

It is important to select detection devices that are appropriate for the type of incidents most likely to occur. Failure to do so will result in either a very slow response or the possibility of a large number of spurious alarms. The latter should be particularly avoided where the detection system is used to activate a fixed fire extinguishing system or provide automatic shutdown of processes.

Main components are central control station equipment, emergency alarm stations, supervisory devices, and visual and audible alarm services. These systems can be used for all types of in-house emergencies, such as fires, explosions, vapor releases, liquid spills, and injuries.

The type of alarm system should be chosen based on personnel resources available at the facility. For continuously staffed facilities, proprietary supervised systems are preferred. For facilities staffed less than continuously, remote supervised station or central station alarm systems are usually considered. In these systems, alarms are monitored by an outside firm responsible for alerting appropriate personnel or by the local fire department.

It is important that appropriate escape routes and / or emergency response procedures are in place. The following outlines emergency response features:
- Emergency exits and safe evacuation routes to allow exit away from the fire hazards, including two means of egress.
- Emergency action procedures and training on the procedures.
- Mustering stations located away from potential fire hazards for accountability purposes.
- Means to warn of fire and method for informing employees of actions to take.

Emergency alarm system design and installation should be in accordance with NFPA 72 (Ref. 7-36). Electrical aspects of the fire alarm systems should be designed and installed in accordance with NFPA 70 (Ref. 7-3). When devices are located in hazardous areas, they should meet the electrical requirements suitable for that hazardous area.

7.6.1 Plant Emergency Alarm and Surveillance

Alarm and surveillance systems are an important element of protection systems. These alarm and surveillance systems:
- Provide notification of emergency events.
- Can be used manually by people observing the emergency.
- Can automatically activate protection systems.
- Notify those onsite of an emergency and communicate actions to take.
- Provide surveillance of the facility for fire.
- Notify offsite emergency response organizations.

7. PROTECTION LAYERS

Failure to promptly report a fire could result in greater damage and, more importantly, could delay warning affected personnel. The alarm and surveillance element of fire prevention triggers emergency response and has a major impact on the control of property losses, safety of personnel, and community impact. Some key components of alarm and surveillance are:

- A continuously staffed location for receiving and acting on reported incidents and emergencies.
- Automated detection and protection systems to signal at an offsite central alarm station service for continuous monitoring.
- Back-up power to operate emergency alarm systems in the event of main power failure.
- A reporting system for personnel to report incidents and emergencies to the staffed station. This could include an "alarm pull-box" system, plant telephones, or radios.
- An alarm system for notifying personnel of an emergency in progress and for communicating action required, such as information only, shelter-in-place, or evacuate. This could include bells, sirens, whistles, horns, or public address systems.
- A documented procedure for periodically and systematically testing the reporting and alarm systems to confirm their functionality.
- Assurance of an acceptable level of surveillance for the facility by appropriate resources, procedures, and facility design features.

Each type of emergency alarm or signal should clearly inform those onsite of the actions to be taken. This requires training and testing of the alarm so personnel can recognize the alarm and take appropriate action. Some of these alarms may be automatic. For example, detection of a fire may be signaled directly by the protection or detection system rather than by an individual. This alarm signal may alert not only personnel in the immediate area, but all facility personnel and the community fire department.

The alarm and surveillance procedure should also describe how to use the warning and alerting equipment, which may include telephones, alarms, buzzers, lights, horns, public address systems, radios, and pagers. A useful addition to this procedure is a simple flow diagram indicating how information is distributed, an emergency call recording form, and a regulatory reporting requirements form.

For additional information, refer to NFPA 72 (Ref. 7-36) and NFPA 101 (Ref. 7-37).

7.6.1.1 Evacuation

Depending on the severity of the incident, evacuation may be necessary. Evacuation can be for portions of a facility or the entire facility. When evacuation is an option at a facility, procedures should include:

- Means to warn personnel both inside building and outside
- Evacuation plan that directs personnel to designated specified assembly area
- Emergency exits and safe evacuation routes
- Emergency action procedures and training that will facilitate evacuation

- Plan to account for occupants

7.6.1.2 Shelter-in-Place

Shelter-in-place is a concept used when a toxic release occurs and personnel do not have time to evacuate because the incident occurs very quickly. Shelter-in-place is the use of a building, vehicle or other enclosed space to provide protection against exposure to a toxic gas or vapor. Personnel downwind of a release who are in a building or vehicle should, in most circumstances, stay in the building or vehicle (shelter-in-place) and take action to minimize the ingress of vapors by closing doors, windows, fresh air intakes, and other openings between the space and the outdoors.

Where onsite buildings are used for shelter-in-place, written procedures should be developed to guide personnel in entering and securing the shelters. Equipment required to implement the procedures should be maintained. Considerations for the design of shelter-in-place buildings include:

- Signals for entering and leaving temporary shelters
- Communication equipment necessary
- Equipment to be maintained at each location, including communications equipment (radio or telephone), and materials for securing the shelter
- Heating, Ventilation and Air Conditioning (HVAC) systems capable of shutdown of the system or placement in recirculation mode, whichever is more appropriate
- Seals for all windows, doors and penetrations

The emergency response plan should include a list of all approved shelter-in-place buildings, procedures for accounting for personnel in shelter-in-place buildings, and methods to evaluate the situation if conditions worsen.

7.6.2 Gas / Fire Detection

Where releases of flammable or toxic chemicals are possible, a flammable or toxic gas alarm system is often established as part of the plant emergency alarm system. Best practices require different types of alarms to be annunciated differently, both audibly and visually. For example, the toxic alarm stations may be provided with a blue light to distinguish them from fire alarm stations that are red. A consistent color system for lights should be adopted.

Gas detection systems are typically utilized for one or more reasons including:

- *Personnel Safety* - These systems typically go beyond those required by regulatory agencies and are intended to:
  - Assure that personnel exposure to ambient concentrations of airborne contaminants remain within health based exposure indices.
  - Alert personnel to the presence of a release.
- *Property Protection* - This protection is provided to detect and avert situations that could lead to fires or explosions. In some cases the intent may be to control other means of property loss, such as excessive corrosion due to airborne contaminants.

7. PROTECTION LAYERS

- *Regulatory Monitoring* - To ensure or prove that the concentrations of airborne contaminants are being kept below regulated levels for personnel safety or community impacts.
- *Community Impacts* - It may be useful to put detection equipment in place to detect or avert large releases that could pose a hazard to a large portion of the facility, environment, and / or the public outside the fence line.

Taking the time to proactively address the need up front will allow the company or facility to make its gas detection decisions based on objective technical merits. If used properly, this information will help to:

- Verify that the releases of concern can actually be detected with a relatively high probability of success. If this is not the case, it will help everyone feel more comfortable in moving towards preventive or other detection technologies to better address the situation.
- Assure that the planned response to the release of concern is appropriate. As an example, it may be totally acceptable and safe for properly equipped personnel to enter a toxic atmosphere to manually isolate a small release. However, it is probably not acceptable to send personnel into a 10-ton release of flammable gas in hopes of isolating it before it ignites.
- Understand the limits of the detection provided by the gas detection systems. This will minimize the potential for over reliance on the gas detection system and foster an atmosphere where personnel apply appropriate degrees of awareness and precaution.

There are a variety of sensors used in gas detection systems. Typical sensors include:

- Electrochemical Sensors
- Infrared (IR) Sensors
- Catalytic Bead Sensors
- Photoionization Detector (PID)
- Thermal Conductivity (TC)
- Colorimetry

For additional information on sensor types, see *Continuous Monitoring for Hazardous Material Releases* (Ref. 7-38).

7.6.2.1 Installation of Sensors

Gas detection location is important in detecting releases, however, obvious placement such as high up for lighter-than-air gases and down low for heavier-than-air gases may not always work.

Some important gas and vapor detection system features to consider when selecting a system are: self-checking, auto-calibrating or self-diagnosing capabilities with sensor out-of-service, or malfunction alarming.

Arrangement of detectors generally falls within the following configuration:

- *Open Path Gas Detectors* (OPGDs) produce an IR beam that is directed across the area to be monitored. The received light is analyzed at two or more frequencies, some of which is absorbed by the target gas or gases; the reference

frequency is not. Given the initial and final intensities, the average concentration of gas in the path is calculated and transmitted.
- *Point Combustible Gas Detectors* (IR) are used to indicate the presence of gas at a particular location (e.g., in a congested area of the plant or in small ducts).

7.6.2.2 Guidance for Fire Detectors

Fire detectors generally fall within one of three categories:
- *Heat* - Heat detectors work by sensing the heat from a fire.
- *Smoke* - Smoke detectors sense the combustion products from the fire.
- *Flame* - Flame detectors identify flame by sensing the IR or UV light it emits.

Where process equipment is provided with fixed-temperature detectors, these should be located as near as possible to the potential fire source, for example, above flammable liquid pump seals, immediately over a solvent draw-off point, or mounted above a crude tank mixer stuffing box. As a general rule, fixed-temperature detectors directed at a potential hazard should be considered only for process equipment where specific fire problems are anticipated. Fixed-temperature detectors are preferred because they require less calibration and maintenance.

Smoke detectors are primarily used where smoldering fires can be expected and where electrical equipment is located indoors. Examples of their use are in offices and sleeping quarters, computer rooms, control rooms, electrical switchgear rooms, etc.

The available types of flame sensing detectors are Ultraviolet (UV), Infrared (IR), combination of these, and monitored Closed Circuit Television (CCTV). These devices operate on the detection of certain wavelengths of light emitted by flames. They are used when there is a potential for fires that rapidly produce flame such as flash fire.

7.6.2.3 Guidance for Flammable Gas Detectors

The objective of an area gas detection system is to detect gas clouds of sufficient size that, if ignited, could cause damage as a result of a flash fire or explosion overpressure. The size of a gas cloud requiring detection is based on the volume of the area, the level of building confinement, and equipment congestion. In assessing the need and value of installing combustible gas detection, the following should be considered: the nature of the hazard, consequence of ignition, gas composition, confinement of the area, equipment congestion, and required response time.

The primary consideration in evaluating the need for combustible gas detection is to determine if the area in question encompasses any sources of flammable vapors or gases. With this said, there are several types of operations that can usually be eliminated from consideration:

- Areas handling combustible liquids below their flash points, as there will be no vapors to detect.
- Areas handling combustible liquids, flammable liquids, or flammable gases at or above their autoignition temperatures, because these materials will ignite spontaneously upon being exposed to air.

7. PROTECTION LAYERS 355

- Areas handling lighter than air gases where no roof or deck is present above the release sources, as the footprint of any release is likely to be too small to practically detect.

The primary challenge in using fixed combustible gas detection is that it is impractical, if not impossible, to detect minor releases, for the same reasons discussed under toxic detection. In addition, flammable releases will have much smaller detectable footprints than similarly sized toxic releases, because their concentrations of concern (essentially measured in parts per hundred) are much higher than those of toxic releases (measured in parts per million). Moderate to large releases also pose a detection challenge because:

- The rate of area engulfment and vapor travel leave very little time for intervention by control systems or operators.
- The threat of flash fires and explosion make operations in the release area particularly dangerous for operating personnel and emergency responders.
- Explosions emanating from enclosed spaces or highly congested or obstructed portions of the plant can be quite powerful and result in widespread damage throughout the facility, thus spreading the incident.

Due to these challenges, the objectives of most flammable detection programs in the process industries are limited to:

- Alerting personnel to the accumulation of combustible gases in buildings due to releases within the space or the ingress of exterior releases.
- Initiating the shutdown of internal process streams / equipment and ventilation systems where combustible gases have accumulated in an enclosed space.
- Alerting personnel to releases that may affect highly congested or obstructed areas of the plant from which powerful explosions may propagate.
- Alerting personnel to large releases in the area of high potential release sources.
- Alerting personnel to releases that may affect commonly used access routes, normally occupied areas, emergency marshalling points or the public.
- Alerting personnel to releases that are affecting their immediate location.

Providing combustible gas detection in elevated locations, such as process unit decks and on offshore docks, is not practical in most cases, because the number of factors at play precludes a reasonable degree of detection success.

7.6.2.4 Guidance for Toxic Gas Detectors

A toxic material is defined as:

Toxic is defined as a chemical having a median lethal concentration (LC50) in air of more than 200 parts per million but not more than 2,000 parts per million by volume of gas or vapor, or more than two milligrams per liter but not more than 20 milligrams per liter of mist, fume, or dust, when administered by continuous inhalation for one hour (or less if death occurs within one hour) to albino rats weighing between 200 and 300 grams each. (ref OSHA 1910 Part z)

The facility should also consider how it intends to manage and physically respond to toxic releases. By doing so, the facility will be able to develop coordinated policies and procedures concerning warning systems, evacuation routes, responder access routes, and mechanisms of control. It may also be helpful to consider the industrial hygiene monitoring requirements that an accidental release may trigger, as it may be impractical or unsafe to utilize portable monitoring equipment during such an event.

Toxic releases are problematic because their low concentrations of concern, as compared to flammable releases, can make even the most minor release a hazard. The footprint of these minor releases can be very small, making it difficult to reliably detect them. It is common in some industries to use a large number of detectors strategically located to provide earlier warning of a likely release, as well as to aid in profiling the location and migration of toxic releases.

For example, chlorine detectors are frequently located in production, processing, and loading areas set to alarm at the lowest detection limits to provide early warning for small leaks to allow detection before they increase in size. Additional detectors are located throughout the facility to aid in tracking the source and direction of chlorine leaks, determining relative quantities involved, and forecasting potential offsite impact.

On the other hand, it is comparably easy to predict / control where personnel will normally enter an area, how they will traverse the area while performing rounds / assignments, and where they may spend significant amounts of time. As a result, the use of toxic gas detection is best focused on protecting frequently and normally occupied locations in order to meet the objective of alerting personnel to:

- Toxic and oxygen-deficient atmospheres in buildings due to internal releases or the ingress of exterior releases. These systems may also be utilized to initiate the shutdown of the HVAC system and the process equipment that the buildings may house.
- Releases that may affect commonly used access routes, normally occupied areas, emergency marshalling points, or the public.
- Large releases in the area of high potential release sources.
- Releases that are affecting their immediate location.

7.6.3 Leak Detection

Most leak detection alarms are either fire or gas; however, there are some that do not fall into those categories.

- *High Temperature Detection* - These systems utilize temperature sensors placed in containment areas, trenches, and sewer systems where hot process fluids may run to or accumulate upon release.
- *Low Temperature Detection* - These systems utilize temperature sensors placed in containment areas, trenches, and sewer systems where cryogenic materials such as LNG may run to or accumulate upon release.
- *Oil-on-Water Detection* - These systems measure electromagnetic energy absorption to identify when the layer of hydrocarbon product on the surface of accumulated water has exceeded threshold thicknesses. These devices could be placed in containment areas, trenches, and sewer systems.
- *Liquid /Liquid Concentration* - These systems measure microwave absorption of a liquid to identify the percentage of hydrocarbon within a water stream or vice versa. These devices could be placed in containment areas, trenches, piping traps, and sewer piping.

7. PROTECTION LAYERS

- *Liquid Level Detection* - There are a multitude of technologies available for monitoring liquid levels.

7.7 FIRE PROTECTION

There are different types of fire and many different firefighting agents for combating them. An understanding of how these different types of firefighting agents are used in fire protection is important because their effectiveness can vary widely when applied to different types of fires.

It is important to analyze all materials and processes associated with a particular process including production, manufacturing, storage, or treatment facilities. Each process requires analysis of the potential for fire.

Hydrocarbon fires are a principal concern in many processing facilities. There are many different types of hydrocarbon fires. The mode of burning depends on characteristics of the material released, temperature and pressure of the released material, ambient conditions, and time to ignition.

Other fires that can occur in specific areas within a processing facility include:
- Solid material fires, e.g., fires involving wood, paper, dust, plastic, etc.
- Electrical equipment fires, e.g., transformer fires
- Fires involving oxygen, e.g., systems for oxygen addition to a Fluid Catalytic Cracking (FCC) unit
- Fires involving combustible metals, e.g., sodium
- Fires involving pyrophoric materials, e.g., aluminum alkyls

Fire protection systems should be designed to accomplish a combination of the following objectives:
- Extinguishment of fire
- Control of burning
- Exposure protection
- Prevention of fire

For additional information, see *Guidelines for Fire Protection in Chemical, Petrochemical, and Hydrocarbon Processing Facilities* (Ref. 7-39).

7.7.1 Structural Fireproofing

This section discusses basic design guidelines for fireproofing or passive fire protection in areas where flammable liquids and gases are processed, handled, and stored. API 2218, *Fireproofing Practices in Petroleum and Petrochemical Processing Plants* (Ref. 7-40), can be referenced for additional information.

Fireproofing is a fire-resistant material or system that is applied to a surface to delay heat transfer to that surface. Fireproofing, a form of passive fire protection, protects against intense and prolonged heat exposure that can cause the weakening of steel and eventual collapse of unprotected equipment, vessels, and supports and lead to the spread of burning liquids and substantial loss of property. The primary purpose is to improve the capability of equipment / structures to maintain their integrity until the fire is extinguished by either stopping the fuel source or active fire protection methods.

The principal value of fireproofing is realized during the early stages of a fire when efforts are primarily directed at shutting down units, isolating fuel flow to the fire, actuating fixed suppression equipment, and setting up portable firefighting equipment. During this critical period, if non-fireproofed equipment and pipe supports fail due to fire-related heat exposure, they could collapse and cause gasket failures, line breaks, and equipment failures, resulting in expansion of the fire. Fireproofing may be applied to control or power wiring to allow operation of emergency isolation valves, vent vessels, or actuate water spray systems during a fire.

Determining fireproofing requirements involves experience-based or risk-based evaluation (Ref. 7-39). An approach for selecting fireproofing includes the following steps:

- Conducting a hazard evaluation, including quantification of inventories of potential fuels.
- Developing fire scenarios, including potential release rates and determining the dimensions of fire-scenario envelopes.
- Determining fireproofing needs based on the probability of an incident considering industry experience, the potential impact of damage for each fire-scenario envelope, and technical, economic, environmental, regulatory, and human risk factors.
- Choosing the level of protection (based on appropriate standard test procedures) that should be provided by fireproofing material for specific equipment, based on the needs analysis.

7.7.2 Firefighting Agents

Before fire protection systems can be adequately discussed, it is important to understand the agents available for fire protection and their particular application. These include:

- Water
- Foam
- Carbon dioxide
- Dry chemical
- Clean agents

Table 7.4 highlights the advantages and limitations of the various extinguishing agents (Ref. 7-39).

7.7.3 Fire Water Systems

Normally, fire water demands a range between 2,000 and 10,000 gpm (7,600 to 38,000 lpm). The design capacity of the fire water system should be at a minimum 4-hours of continuous operation of the largest fire water demand. The capacity is based on a number of factors, including:

- Sources of water available
- Reliability of make-up water supply
- Potential for escalation to other areas of the facility
- Isolation philosophy and the ability to depressure high pressure units

Table 7.4 Advantages and Limitations of Various Extinguishing Agents

Agent	Type Extinguishment	Advantages	Limitation
Water	Cooling Smothering Dilution Exposure	Available Very low cost	Not for Class C electrical fires Freezes at 32°F (0°C) Reactive with some material, e.g., sodium, magnesium Cannot extinguish low flash point materials
Foam	Smothering	Best for Class B Pool Fires (Two-dimensional fires)	Not for electrical fires Foam blanket may break-up Not applicable for LPG
CO_2	Smothering Reduction Some cooling	Non-reactive No residue Class C	Reduces O_2 level Toxic to people (asphyxiant) Not applicable for oxidizers Not practical for outdoor use
Dry Chemical	Chain breaking	Classes B and C	Fire reflashes if not completely extinguished or hot surfaces are present (especially flammable / combustible liquids)
Clean Agent	Chain breaking Inerting	Good for Classes A, B, C	Not for outdoors May produce toxic gases

The reliability of the fire water supply should be such that the loss of any one source does not result in a loss of more than 50% of the flow requirements of the system. For example:

- A large facility connected to the city water supply should have two independent connections off different branches of the city underground piping.
- A facility drawing water from a stream or lake should either have independent locations from which to draw from or have a back-up supply from the city system or a private well. Care should be taken when using potable and non-potable sources so that cross-contamination does not occur.
- Fire water pumping capacity (flow rate) should be sufficient to provide the required amount of water at required pressure to the fire areas having the greatest demand. At least 50% of the pumping capacity should be from diesel-driven pumps. Fire water pumps should have a minimum capacity of 1,500 gpm (5,700 lpm) and can range up to 5,000 gpm (18,930 lpm).

It is common practice to provide pumping capacity so that when the largest fire water pump is out of service, the total fire water demand can still be met. In situations where the demand does not exceed 1,500 gpm (5,700 lpm), it may be acceptable to use a single pump.

The reliability of the power supply should be determined, taking into account the frequency of power outages and extent of interruption. Consideration should be given to connecting electrically driven fire water pumps to the emergency power system, where one exists.

7.7.4 Mitigation Systems

7.7.4.1 Water Spray Systems

The term "water spray" refers to the use of water discharged from nozzles having a predetermined pattern, droplet size, velocity, and density. While deluge systems are for the overall protection of a given area in accordance with NFPA 13, *Installation of Sprinkler Systems* (Ref. 7-41), water spray systems can be installed for the protection of a given area or specific equipment / hazards. Design guidance for water spray systems can be found in NFPA 15, *Standard for Water Spray Fixed Systems* (Ref. 7-42).

Water spray systems are used for protection against hazards involving gaseous and liquid flammable materials and combustible solid materials. These systems are used to:

- Cool metal
- Control fire intensity
- Prevent ignition
- Protect vital instruments runs
- Prevent formation of flammable vapor clouds
- Control toxic vapor clouds

Fixed water spray systems are most commonly used to protect flammable liquid and gas vessels, piping and equipment, process structures and equipment, electrical equipment such as transformers, oil switches, rotating electrical machinery, and openings through which conveyors pass. The type of water spray will depend on the nature of the hazard and the purpose for which the protection is provided.

The design of water spray systems should be in accordance with NFPA 15 (Ref. 7-42).

7.7.4.2 Water Mist Systems

Water mist systems are intended for rapid suppression of fires using water discharged into completely enclosed limited volume spaces. Water mist systems are desirable for spaces where the amount of water that can be stored or that can be discharged is limited. In addition, their application and effectiveness for flammable liquid storage facilities and electrical equipment spaces continues to be investigated with optimistic results. Water mist systems are also used for gas turbine enclosure protection.

A water mist system is a proprietary fire protection system using very fine water sprays. The very small water droplets allow the water mist to control or extinguish fires by cooling of the flame and fire plume, oxygen displacement by water vapor, and radiant heat attenuation.

7. PROTECTION LAYERS

The design of water spray systems should be in accordance with NFPA 750 (Ref. 7-43).

7.7.4.3 Foam Systems

Foam is primarily used for extinguishment of two-dimensional surface fires involving liquids that are lighter than water. Foams may be used to insulate and protect against exposure to radiant heat. They also act to prevent ignition of flammable liquids that are inadvertently exposed to the air (typically due to a spill), by separating them from air by spreading foam completely over the exposed surface. However, progressive foam breakdown can render the protective foam coating useless; thus, frequent reapplication may be necessary.

Because of water content, foams may be used to extinguish surface fires in ordinary combustible materials, such as wood, paper, rags, etc. Foams are arbitrarily subdivided into three ranges of expansion roughly corresponding to certain types of usage:

- *Low Expansion Foam* - expansion up to 20 times foam to solution volume
- *Medium Expansion Foam* - expansion from 20 to 200 times foam to solution volume
- *High Expansion Foam* - expansion from 200 to 1,000 times foam to solution volume

The design of foam systems should be in accordance with NFPA 11 (Ref. 7-44) and NFPA 16 (Ref. 7-45).

7.7.4.4 Clean Agents

Clean agents are electrically nonconductive, volatile, or gaseous fire extinguishing agents that do not leave a residue. Clean agents fall within two categories: halocarbons and inert gases. Typical halocarbons include Hydrofluorocarbons, (HFCs), Hydrochlorofluorocarbons (HCFCs), Perfluorocarbons (PFCs or FCs), and Fluoroiodocarbons (FICs). Typical inert gases include argon, nitrogen, carbon dioxide, or combinations of these agents.

Clean agent fire extinguishing systems are used primarily to protect enclosures. Clean agents can be used to protect enclosures containing:

- Electrical and electronic equipment
- Sub-floors and other concealed spaces
- Flammable and combustible liquids
- Telecommunication equipment

Clean agent systems can also be used for explosion prevention and suppression where flammable materials are confined. Clean agents should not be used for fires involving:

- Materials that are capable of rapid oxidation such as cellulose nitrate and gunpowder
- Reactive metals, such as lithium, sodium, potassium, magnesium, titanium, zirconium, uranium, and plutonium
- Metal hydrides
- Chemicals capable of decomposition such as organic peroxides and hydrazine

The design of clean agent systems should be in accordance with NFPA 2001 (Ref. 7-46).

7.7.4.5 Carbon Dioxide Systems

Carbon dioxide systems should be designed and installed and tested in accordance with NFPA 12. Fixed CO_2 systems may be total flooding or local application systems as described in the following sections. CO_2 systems should not be used where personnel may be present due to the asphyxiation hazard with CO_2.

Total flooding carbon dioxide system may be used where there is a permanent enclosure around the area or equipment to be protected that is adequate to enable the required concentration to be built up and to be maintained for the required period of time to ensure complete and permanent fire extinguishment. The minimum inert gas concentrations for the suppression of flammability in air are provided in NFPA 12 (Ref. 7-47).

The design of carbon dioxide systems should be in accordance with NFPA 12 (Ref. 7-47).

7.7.4.6 Dry Chemical Systems

Dry chemicals are recognized for their unusual efficiency in extinguishing two-dimensional fires involving flammable liquids. Fast extinguishing action is achieved provided the agent engulfs the fire without interruption of the application. The finely divided powder acts with a chain-breaking reaction by inhibiting the oxidation process within the flame itself.

These agents are effective on small spill fires. If there is risk of reignition from embers or hot surfaces, these ignition sources should be quenched or cooled with water and secured with foam, or the source of fuel should be shut off before attempting extinguishment.

The design of dry chemical systems should be in accordance with NFPA 17 (Ref. 7-48).

7.7.5   Portable Fire Suppression Equipment

Based on the physical layout of the site, the hazards of the process, and the fixed fire protection systems, additional fire suppression equipment may be required to effectively manage a fire emergency. Portable fire suppression equipment assists in providing fire protection for the equipment involved in an emergency. The use of portable equipment provides protection, where the cost of a fixed system may not be acceptable or the fire water supply may be limited.

7.7.6   Fire Extinguishers

Fire extinguishers purchased in the United States should be listed by Underwriters Laboratories (UL). Extinguishers for marine use should bear the label of the U.S. Coast Guard or other Authority Having Jurisdiction (AHJ). Extinguishers and agents purchased outside the United States should be approved by the AHJ, such as the governmental authority. For more information on fire extinguishers, see NFPA 10 (Ref. 7-49).

7. PROTECTION LAYERS

7.8 DEFLAGRATION/DETONATION ARRESTERS

Flame arresters are broadly divided into two major types: deflagration and detonation flame arresters. A number of essential points about flame arresters are as follows (Ref. 7-50):

- A flame arrester is a device permeable to gas flow but impermeable to any flame it may encounter under anticipated service conditions. It should both quench the flame and cool products sufficiently to prevent reignition at the arrester outlet.
- Proper application of a flame arrester can help avoid catastrophic fire and explosion losses by providing a flame barrier between at-risk equipment and anticipated ignition sources.
- Flame arresters have often failed in practice. Plant inspections have shown that misapplication of flame arresters continues to be common.
- Flame arresters can only be proven by tests simulating the conditions of use. The user should ensure that a flame arrester has been properly tested to meet the intended purpose and should be prepared to stipulate the required performance standard or test protocol to be followed.

In almost all cases, if a flame arrester is placed in-line rather than at (or close to) the open end of a vent pipe, a detonation flame arrester is needed. Detonation flame arresters should be able to stop both deflagrations and detonations. They require extensive testing and mandatory testing protocols may apply.

Flame arresters are classified according to certain characteristics and operational principles, as follows:

- *Location in Process* - Flame arresters are classified according to their location with respect to the equipment they are protecting. When a flame arrester is located directly on a vessel / tank vent nozzle or on the end of a vent line from the vent nozzle, it is called an end-of-line arrester and is usually a deflagration flame arrester. These are commonly installed on atmospheric pressure storage tanks, process vessels, and transport containers. If the arrester is not placed at the end of a line, it is known as an in-line arrester. In-line arresters can be of the deflagration or detonation type, depending on the length of piping and pipe configuration on the unprotected side of the arrester and the restrictions on the protected side of the arrester. A detonation flame arrester is used in all cases where sufficient "run-up" distance exists for a detonation to develop.
- *Combustion Conditions* - Deflagration flame arresters on tanks are designed to stop a flame from propagating into a tank from an unconfined atmospheric deflagration or to prevent a flame generated from a confined volume deflagration in a vessel from escaping to the outside of the vessel (Ref. 7-51). They normally cannot withstand significant internal pressure and cannot stop detonations. Typical flame speeds in a deflagration occurring in piping range from 10 to 200 ft/s. Detonation flame arresters are devices designed to withstand and extinguish the high speed and high pressure flame front that characterizes a detonation propagating through a piping system. Therefore, a detonation arrester should be able to withstand the mechanical effects of a detonation shock wave while quenching the flame. Some designs have a "shock absorber" in front of the flame arresting element to reduce both the high

pressure shock wave and the dynamic energy and to split the flame front before it reaches the flame arrester element (Ref. 7-51).

- *Arrester Element (Matrix) Construction for Dry-Type Arresters* - Dry-type deflagration and detonation flame arresters have an internal arrester element (sometimes called a matrix) that quenches the flame and cools the products of combustion. A great number of arrester elements have been developed and used. The most common types currently available are as follows:
  - Crimped metal ribbon
  - Parallel plate
  - Expanded metal cartridge
  - Perforated plate
  - Wire gauze and wire gauze in packs
  - Sintered metal
  - Metal shot in small housings
  - Ceramic balls
  - Oxidant concentration reduction

7.8.1 Selection and Design Criteria

7.8.1.1 Flame Propagation Direction

The flame propagation direction affects the type of flame arrester selected. An end-of-line or in-line deflagration flame arrester used for the protection of an individual tank may be of a unidirectional design because the flame will only propagate from the atmosphere towards the tank interior. A bidirectional flame arrester design, however, is needed for an in-line application in a vapor recovery (vent manifold) system because the vapors should be able to flow from the tank interior into the manifold or from the manifold into the tank interior. Consequently, flame may propagate in either direction.

7.8.1.2 Material Selection Requirements

When the materials handled are noncorrosive, the flame arrester vendor's standard materials of construction for the housing and arrester element are commonly used. For noncorrosive service housings are normally available in aluminum, carbon steel, ductile iron, and 316 stainless steel; the elements are commonly available in aluminum or 316 stainless steel.

7.8.1.3 System Constraints

The length and configuration of the piping system on the run-up (unprotected) side of the arrester can determine whether a detonation will occur. This includes the positioning of any turbulence-promoting flow obstructions such as tees, elbows, and valves, which can significantly increase the flame speed. Thus, it is of great importance to establish where ignition might occur in the system and how this will affect the flame path to the arrester. Pipe diameter also affects the distance required for Deflagration-to-Detonation Transition (DDT); larger pipe diameters typically require longer run-up distances for detonation. Testing has shown that reductions in pipe diameter along the pipe run dramatically increase the flame speed and pressure. The opposite effect occurs with

7. PROTECTION LAYERS

increased pipe diameter. Although detonations may fail on encountering branches into smaller pipe diameters, run-up to detonation may reoccur (Ref. 7-52).

7.9 EXPLOSION SUPPRESSION

In many cases, deflagration flames can be extinguished before unacceptable pressures occur within process equipment and piping if the onset of combustion can be detected early and an appropriate extinguishing agent can be delivered to the proper location within equipment or piping. The technique of deflagration suppression is applicable to most flammable gases and vapors, combustible mists, or combustible dusts that are subject to deflagration in a gas-phase oxidant. Suppression systems are active systems that include components for detection, suppressant delivery, and electrical supervision to assure readiness to operate and interlock functions to shut off or isolate other process equipment connected to the equipment to be protected.

Deflagration suppression systems can be applied to a large number of types of process equipment, rooms, and piping systems (including vent manifolded systems) (Ref. 7-35). Deflagration suppression is a competitive process between a rising rate of combustion heat release and a delayed, but rapid, delivery of extinguishing agent. The deflagration will be suppressed when the unburned fuel-air mixture has been rendered noncombustible due to the addition of an extinguishing agent, or the combustion zone has been cooled to the point of extinguishment, or the reaction kinetics are impeded. The time required for a suppression system to stop a flame front from propagating is dependent on the equipment or piping volume (time increases with increasing volume) and the flame speed of the material being handled (time decreases with increasing flame speed). For example, the time to suppress a deflagration in a 1.9-cubic-meter vessel takes about 100-milliseconds, while it takes 250-milliseconds for the same event in a 25-cubic-meter vessel. Each application requires experimental validation of suppression system design.

A suppression system consists of three subsystems for:

1. Detection
2. Extinguishment
3. Control and supervision

Incipient deflagrations are detected using pressure detectors, rate of pressure rise (or "rate") detectors, or optical flame detectors. Pressure detectors are employed in closed process equipment or piping, and particularly where dusty atmospheres are present. Rate detectors are used in processes that operate at pressures significantly above or below atmospheric pressure. Optical detectors may be Infrared (IR), Ultraviolet (UV) or Hybrid (i.e., both IR and UV) depending on the flame to be detected and the absorbent properties of the operating environment.

The extinguishment subsystem consists of one or more High-Rate Discharge (HRD) suppressant containers charged with suppressing agent and propellant. Normally, dry nitrogen is used to propel the agent out of the container into the equipment or piping. The propellant pressure is normally in the range of 300 to 900 psig, depending on the supplier of the suppression system. An explosive charge is electrically detonated and opens valves providing rapid agent delivery to the equipment or piping being protected.

Common extinguishing agents are water, Halon substitutes, and dry chemical formulations typically based on sodium bicarbonate or ammonium dihydrogen phosphate. The extinguishing mechanism of each agent is often a combination of thermal quenching and chemical inhibition. The selection of the appropriate agent is usually based on several considerations such as effectiveness, toxicity, cost, product compatibility, and volatility. Water is often a very effective suppressant, and should be used whenever possible since it is not toxic and is easier to clean up in comparison to the other types of suppressants. Dry chemical agents have been used for many years in Europe and are being used now more often in the United States. Halons were used for many years as they were very effective suppressants, but they have been outlawed in many countries because of their adverse effect on stratospheric ozone (the "ozone layer"). Numerous substitutes are now available, but none have been found to be as effective as Halon 1301 and other Halons. One new substitute that is being widely used is FM 200™ (a hydrofluorocarbon).

7.9.1 Oxidant Concentration Reduction

One of the most widely used methods of preventing deflagrations and detonations is oxidant concentration reduction. This method can be applied to process equipment and vent manifold systems. The prevention of deflagrations or detonations can be accomplished by either inerting or fuel enrichment.

In the case of inerting, the oxidant (usually oxygen) concentration is reduced by the addition of inert gas to a value below the Limiting Oxygen Concentration (LOC). Values of the LOC for many gases and dusts can be found in NFPA 69 (Ref. 7-35). Some commonly used inert gases used in industry are nitrogen, steam, carbon dioxide, and rare gases.

In the design of inerting systems, sufficient inerting gas should be provided to assure not only that the normal process conditions are rendered nonflammable, but also that any credible alteration of the process environment remains outside the combustible limits.

The minimum concentration of oxygen that can support a flame is known as the "Limiting Oxygen Concentration," or LOC, which is a singularity appearing at a fuel concentration marginally above the LFL.

A safety margin between the LOC and the normal oxidant concentration in the process equipment or piping system is mandated by NFPA 69.

Design and operating criteria for inerting systems are presented in NFPA 69. Another excellent reference on inerting systems is provided in the Expert Commission for Safety in the Swiss Chemical Industry (ESCIS) Booklet No. 3 (Ref. 7-53).

7.9.2 Deflagration Pressure Containment

Deflagration pressure containment is an approach for selecting the design pressure of a vessel so that it is capable of withstanding the maximum pressure resulting from an internal deflagration. Vessels or process equipment can be designed to either:
- Prevent rupture but allows deformation (known as "shock-resistant" design in Europe)
- Prevent rupture and deformation (requires a thicker vessel wall)

NFPA 69 (Ref. 7-35) provides equations for calculating the required design pressures for both types of containment design. It also discusses the limitations of deflagration pressure containment design.

7. PROTECTION LAYERS

In Europe process equipment such as spray dryers, fluid-bed dryers, and mills are available in "shock-resistant" designs for pressures up to 145 psig (10 bars).

Pressure containment can also be provided by using piping systems with a pressure rating above the anticipated maximum pressure generated during a deflagration.

More information about pressure containment design is available in NFPA 69 (Ref. 7-35).

7.9.3 Explosion Venting

Explosion venting can be used to lower the maximum pressure developed by a deflagration, for example, where the flame speed is lower than the speed of sound, but is not effective against detonations. The vent area required to prevent the explosion pressure from exceeding a desired value is dependent on several variables, including:

- The properties of the materials present, which are characterized by their deflagration index
- The vent opening pressure
- The length-to-diameter ratio (L/D) and volume of the enclosure being vented
- The weight of the vent panel (which should normally be below 2.5 lb/ft^2)

Other special circumstances should also be considered:

- If the enclosure is vented through a duct the vent area requirements will increase.
- If the vessels are interconnected, pressure piling can significantly increase the vent area.
- The explosion panel should be provided with a restraint to prevent the panel from becoming a hazard.

Additional details on this subject are provided by NFPA 68 (Ref. 7-54).

7.9.4 Equipment and Piping Isolation

It is common practice in the chemical process industries to provide isolation devices for stopping flame fronts, deflagration pressures, pressure piling, and flame-jet ignition between process equipment interconnected by pipes or ducts. There are several devices for providing this isolation:

- Suppressant barriers
- Fast-acting valves
- Material chokes
- Hydraulic (liquid seal) flame arrester

7.9.4.1 Suppressant Barriers

This type of isolation device (also called a chemical barrier) is similar to deflagration suppression systems used on process equipment. This barrier system consists of an optical sensor installed in the pipeline or duct between two items of equipment that detects an oncoming deflagration flame and emits a signal to a control unit. The amplified signal triggers the detonator-activated valve in a suppressant bottle, which injects an extinguishing agent into the pipeline through suitable nozzles. Pressure sensors are not normally used for pipeline barriers since there is no clear correlation between the front of the pressure wave and the flame front, and pressure sensor response times often are too slow for use in this application.

7.9.4.2 Fast-Acting Valves

A variety of fast-acting valves are available, including slide gate, flap (butterfly), and float (poppet) valves. Slide gate and flap-type valves are actuated (closed) upon a signal from a detector (sensor) in the pipeline between two items of interconnected process equipment. The detector sends a signal to a compressed air cylinder which then discharges the compressed air to a mechanism at the top of the valve, thereby closing the valve. A typical closing time for a fast-acting valve is about 25 milliseconds. The deflagration detector is located about 1 meter away from the source of ignition (equipment), and the fast-acting valve is usually installed 5 to 10 meters along the connecting pipeline, so that by the time the flame front reaches it, the valve is fully closed.

7.9.4.3 Material Chokes

Flame propagation can also be stopped between process equipment handling bulk solids and powders by judicious selection and design of bulk solids / powders conveying equipment such as rotary valves (rotary airlocks) and screw conveyors. The mass of bulk solids / powders contained in these items of equipment provide a tortuous path through which the gas and flame have to pass, and so can act as a "material choke" when certain design features are implemented.

Rotary valves will generally prevent flame propagation if the following criteria are followed (Ref. 7-35 and 7-55).

- Two vanes per side are always in contact with the housing,
- The vanes or tips are made out of metal (no plastic vanes),
- The gap between the rotor and housing is ≤0.2 mm,

In screw conveyors the removal of part of the screw will ensure that a plug of bulk solids / powder will always remain as a choke.

Several considerations have to be taken into account when using a rotary valve as a material choke. When a deflagration occurs in the equipment upstream of the rotary valve, the rotary valve has to be stopped immediately by a suitable detector in order not to pass burning or glowing solids into downstream equipment, which could then cause a second fire or deflagration. Rotary valves should be tested for their suitability as flame arresters as well as for their pressure rating with appropriate explosion tests (Ref. 7-55). These devices should properly be maintained to ensure that normal wear and tear not result in a loss of seal between the rotor blades and the housing.

7.9.4.4 Hydraulic (Liquid Seal) Flame Arrester

While all the flame arrester types discussed above have a solid arresting element (matrix), the hydraulic (liquid seal) flame arrester contains a liquid, usually water, to provide a flame barrier. It operates by breaking up the gas flow into discrete bubbles by means of an internal device to quench the flame. A mechanical non-return valve (check valve) is sometimes incorporated to prevent the displacement of liquid during or after a flame event (deflagration or detonation).

7. PROTECTION LAYERS

This arrester is usually designed to be effective in one direction only. However, hydraulic arresters exist that are reported to be effective in preventing flame propagation in both directions.

Proper design against flashback should ensure mechanical integrity of the vessel and internals during the flame event and prevent loss of the liquid seal. Suitable testing should also be performed to ensure that a hydraulic flame arrester design will work for a specific application.

ANSI / API Std 521 (Ref. 7-22) discusses the design of hydraulic flame arresters (liquid seal drums) for flares.

7.10 SPECIALTY MITIGATION SYSTEMS

7.10.1 Water / Steam Curtain

A water / steam curtain system uses an array of spray nozzles to create a barrier of water to contain a fire inside a small area. The nozzles can be mounted to shoot water either upwards or downwards. The goal of a water / steam curtain is to provide exposure protection or to stop the flow of flammable vapors from migration into a hazardous area.

7.10.2 Steam Snuffing

Steam snuffing is a method to assist in controlling a fire in a confined space. The snuffing occurs when steam is discharged into the confined space. The confined space can be a firebox for a furnace, boiler, or other heating mechanism. Snuffing is used when the application of fire water into the confined space cannot be done safely or the application of water could cause the loss of the mechanical integrity of the equipment or piping. Special care should be taken to ensure that the chemicals in the space will not react with the steam or steam condensate and cause additional damage to the equipment or create safety hazards for firefighting personnel.

7.10.3 Mechanical Interlocks

Mechanical interlocks are used for the following:
- One of the simplest forms of mechanical interlock is lockout / tagout. Lockout / tagout procedures have been implemented in most facilities.
- For electrical switches a mechanical interlock is arranging forward and reverse contacts so that it is physically impossible for both sets of contacts to close at the same time.

Actuated valves cannot be locked by preventing the valve from turning. Instead, the valve is locked by controlling access to the control equipment that energizes the actuator. For pneumatic and hydraulic actuators, this is usually a small diameter valve in the air or oil supply line to the actuator. Closing and locking this valve will disable the actuator. When interlocking motor-actuated valves, there are three considerations to be observed:

1. The switch buttons can only be operated within the permitted interlock sequence.
2. The exact position of the valve is confirmed before progressing / proceeding.
3. The manual override facility is incorporated within the interlock scheme.

Process interlocking is a long used principle to guide the operator safely through an operating sequence. Once the proper steps have been identified mechanical interlocks can be installed that prevent continuing until a key is inserted. An interlock guides the operator through the sequence with unique keys for each step. It is only when a mistake is made that the operator will not be allowed to proceed: a key will not fit or a valve will be locked in position. The principle of mechanical key interlocking is the transfer of keys. Each lock is executed with two keys, One for the locked open position and one for the locked closed position. When the valve is open the "open key" can be released and transferred to another lock with the same code. All keys are unique and depend on the sequence.

A three-way valve can be considered a type of interlock since it ensures the flow is directed in only the desired direction.

7.10.4 Inhibitor Injection

Inhibitor injection systems are primarily used with polymerizing materials such as vinyl acetate. If the material begins to self-react in an uncontrolled manner, then injection of a polymerization inhibitor can interfere with the reaction before sufficient pressure and temperature are built up to cause a release from the storage / handling containment. The type of inhibitor needed will depend on the nature of the polymerization reaction; for example, a free radical scavenger may be used as an inhibitor for a material that reacts by free-radical polymerization. The inhibitor is often the same inhibitor used for normal storage stability requirements but injected in a much larger quantity. If a different inhibitor is used that is designed to quickly kill the reaction, it is generally called a "short-stop" system. An inhibitor injection system consists basically of a supply of inhibitor, a means of delivery of the inhibitor to the reactive material, and a means of sensing the need for the inhibitor injection system and actuating the inhibitor delivery.

Provision should also be made for ensuring the inhibitor is adequately mixed. The potential for common mode failures should be carefully evaluated in this regard; for example, in a polymerization process, loss of agitation might both initiate an out-ofcontrol situation and prevent adequate mixing of added inhibitor. A backup mixing system such as a gas sparge may be required in such cases. Other means of inhibitor mixing include injection into a recirculating loop (requiring the proper functioning of the recirculation pump), using a mixing eductor, directly sparging into the storage tank, or by manual agitation in smaller, atmospheric-pressure operations with adequate safety precautions.

Inhibitor injection systems need to be carefully designed and maintained to provide a highly reliable last-resort safety system. Since the inhibitor injection system is on standby and may not be used for months, attention should be paid to how the system components can be functionally and effectively tested on a periodic basis, such as once a month, without excessive disruption of normal operations. This functional testing is important not only for the checking of adequate inhibitor supply and properly functioning delivery system, but also for the means of detecting an out-of-control situation and actuating the inhibitor injection system. If the system is actuated manually, the system should be part of periodic operator drills or simulation training. Most inhibitor injection systems are designed so that they can be actuated either automatically, such as upon sensing of high-high temperature or pressure, or manually, by the control room or field operator.

7. PROTECTION LAYERS

7.10.5 Quench System

Quench systems are used for essentially all types of reactive chemicals. A quench system involves the addition of flooding quantities of water or other quenching medium to the reactive material; the quenching medium might be a subcooled material such as liquid nitrogen or dry ice in special applications.

The means by which a quench system works depends on the nature of the reactive material; for example, for water-reactive materials, a quench system will destroy the material in a last-resort situation and generally form less-hazardous products and will at the same time absorb some of the heat of reaction. Most quench systems are designed to both cool down and dilute a material that may be reacting uncontrollably; the quenching medium may also actually interfere with the chemical reaction or deactivate a catalyst.

In a quench system, the water or other quenching medium is generally added to the storage / handling facility in significantly greater quantities than the inhibitor in an inhibitor injection system. For this reason, the system design should take the extra volume of material into account. Also, as for inhibitor injection systems, a means of mixing the quenching medium with the reactive material may be needed for effective quenching action.

7.10.6 Dump System

For an inhibitor injection or quench system, the inhibitor or quenching medium is transferred from an external supply to the reactive material; in a dump system, the reactive material is transferred from the storage / handling facility to a safer location that is the same size or, more commonly, larger than the normal capacity of the facility. This allows depressuring and deinventory of the reactive material from the facility in an out-of-control situation, such as an incipient thermal decomposition.

Detection and actuation of a dump system is similar to an inhibitor injection or quench system, with the dumping typically being started by actuating one or more quick-opening dump valves. Containment of the dumped material may be in an above-ground, below-grade, or "piggyback" tank; a sump; or a larger basin.

Design considerations include:
- Time required to dump the system, which should be carefully evaluated and take full consideration of back pressure and two-phase flow.
- Reliability of the detection and actuation system and the dump valves, including protection against inadvertent dumping.
- Ensuring adequate volume in the dump tank if one is used.
- Possible leakage of water or introduction of an incompatible material into the dump tank when not in use.
- Overpressure protection on the dump tank as well as the reactive material storage tank.

It is common to combine the effects of both dumping and quenching by dumping the reactive material to a location where a quenching medium is already located, such as a pond, basin, or large tank. To maintain an adequate level of quenching medium, periodic checks should be part of the facility's operating instructions and routine patrols. An alternate configuration is an eduction system with a high rate water supply, such as a fire water system, with the educted material being diverted to an empty system.

7.11 EFFLUENT HANDLING/POST-RELEASE MITIGATION/WASTE TREATMENT ISSUES

Large quantities of water may be used to fight fires in facilities handling chemicals. Since most flammable materials float on water, there is the potential for fire protection water to spread the fire. In addition, many chemicals have the potential to contaminate the soil and groundwater. Water used for firefighting can disperse these chemicals spreading the pollution. As a result of all these factors, there is a need for drainage systems to control water runoff. An equally important point to make is that drainage systems can control flammable liquid spills.

Drainage facilities should be designed to simultaneously carry flammable liquids and fire protection water away from buildings, structures, storage tanks, pipe racks and process equipment. Drainage systems should not expose adjacent plant facilities to burning or toxic materials during an incident. This may require diversionary curbs, trenches, collection sumps, skimmers, and holding ponds or basins.

In many cases, the water and chemicals collected during an incident will need to be "pretreated" prior to disposal in a waste water treatment facility, or the rate that these materials are introduced to the treatment process controlled. Small holding ponds for specific process areas should be sized to hold 30 to 60 minutes of discharge as a minimum. Where there is the potential for fires of long duration, such as in petrochemical or refining facilities, special precautions will be necessary. It may be possible to separate organics from fire water prior to disposal, thus reducing the size of the required holding area. The potential for soil and water contamination should not be used as a reason to avoid providing fire protection or drainage systems.

Drainage of liquids may not be desirable in some cases, since this may actually result in more serious concerns (such as large vapor clouds or contamination of drinking water), and in other cases, drainage and remote containment may not be practical due to space limitations or other reasons. In such cases, additional automatic fixed fire protection measures such as foam or water spray may be necessary to ensure adequate protection is afforded.

Drainage / spill control systems typically consist of combination of features to achieve the necessary spill control depending on the particular situation. These features include the following:
- Curbs or dikes
- Drains (with traps to prevent flashback)
- Flumes or sluiceways
- Remote containment basins or tanks
- Separators or skimmers (to separate contaminants from water)
- Sloped surfaces (typically minimum 1%)
- Sumps or pits (to contain small spills)
- Trenches

7. PROTECTION LAYERS

The design of drainage / spill control systems can be complex. Some of the factors that should be considered in the design of drainage / spill control systems include the following:

- Expected duration of fire (or time to implement contingency plans)
- Expected flow from water-based fire extinguishing systems, such as sprinklers, foam systems, hoses, and monitor nozzles
- Local codes and regulations
- Properties of the liquid which could be released, including extinguishability, viscosity, water solubility, specific gravity, volatility, etc.
- Rainfall (containment facilities should normally contemplate some rainfall in capacity design)
- Reactivity of chemicals with water or other chemicals in the drainage system
- Risk of environmental contamination (proximity to water supplies, geology, etc.)
- Separation of organics from water to prevent drainage to rivers
- Spacing and location of facilities
- Surface type (earth, gravel, concrete, etc.)
- Topography
- Volume of liquid which could be released as well as the rate and mode of release

NFPA 30 (Ref. 7-2) and the appendix of NFPA 15 (Ref. 7-42) should be consulted for details on the design of drainage / spill control systems.

Drainage / spill control systems should be inspected on a regular basis to ensure they are in good condition. In particular, drains and trenches should be examined to ensure they do not contain any blockages. Rainwater should be drained or pumped out of containment facilities following each rainfall.

7.12 REFERENCES

7-1. NFPA 70B. *Recommended Practice for Electrical Equipment Maintenance.* National Fire Protection Association. Quincy, Massachusetts. 2010.

7-2. NFPA 30. *Flammable and Combustible Liquids Code,* National Fire Protection Association. Quincy, Massachusetts. 2008.

7-3. NFPA 70. *National Electrical Code (NEC), 2011 Edition, Article 500,* National Fire Protection Association, Quincy, Massachusetts. 2011.

7-4. API RP 500 (R2002). *Recommended Practice for Classification of Locations for Electrical Installations at Petroleum Facilities Classified as Class I, Division 1 and Division 2, Second Edition*, American Petroleum Institute. 1997.

7-5. NFPA 496. *Standard for Purged and Pressurized Enclosures for Electrical Equipment, 2008 Edition*, National Fire Protection Association. Quincy, Massachusetts. 2008.

7-6. ANSI / UL 913. *Intrinsically Safe Apparatus and Associated Apparatus for Use in Class I, II and III, Division 1, Hazardous (Classified) Locations*, Underwriters Laboratory. 2002.

7-7. Static Electricity. FM Global 5-8, Factory Mutual Insurance Company. 2001.

7-8. API RP 2003. *Protection Against Ignitions Arising out of Static, Lightning, and Stray Currents*, American Petroleum Institute. 2008.

7-9. CCPS. *Avoiding Static Hazards in Chemical Operations*, Center for Chemical Process Safety of the American Institute of Chemical Engineers. New York, New York. 1999.

7-10. NFPA 780. *Standard for the Installation of Lightning Protection Systems*, National Fire Protection Association. Quincy, Massachusetts. 2011.

7-11. ISA 84.91.01. *Identification and Mechanical Integrity of Instrumented Safety Functions in the Process Industry*, International Society of Automation. Research Triangle Park, North Carolina. 2011.

7-12. IEC 61511. *Functional Safety: Safety Instrumented Systems for the Process Sector*, International Electrotechnical Commission, Geneva, Switzerland. 2003.

7-13. ANSI / ISA 84.00.01-2004 (IEC 61511 modified). Part 1, *Functional Safety: Safety Instrumented Systems for the Process Industry Sector - Part 1: Framework, Definitions, System, Hardware and Software Requirements*, International Society of Automation, Research Triangle Park, North Carolina. 2004.

7-14. ANSI / ISA 84.00.01-2004 (IEC 61511 modified). Part 2, *Functional Safety: Safety Instrumented Systems for the Process Industry Sector - Part 2: Guidelines for the Application of ANSI / ISA-84.00.01-2004 Part 1 (IEC 61511-1 Mod) – Informative*, International Society of Automation, Research Triangle Park, North Carolina. 2004.

7-15. ANSI / ISA 84.00.01-2004 (IEC 61511 modified). Part 3, *Functional Safety: Safety Instrumented Systems for the Process Industry Sector - Part 3: Guidance for the Determination of the Required Safety Integrity Levels - Informative*, International Society of Automation, Research Triangle Park, North Carolina. 2004.

7-16. ISA TR84.00.04. *Guidelines on the Implementation of ANSI / ISA 84.00.01-2004 (ISA 61511 Modified)*, International Society of Automation, Research Triangle Park, North Carolina. 2006.

7-17. ISA TR84.00.02. *Safety Integrity Level (SIL) Verification of Safety Instrumented Functions*, International Society of Automation, Research Triangle Park, North Carolina. 2002.

7-18. ISA TR84.00.03. *Mechanical Integrity of Safety Instrumented Systems (SIS)*, International Society of Automation, Research Triangle Park, North Carolina. 2002.

7. PROTECTION LAYERS

7-19. CCPS. *Guidelines for Hazard Evaluation Procedures, Third Edition,* Center for Chemical Process Safety of the American Institute of Chemical Engineers. New York, New York. 2008.

7-20. CCPS. *Layer of Protection Analysis: Simplified Process Risk Assessment,* Center for Chemical Process Safety of the American Institute of Chemical Engineers. New York, New York. 2001.

7-21. CCPS. *Guidelines for Independent Protection Layers and Initiating Events,* Center for Chemical Process Safety of the American Institute of Chemical Engineers. New York, New York. 2011.

7-22. ANSI / API Std 521. *Pressure-Relieving and Depressuring Systems, Fifth Edition,* American Petroleum Institute. 2007.

7-23. API Std. 2000. *Venting Atmospheric and Low-pressure Storage Tanks, Sixth Edition,* American Petroleum Institute. 2009.

7-24. ASME Section VIII-DIV 1. *2010 ASME Boiler and Pressure Vessel Code, Section VIII, Division 1: Rules for Construction of Pressure Vessels.* American Society of Mechanical Engineers. 2010.

7-25. API Std 520. *Sizing, Selection, and Installation of Pressure-relieving Devices in Refineries, Part I - Sizing and Selection, Eighth Edition,* American Petroleum Institute. 2008.

7-26. Glinos, K. and R.D. Myers. *Sizing of Vacuum Relief Valves for Atmospheric Distillation Columns,* Journal of the Loss Prevention in the Process Industries, Vol. 4, No. 3, pp. 166-169. 1991.

7-27. Simpson, L.L. E*stimate Two-Phase Flow in Safety Devices,* Chemical Engineering, Vol.98, No. 8, pp. 98-102. 1991.

7-28. Leung, J.C. *Size Safety Relief Valves for Flashing Liquids,* Chemical Engineering Progress, Vol. 88, No. 2, pp. 98-102. 1992.

7-29. DIERS. *Emergency Relief System Design Using DIERS Technology, DIERS Project Manual.* American Institute of Chemical Engineers. New York, New York. 1992.

7-30. CCPS. *Safe Design and Operation of Process Vents and Emission Control Systems,* Center for Chemical Process Safety of the American Institute of Chemical Engineers. New York, New York. 2006.

7-31. DIERS. *Systems Analysis for Integrated Relief Evaluation (SAFIRE) User's Manual, SAFIRE Computer Program and Documentation.* American Institute of Chemical Engineers. New York, New York. 1986.

7-32. CCPS. *Guidelines for Pressure Relief and Effluent Handling Systems, 2nd Edition,* Center for Chemical Process Safety of the American Institute of Chemical Engineers. New York, New York. 2011.

7-33. Scandpower. *Guideline for Protection of Pressurized Systems Exposed to Fire,* Scandpower. 2002.

7-34. ANSI / API Std 537. *Flare Details for General Refinery and Petrochemical Service,* American Petroleum Institute. 2008.

7-35. NFPA 69. *Standard of Explosion Prevention Systems*, National Fire Protection Association. Quincy, Massachusetts. 2008.

7-36. NFPA 72. *National Fire Alarm and Signaling Code*, National Fire Protection Association. Quincy, Massachusetts. 2010.

7-37. NFPA 101. *Life Safety Code*, National Fire Protection Association. Quincy, Massachusetts. 2009.

7-38. CCPS. *Continuous Monitoring for Hazardous Material Releases,* Center for Chemical Process Safety of the American Institute of Chemical Engineers. New York, New York. 2009.

7-39. CCPS. *Guidelines for Fire Protection in Chemical, Petrochemical, and Hydrocarbon Processing Facilities*, Center for Chemical Process Safety of the American Institute of Chemical Engineers. New York, New York. 2003.

7-40. API Publ 2218. *Fireproofing Practices in Petroleum & Petrochemical Processing Plants,* American Petroleum Institute. 1999.

7-41. NFPA 13. *Installation of Sprinkler Systems*, National Fire Protection Association. Quincy, Massachusetts. 2010.

7-42. NFPA 15. *Standard for Water Spray Systems for Fire Protection*, National Fire Protection Association. Quincy, Massachusetts. 2007.

7-43. NFPA 750. *Standard on Water Mist Fire Protection Systems*, National Fire Protection Association. Quincy, Massachusetts. 2010.

7-44. NFPA 11. *Standard for Low-, Medium-, and High-Expansion Foam*, National Fire Protection Association. Quincy, Massachusetts. 2010.

7-45. NFPA 16. *Standard for the Installation of Foam-Water Sprinkler and Foam-Water Spray Systems, 2007 Edition.* National Fire Protection Association. Quincy, Massachusetts. 2007.

7-46. NFPA 2001. *Standard on Clean Agent Fire Extinguishing Systems,* National Fire Protection Association. Quincy, Massachusetts. 2008.

7-47. NFPA 12. *Standard on Carbon Dioxide Extinguishing Systems*, National Fire Protection Association. Quincy, Massachusetts. 2008.

7-48. NFPA 17. *Standard for Dry Chemical Extinguishing Systems*, National Fire Protection Association. Quincy, Massachusetts. 2009.

7-49. NFPA 10. *Standard for Portable Fire Extinguishers*, National Fire Protection Association. Quincy, Massachusetts. 2010.

7-50. CCPS. *Deflagration and Detonation Flame Arresters*, Center for Chemical Process Safety of the American Institute of Chemical Engineers. New York, New York. 2002.

7-51. Halstrick, V. *Technical Report Part 1*, Protego Fundamentals. Braunschweiger Flammenfilter GmbH. Braunschweig, Germany. 1995.

7. PROTECTION LAYERS

7-52. Frobese, D.H. and Forster, H. *Propagation of Detonations Through Pipework Junctions*, Proc, Seventh International Symposium on Loss Prevention and Safety Promotion in the Process Industries. Taormina, Italy. 1992.

7-53. ESCIS. *Inerting*, Booklet No. 3, *Expert Commission for Safety in the Swiss Chemical Industry*. Basel, Switzerland. 1994.

7-54. NFPA 68. *Standard on Explosion Protection by Deflagration Venting*, National Fire Protection Association. Quincy, Massachusetts. 2007.

7-55. Bartknecht, W., *Dust Explosions: Course, Prevention, Protection*. Springer-Verleg, New York. 1989.

8

DOCUMENTATION TO SUPPORT PROCESS SAFETY

Documentation is important to long term management as well as the day-to-day safe operation of a process facility. As the regulatory mandate for documentation evolves, failure to maintain accurate and complete records can become a legal liability. Documentation is frequently the means to implement a corporate process safety management program and to verify plant compliance to its provisions. In addition, quality documentation can facilitate continuous improvement. The primary elements of a document management program are information infrastructure, procedures, retention, and control. Access to necessary information during emergency conditions is essential, as well as to support Management of Change.

A robust document management system ensures version control of procedures and other process safety information. This is particularly challenging when process safety information is computer based and managed over an intra-net system.

8.1 PROCESS KNOWLEDGE MANAGEMENT

The importance of clear and comprehensive documentation of the design basis cannot be over-emphasized. Often referred to as Process Safety Information (PSI), Process Knowledge Management is essential throughout the life cycle of a facility. Table 8.1 contains commonly required types of Process Safety Information. It is important to note that all information contained in this table is not appropriate for every design project.

Designing and building an efficient and safe process unit requires significant investment of resources. In addition to this initial investment, almost all units are modified over time to increase throughput and / or efficiency. The information required to design, construct, and optimize a unit represents a significant and valuable corporate asset. Because process knowledge and information provide the foundation for long-term viability and continued success of the business, a management system should be established to protect and promote the use of this information. Establishing an efficient and dependable means to collect, maintain, and protect a company's process knowledge helps protect an important asset which simply makes good business sense.

Table 8.1 Typical Types of Process Safety Information

Chemical Hazards Information	

- Calorimetric data
- Corrosivity data
- Hazardous effects of inadvertent mixing of typical contaminants (e.g., air, water) with different materials contained in process streams and utility systems
- Industrial hygiene data
- Measure of dust explosibility (K_{St})
- Maximum deflagration or detonation pressure and flame speed
- Minimum ignition energy
- Permissible exposure limits
	- Physical data
	- Reactivity data
	- Special hazards:
	  - Shock sensitivity
	  - Pyrophoric properties
	  - Chemical stabilizers, including effects of purification (removal of a stabilizer or other chemical species)
	- Material compatibility
	- Thermal and chemical stability data
	- Thermodynamic data
	- Toxicity information

Process Technology Information	
- A description of control system logic in narrative format and / or simple figures - Adiabatic reaction temperature and the corresponding system pressure, based on both intended and worst credible case material composition - Cause-and-effect charts - Consequences of deviations from safety limits - Hazards related to credible undesired chemical reactions - Map and / or tables showing zones / distances of concern for overpressurized or toxic exposure hazards based on consequence analysis	- Material and energy balances - Maximum intended inventory - Process chemistry, including laboratory notebooks that provide information developed during the early stages of product or process development - Basis for and values of safe upper and lower limits - Separation equipment design information and design bases - Simplified process flow diagram or block flow diagram
Process Equipment Information	
- Control system logic diagrams, loop sheets, and interlock tables - Electrical classification diagrams - Electrical data, including one-line diagrams, a motor database, and grounding / bonding drawings - Facility data, including plot plans that document the location of underground utility and process piping, structural drawings and structural analysis, design and design basis information for fixed fire protection systems, and information on heat / blast loads and fire / blast walls - Instrument data, including a register or database of key parameters for field instruments, alarms, interlocks, etc. - Isometric drawings	- List of design codes and standards applicable to the process - Location of safety showers / eye wash stations, fire extinguishers, and other safety equipment - Materials of construction - Mechanical data / design basis sheets for process equipment - Piping and Instrumentation Diagrams (P&IDs) - Piping specifications - Portable multi-unit equipment - Relief system design basis and calculations, including any flare system - Safety systems (e.g., interlocks, detection, or suppression systems) - Shop fabrication drawings - Ventilation system design basis and calculations

8. DOCUMENTATION TO SUPPORT PROCESS SAFETY

8.1.1 Importance of Process Knowledge Management

Understanding the hazards and risk requires accurate process knowledge and information. The primary objective of Process Knowledge Management is to maintain current, accurate, complete, and understandable information that can be accessed on demand.

8.1.2 Types of Process Knowledge and Information Documentation

Developing, documenting and maintaining process knowledge represents one of two elements in the Risk Based Process Safety pillar of understanding hazards and risk that is described in *Guidelines for Risk Based Process Safety* (Ref. 8-1). This section provides an overview of documenting process safety knowledge and information and focuses on information that can easily be recorded, such as:

- Written technical documents and specifications
- Engineering drawings and calculations
- Specifications for design, fabrication, and installation of process equipment
- Other written documents such as Materials Safety Data Sheets (MSDS)

Knowing how to access information for use is essential. An efficient system should:

- Make information available in an organized manner
- Protect knowledge from inadvertent loss
- Store calculations, design data, and similar information in an easily accessible manner

Further detail on process safety documentation can be found in:

- *Guidelines for Risk Based Process Safety* (Ref. 8-1)
- *Guidelines for Process Safety Documentation* (Ref. 8-2)

8.1.3 Design Basis

Process safety depends on how a unit is designed, constructed, operated, and maintained. The input of the process engineer is essential in establishing a permanent record of the design basis and operational requirements. Thorough documentation is necessary so that the design basis is not defeated by incorrect fabrication, operation, or maintenance techniques.

One example of design basis information is documenting what specific scenario or basis was used to calculate relief valve sizing and load. Another example is to document the sizing basis of vent and relief headers, giving information as to what cases of simultaneous venting were or were not included when calculating the emergency vent load.

The original design package (and all subsequent revisions), which should include a set of design specifications, standards, and drawings used to construct a chemical facility, is usually the most accurate and complete set of information assembled for a given processing unit. The design documents are used as the basis for all future improvement projects and the need for maximum completeness and accuracy cannot be overemphasized. Therefore it is best to immediately institute a procedure for storage, control, and revision of this information. Design documents typically include those described below.

- *Design Basis Documents* - Process definition and design criteria are usually the initial information assembled. The basic process knowledge includes process chemistry, energy and mass balances, general control philosophy, process hazard analysis, etc. Applicable codes and design standards are identified. Design calculations and research and development reports, which explain the original design bases with their underlying philosophy and define safe operating ranges for process variables, should also be clearly documented. The latter are often a useful place to begin troubleshooting or planning alternative operating conditions. Design basis should also be documented for mitigation systems.

- *Equipment Specifications* - These documents describe all of a plant's equipment in a concise and complete way. The original design basis is clearly stated. Sufficient process and mechanical data are provided to allow procurement of the items required. Changes sometimes occur after the purchase order is awarded. The specifications should be updated to show "as delivered" and installed.

- *Design Standards* - Design standards explain in detail the proper components, fabrication, assembly or construction techniques, or references used for items other than specific equipment.

- *Drawings* - While design standards may go through minor adjustments, engineering drawings are revised frequently to reflect the addition of equipment and instruments or rerouting of lines. Regulatory agencies most often require retention of P&IDs and plot plans; these documents encompass the essence of the facility in a condensed form.

- *Hazard Analysis* - One of the most common elements of industry guidelines and regulations is the performance and documentation of a hazard analysis. This review does not ensure that all hazards have been identified, but it is currently the most effective method to systematically review a process and its components for hazards. The hazard analysis should be thoroughly documented with detailed minutes of meetings and records of decisions and actions taken. Besides serving as a reference against which potential changes may be assessed, the hazard analysis can serve as a case study for similar process units. For more information on hazard analyses, see Chapter 4, Analysis Techniques, and *Guidelines for Hazard Evaluation Procedures* (Ref. 8-3).

- *Vendor Information* - Equipment manufacturers should provide drawings and operating manuals for each piece of equipment. These drawings and manuals are useful because they reflect exact detail or "as built" descriptions and include proper operating instructions intended to ensure safe and trouble-free operation. These documents are particularly useful in establishing the historical background of specific pieces of equipment. Vendor training manuals are useful for ensuring proper and consistent maintenance of equipment. Manuals, drawings, and Material Safety Data Sheets (MSDSs), and all test reports should be retained in the plant maintenance department, the engineering office, or operating department.

- *Quality Control (New Equipment)* - Procedures should be developed to ensure that equipment is purchased, fabricated, inspected, tested, and installed to meet

8. DOCUMENTATION TO SUPPORT PROCESS SAFETY

equipment design specifications and assure process safety. The process engineer may be involved, along with the materials engineer, quality surveillance representatives, and equipment specialists, in developing these specifications.

Original materials, thickness, and construction details must be known for an accurate determination of corrosion rates and equipment life. Chemical composition of alloy piping and pressure components require verification as detailed in the fabrication specification. Records of weld integrity, post weld heat treatment, and testing of material and / or equipment are often required. Original shell and nozzle wall thickness should be verified on pressure vessels and exchangers. Material verification may be required for process piping. For more information on material verification, see Chapter 5.7, Material of Construction.

Pressure testing requirements are described as minimum requirements in ASME / ANSI Codes, API Std 510 (Ref. 8-4), and the National Board Inspection Code (NBIC). Pressure testing of piping repairs should, as a minimum, be in accordance with the ASME B31 Piping Code. Pressure testing of large vessels should be covered by a written procedure defining test pressure; location of pressure and temperature indicators, test fluid temperature, venting, and pressurization / inspection sequencing, and any safety requirements. All test records, both by owner and by third parties, must be preserved for the entire life of the equipment / system.

8.1.4 Managing Change

A formal system for managing changes that occur during the design phase and throughout a process life cycle helps ensure the risk-based design integrity of the system. Changes to physical equipment and other revisions to the design are often made to correct minor deficiencies or to improve operability of a unit once the actual layout comes together. This may occur during design, and construction, and is almost inevitable during operation. Managing changes during the detailed design phase is an important activity that is not easily accomplished. Changes made after the final PHA should be accurately and thoroughly documented.

Proposed modifications to the process or the plant should be subjected to critical analysis and safety assessment; construction, inspection, and testing specifications and codes must be determined. These changes should be reflected in revisions to the original design package. All these documents must be preserved.

The individual overseeing the change (or project) should determine whether other documents are affected and should be responsible for making sure that all complementary documentation is also revised.

An effective Management of Change (MOC) program should be in place to ensure that process safety information is kept up-to-date by providing:
- Notification of the change to the custodian of process safety information.
- Basic data, drawings, and other information to persons assigned to design, review, and approve the change.
- A set of updated process information or a list of changes that need to be made.
- A means to ensure that process safety information has been properly updated with the change implemented. For example, that "as-built" drawings are

produced or that the "approved for construction" drawings be field verified after a change is made.

8.1.5 Other Considerations

As discussed above, process safety depends on how a unit is designed, constructed, operated, and maintained. The input of the process engineer is essential in establishing a permanent record of the design basis and operational requirements. Some examples of documentation that is important in understanding the hazards and risk are:

- *Material Reactivity* – The design team should identify and document potential reactivity hazards. Scenarios that should be considered include cases where a material self reacts, e.g., by polymerizing or decomposing, or when combinations of materials and conditions may cause a runaway reaction. A major concern involving vessels handling reactive materials is the potential for runaway reactions causing vent rates that may be several times greater than the normal process vent flows.

- *Material of Construction Compatibility and Interactions* – Equipment service life is influenced by many factors, such as materials of construction, design details, fabrication techniques, operating conditions, and inspection and maintenance procedures. Material failures, while relatively infrequent, can be extremely severe, resulting in catastrophic accidents. The best way to reduce the risk of material failure is to fully understand the internal process, the exterior environment, and failure modes, select materials for the intended application, apply proper fabrication techniques and controls, and provide good maintenance and inspection and repair techniques.

- *Expected Risk Reduction by Various Layers of Protection* – Changing the design of protection systems, proof-test intervals, or other aspects of the mechanical integrity program may violate assumptions made when determining compliance with a company's risk criteria. To ensure continued compliance, the MOC work process needs to have the initial work available so that the impact of the proposed changes can be properly evaluated.

8.2 ENGINEERING DESIGN PACKAGE

The Engineering Design Package follows a process unit throughout its life cycle, providing valuable information essential for the design, operation, maintenance, and eventual decommissioning of a facility.

The engineering design package documentation should include:
- Equipment specifications, including all assumptions / stipulations used in design and equipment selection.
- Requirement for and documentation of acceptance testing.
- Manufacturer's recommendations for periodic integrity testing (method, acceptance criteria, frequency).
- Initial functional testing of liquid overfill protection and assumptions (strapping tables, density, assumed temperature, and composition of the fluid level being measured.

8. DOCUMENTATION TO SUPPORT PROCESS SAFETY

- Expected composition of each stream and the expected variations that were considered and used in design. This information impacts several important process safety design considerations, e.g., chemical reactions, corrosion, instrumentation calibration and response, material of construction, and pressure relief sizing.
- Document thermochemistry data for expected reaction and credible cases of advertent reactions or inadvertent reaction rates.
- Mechanical integrity program expectations to conduct appropriate inspections and tests according to manufacturer recommendations and industry practice.

8.3 OPERATING / MAINTENANCE PROCEDURES

Procedures are an essential part of every process operation, providing rules to be followed and standardized records of safe and approved operations and maintenance practices. They also provide consistent information across the plant and help minimize guesswork, leading to more efficient and safe operations.

Good procedures also describe the process, hazards, tools, protective equipment, and controls in sufficient detail that operators and maintenance personnel understand the hazards, can verify that any necessary controls are in place, and can confirm that the process responds in an expected manner. Procedures should specify when an emergency shutdown should be executed and should also address special situations, such as temporary operations when equipment is out-of-service. Operating procedures are normally used to control activities, such as transitions between products, periodic cleaning of process equipment, preparing equipment for maintenance, and other activities normally performed by the operations department. To ensure accurate procedures are published and maintained, a facility should have a review and approval process.

> **Part of the project design and engineering design team responsibility is developing initial operating procedures.**

The project design team needs to provide the information necessary for safe operation, including consequences of deviations and steps to correct the deviation. This information is essential for operations training.

There are four major types of procedures in process operations:

- *Operating procedures* are written step-by-step instructions and associated information (cautions, warnings, notes, etc.) for safely performing a task within operating limits. Procedures should cover all modes of operation. Typically, operating procedures are required for:
  - Initial startup
  - Normal startup
  - Startup after a turnaround
  - Normal operations
  - Temporary operations
  - Emergency operations

- Normal shutdown
- Emergency shutdown
- *Emergency or abnormal operating procedures* are written instructions that provide step-by-step actions for operations personnel to ensure the process is in a safe and stable mode following a system upset or when a process is in intentional (startup, shutdown, etc.) or unintentional (upset) transition.
- *Temporary operations* are written instructions that document the steps taken during operations that are not conducted on a daily basis.
- *Maintenance procedures* are written instructions that address material control and maintenance practices needed to ensure system operability and integrity as well as maintenance, testing, and inspection frequency.

8.3.1 Need for Procedures

There should be a clearly defined process in place that identifies the jobs that require procedures. The first step is to conduct a task analysis to develop a comprehensive list of tasks to be performed. From this comprehensive list, each task can be reviewed to determine if a written procedure is necessary. Many routine procedures, such as taking a sample or starting a pump, are covered by training and a procedure may not be necessary. Some companies use hazard and risk analyses to determine the job tasks that require the procedure to be "in hand" during their execution and the job tasks where it is appropriate for the procedure to be "available for reference."

There may be regulatory requirements for certain procedures, particularly OSHA PSM Standard and the Food and Drug Administration. Additionally, procedures may be necessary for voluntary programs such as Responsible Care and ISO quality certification.

The need for procedures is best determined by the:
- Complexity of the tasks
- Frequency that the tasks are executed
- Consequences of possible errors made while executing the job tasks

The level of detail needed in the procedures is a function of:
- The risk associated with the task
- The knowledge, skill, and abilities of the person assigned to perform the task

Procedures should include sufficient detail so that the newest or least qualified person can successfully perform the task. If unnecessary detail is provided, the person performing the task may not use the procedure.

8.3.2 Developing Procedures

As previously discussed, the engineering design team is responsible for developing the initial operating procedures. To develop effective procedures, the following guidelines should be used:
- Ensure that the procedure reflects the way the work is done and is technically accurate.
- Encourage procedure users to actively participate in the development and review of procedures.

8. DOCUMENTATION TO SUPPORT PROCESS SAFETY

- Ensure enough information for the user to perform the task safely and correctly.
- Ensure that the level of detail considers the experience and capabilities of the users, their training, and their responsibilities.
- Develop processes to ensure that users are able to quickly and accurately locate the correct procedure for the job.
- Write all procedures in a standard format which is set by a "style guide."
- Instruct users in the use of their procedures.

8.3.2.1 Upper and Lower Safe Operating Limits

Safe operating limits are normally set for critical process parameters such as temperature, pressure, level, flow, or concentration based on a combination of equipment design limits and the dynamics of the process. Safe operating limits are most often specified when the system response may be so severe that the risk of continued operation is unacceptable. In this case, the procedures should include clear, simple instructions for responding to the situation. For example, a reactor may normally operate at 90°F (32°C) with a safe operating limit of 120°F (49°C), based on the potential for runaway reaction.

For additional detail on safe operating limits, see Chapter 5, Section 5.5.1, Process Equipment Safe Operating Limits.

8.3.2.2 Consequences of Deviations

In the above example, if the temperature were to exceed the operating limit of 120°F, there is the potential for a runaway reaction. The consequences of the runaway reaction could be increased temperature and pressure, if steps are not taken immediately, which could result in vessel rupture with corresponding fire and explosion.

The operating procedures should include:
- Consequence of exceeding the safe operating limit
- Immediate steps to be taken (automatically or manually) to stop the deviation from continuing to avoid exceeding the safe operating limit

For the above example, the steps could be to stop the reactant flow, shut down the heat input to the reactor, set the system to maximum cooling, or add quench or inhibitor to stop the reaction.

For additional detail on consequences of deviations, see Chapter 5, Section 5.5.2, Consequences of Deviation.

8.3.2.3 Considerations for Batch Procedures

The basic principles of effective procedure writing previously presented apply equally well to batch processes, but batch processes also have specific procedure considerations. Batch equipment is often used to produce multiple products with the same components. Typical batch operations are:
- Staging
- Charging
- Processing
- Monitoring
- Transferring

- Cleaning / Decontaminating
- Emergencies

For batch processes, there may be two types of documents necessary to conduct operations because the same equipment is often used in different configurations for different products. First, there are operating procedures that contain the steps and safety information for performing each task. Second, there are the "batch" or "recipe" or "process" sheets that contain operating parameters such as temperature, material amounts, and sequencing. These batch sheets may change with each run, although the actual operating procedures remain the same. The batch sheets may change several times a week, but the operating procedures are always applicable to the equipment. Together, batch sheets and operating procedures provide the necessary information for safe operation.

8.3.2.4 Considerations for Maintenance Procedures

Maintenance activities at a site are usually considered "crafts," and it can be a new experience for maintenance personnel to write and use written procedures. Maintenance procedures require special consideration depending upon the type of maintenance force the facility maintains. If the facility uses cross-trained maintenance personnel, maintenance procedures that are written to a very high level of detail may be required. The increased use of contract maintenance personnel at facilities presents a similar problem.

Referencing vendor manuals is a choice that maintenance managers can use to keep numbers of procedures manageable, but it implies another level of document control. If vendor manuals are referenced in maintenance procedures, the facility must possess these documents and ensure they are accessible, up-to-date, and accurate. Sometimes vendor-supplied generic equipment manuals or sections of catalogs are included as part of the project design records. In such cases, it is necessary for the project design team to clearly indicate which specific part or section is applicable to the actual system that was installed. This critical design information can easily be lost. Absence of this documentation or erroneous assumptions by later users of the equipment information files can result in adverse safety and operating consequences, for example, that a specific gas detector head illustrated in a vendor manual is indeed the one that was installed in the field. Vendor manuals often do not provide the application-specific cautions, warnings, and level of detail that the facility may require. A facility's maintenance procedures may need to augment this information in order to reflect site needs accurately.

8.3.3 Maintaining Procedures

The process industry is a dynamic operating environment that is constantly changing and requires the need for accurate procedures. Administrative controls for maintaining procedures should be established and normally consist of the following:

- A review process to ensure that all procedures are reviewed thoroughly before initial use and periodically thereafter
- Methods to ensure that procedures are reviewed and updated when the task, equipment, or process changes (Management of Change)

8. DOCUMENTATION TO SUPPORT PROCESS SAFETY

- A system to ensure that shared knowledge between operations, maintenance, and engineering is routinely incorporated into procedure updates
- A system to advise users when a procedure changes and instruct them on the changes
- A system to learn from accidents, abnormal situations, near misses, behavior observations, and simulations and exercises and use the information to update procedures

8.4 ASSET INTEGRITY / RELIABILITY / PREDICTIVE MAINTENANCE DATA

In general, an asset integrity program is a management function, not a design function. Therefore, its design, execution and maintenance are beyond the scope of this publication. However, key data developed or obtained during the design phase of a project are critical to the efficient specification and recordkeeping functions of the program, and should therefore be addressed as part of the design effort.

The engineering design project team plays a key role in providing the foundation data for an effective asset integrity program. Incomplete or inaccurate design information can adversely impact the functional operation and integrity of the equipment, which could result in significant adverse consequences, premature equipment failure, possible injuries, and financial impact.

Engineering design documentation permits maintenance personnel to:
- Ensure correct maintenance procedure(s) are understood and performed.
- Ensure maintenance is performed on the desired frequency / interval.
- Preserve historical data and trends on equipment to optimize maintenance work.
- Evaluate potential repairs and modifications.
- Determine necessary spare parts inventory to minimize unplanned shutdowns and outages.

The objective of maintenance is to ensure the integrity of the process equipment. The maintenance department of a process facility should be responsible for archiving maintenance records and preserving other documents, such as service manuals. Maintenance procedures should be reviewed periodically and updated as required.

One of the purposes of an asset integrity program (also referred to as mechanical integrity program) is to identify and correct equipment deficiencies. There is also the additional expectation of using knowledge gained from these inspections and tests to understand why failures occur and seek improved reliability and safety. To realize such an improvement requires a mechanical integrity program engineered to measure performance in a manner that facilitates the eventual fundamental understanding (through statistical analysis) of the equipment's reliability as a function of many factors: design, operation, process conditions, ambient environment, and maintenance practices.

For example, equipment specifications, mechanical tolerances, and even design basis assumptions are important information needed to establish initial inspection intervals as well as pass / fail inspection criteria. These observations and measurements then become the basis for a preventive / predictive maintenance and reliability program, which can be adjusted on the basis of a series of inspections and measurements. It is therefore critical that key information from the design effort be passed along in a consistent and useful manner to facilitate the initiation of the reliability effort.

Non-destructive testing findings, details of construction, repairs, alterations, or other conditions may also affect the future evaluation of the equipment's integrity. From the point of view of tracking the service history of equipment, the following initial engineering design project records in addition to the equipment specifications listed earlier are useful:

- ASME Code Data Reports for pressure vessels.
- Field-verified inspection drawings for major equipment with reference inspection points. Wall thickness measurements (including original measurements) and other non-destructive examination findings, both past and present, should be on the drawings or a separate sheet.
- A copy of jurisdictional reports and permits which are required to operate boilers or pressure vessels (for the duration of the permit).
- Repair and alteration documentation for major equipment and process piping.

In addition to the transmittal of initial design information, the design process must also be enlisted to support the ongoing collection of equipment performance data, both to validate (or, if necessary, correct) the initial design basis, and to ensure that equipment integrity is maintained in an efficient and effective manner. Collection of asset integrity and equipment performance data into meaningful data sub-sets during analysis and "as found, as left" data in a format designed to make analysis efficient is required. The CCPS Process Equipment Reliability Database Project (PERD) is one example of a structured method for documenting and recording equipment data which may be employed as a part of a structured design information transfer effort that facilitates accurate and meaningful data collection, based on design information, to ensure the ongoing integrity of process equipment.

By making the data collection factual (requiring little or no interpretation on the part of field personnel), easy (automated wherever practical to minimize both tedium and transcription errors), and specific [providing information on the failure(s) of interest], the ability to extract useful trends is greatly enhanced. These qualities are best built into the system with the input of the initial designers, while design intent is fresh, and all assumptions and key variables are clearly understood and remembered.

Testing intervals are also a key component for equipment integrity maintenance. Intervals may be based on regulatory requirements or determined through a facility's Risk Based Inspection (RBI) or Reliability Centered Maintenance (RCM) program. Initial test intervals should be a routine part of the design basis, which may then be modified based on actual field experience. Refer to *Guidelines for Mechanical Integrity Systems* for additional information (Ref. 8-5).

8.5   REFERENCES

8-1.   CCPS. *Guidelines for Risk Based Process Safety*, Center for Chemical Process Safety of the American Institute of Chemical Engineers. New York, New York. 2007.

8. DOCUMENTATION TO SUPPORT PROCESS SAFETY

8-2. CCPS. *Guidelines for Process Safety Documentation,* Center for Chemical Process Safety of the American Institute of Chemical Engineers. New York, New York. 1995.

8-3. CCPS. *Guidelines for Hazard Evaluation Procedures*, Center for Chemical Process Safety of the American Institute of Chemical Engineers. New York, New York. 2008.

8-4. API Std 510. *Pressure Vessel Inspection Code: Maintenance Inspection, Rating, Repair, and Alteration, 8th Edition.* American Petroleum Institute. June 1997, Addendum 1 (December 1998), Addendum 2 (December 2000), Addendum 3 (December 2001), and Addendum 4 (August 2003).

8-5. CCPS. *Guidelines for Mechanical Integrity Systems*, Center for Chemical Process Safety of the American Institute of Chemical Engineers. New York, New York. 2006.

INDEX

Active Design Solutions, 125
Agitation, 192
American Conference of Governmental Industrial Hygienists. (ACGIH), 52
Asset Integrity/Reliability/Predictive Maintenance Data, 389
Atmospheric Storage Tanks, 179
Autoignition Temperature, 43
Baker Panel Report, 2
Basic Process Control System, 132-134
    Alarm Management, 133
    Testing Instrumentation, 134
Batch Reaction Systems, 29
Below Grade Structures, 150
BLEVE, 74
Blowdown Systems, 346-347
    Disengaging Facilities, 347
    Equipment Drainage Systems, 347
    Quench Drums, 347
Buffer Zone, 135
Building Damage Levels, 92
Buncefield Incident, 128
Calorimetric Data, 50
Catalysts, 142
Cathodic Protection and Anodic Protection, 146
Centrifuges, 244
Checklist Analysis, 97
Chemical / Material Hazards, 72
Chemical Hazard Response Information System (CHRIS), 55
Chemical Incompatibility Charts, 51
Chemical Interaction Matrix, 53
Chemical Reactivity Hazard, 46, 47
Civil/Structural/Support Design, 146

Columns, 203
Compressors, 234
Consequence/Impact Assessment, 109
Consequences of Deviation, 137
Consequences, 25
Contaminants, 142
Cooling Water, 310
Corrosion Allowance, 146
Corrosion Fatigue, 144
Corrosion, 143, 265
Corrosive Environments, 142
Corrosion Under Insulation, 146, 153
    Contributing Factors, 153
    Material Stress Conditions, 154
    Prevention of Corrosion, 155
Corrosivity / pH Hazards, 75
Crevice Corrosion, 145
Critical Task Analysis, 105
Culture, 35
Deflagration, 57
Deflagration/Detonation Arresters, 363
    Selection and Design Criteria, 365
Depressurization, 340
Design Alternatives, 78
Design Basis, 381
Design Considerations for Flare, 344
    Flare Header Design, 344
    Flare Stack, 345
    Flare System Safety, 346
    Knockout Drums, 345
Design Considerations, 145
Design Institute for Physical Properties (DIPPR®), 16, 39

Design of Relief Devices: Other Considerations, 338
Detonation, 57
Detonations and Deflagrations, 49
Documentation, 379
Dump System, 372
Dryers, 214
    Design Considerations, 223
    Design Solutions, 217
    Failure Scenarios, 216
    Past Incidents, 215
Dust Deflagration Index - K_{st}, 45
Effects on Environment, 94
Effluent Disposal Systems, 342
Effluent Handling, 373
Electrical Classification, 2, 316
Electricity, 308
Emergency Exposure Guidance Levels (EEGLs), 88
Emergency Power Supply, 308
Emergency Response Alarm Systems, 350
Emergency Response Planning Guidelines (ERPGs), 86
Endpoints, 85
Engineering Aspects of Instrumented Safety Systems, 328
    Hazard and Risk Analysis, 328
    ISS Documentation, 332
    Process Requirements Specification, 330
Engineering Aspects of Instrumented Systems Safety Requirements Specification, 331
Engineering Design Package, 384
Equipment Design, 165
Equipment Isolation, 340
Erosion, 146
Evacuation, 352
Event Tree, 104
Exothermic Reactions, 47
Expansion Joints, 282

Explosion Suppression, 365
    Deflagration Pressure Containment, 367
    Equipment and Piping Isolation, 368
    Explosion Venting, 367
    Oxidant Concentration Reduction, 367
Exposure to Smoke and Gas, 73
Failure Modes and Effects Analysis, 101
Failure Modes, Effects and Criticality Analysis, 118
Failure Modes, Effects and Diagnostic Analysis, 118
Fault Tree, 103
Filters, 244
Fire Hazards, 73
Fire Impact, 82
Fire Point, 43
Fire Protection Handbook, 40
Fire Protection, 357
    Carbon Dioxide Systems, 362
    Clean Agents, 362
    Dry Chemical Systems, 363
    Fire Extinguishers, 363
    Fire Water Systems, 360
    Firefighting Agents, 359
    Foam Systems, 361
    Mitigation Systems, 360
    Portable Fire Suppression Equipment, 363
    Structural Fireproofing, 358
    Water Mist Systems, 361
    Water Spray Systems, 360
Fired Equipment, 255
    Design Considerations, 265
    Design Solutions, 258
    Failure Scenarios, 258
    Past Incidents, 257
Flammability Data, 40
Flammability, 15, 17
Flammable Limits, 44
Flanges, 282
Flare Radiation Levels, 84
Flares, 342
Flash Point, 40, 41, 42

INDEX

Fluid Transfer Equipment, 224
    Design Considerations, 231
    Design Solutions, 226
    Failure Scenarios, 225
    Past Incidents, 225
FMEA, 118
Foundations, 147
Frequency Assessment, 112
Fuel, 313
Galvanic Corrosion, 145
Gas Deflagration Index - K_g, 45
Gas/Fire Detection, 353
    Fire Detectors, 354
    Flammable Gas Detectors, 355
    Installation of Sensors, 354
    Toxic Gas Detectors, 356
Gas/Liquid Separators, 178
Gas/Vapor-Liquid Mixtures, 286
Geotechnical Studies, 147
GESTIS-DUST-EX, 46
Glass-Lined Reactors, 193
Grade Level Structures, 150
Grounding and Bonding, 323
Hazard Analysis Techniques, 63, 94
Hazard and Operability Study, 99
Hazard Evaluations, 2
Hazard Identification, 63, 64, 96
Heat Tracing, 283
Heat Transfer Equipment, 204
    Design Considerations, 212
    Design Solutions, 206
    Failure Scenarios, 206
    Past Incidents, 205
Heat Transfer Fluid, 314
Heating and Cooling Systems, 193
High Pressure, 19
High Temperature, 19
Higher Pressures, 70
Higher Temperatures, 70
Human Factors, 30, 104

Human Factors in Design, 156
    Culture and Working Environment, 156
    Tools for Project Management, 158
Human Impact Data, 79
Human Reliability Analysis, 106
Hydrogen Induced Attack, 145
Ignition Control, 316
Ignition Sources, 76
Immediately Dangerous to Life or Health (IDLH), 88
Inadvertent Mixing, 75
Incident and Near-Miss Investigations, 34
Incineration Systems, 348
Incompatible Materials, 30
Individual Risk, 115
Inert Gas, 311
Inherent Safety, 123
Inherently Safer Design, 124, 128
    Dilution, 131
    Minimize, 129
    Moderate, 131
    Simplify, 131
    Substitute, 130
Inherently Safer Technology, 128
Inhibitor Injection, 371
Instrument Air, 312
Instrumented Safety Functions, 135
Instrumented Safety Systems, 135, 325
Integrated Risk Information System (IRIS), 52
Intergranular Corrosion, , 144
Intrinsically Safe and Nonincendive Equipment, 321
Inventory, 20
Layer of Protection Analysis, 100, 329
Leak Detection, 357
Life Cycle, 33, 94
Lightning, 324
Likelihood, 25
Liquid Seals, 336
Loading/Unloading, 300
Low Energy Electrical Equipment, 320
Low Pressure, 19, 70
Low Temperature, 19

Management of Change, 34
Managing Change, 383
Manual Handling, 301
Mass Transfer Equipment, 194
    Design Considerations, 202
    Design Solutions, 197
    Failure Scenarios, 196
    Past Incidents, 195
Material Characteristics, 15
Material Handling and Warehousing, 292
    Design Considerations, 300
    Design Solutions, 295
    Failure Scenarios, 295
    Past Incidents, 293
Material Safety Data Sheets (MSDSs), 15, 40, 50
Materials of Construction, 30, 71, 141
Mechanical Interlocks, 370
Metallurgical Changes, 143
Methods of Overpressure Protection for Two-Phase Flows, 339
Milling Equipment, 256
Minimum/Limiting Oxygen Concentration, 45
National Institute for Occupational Safety and Health (NIOSH), 52, 88
National Oceanic and Atmospheric Administration (NOAA), 55
Normal Operating Zone, 135
Occupational Safety and Health Administration (OSHA), 52
Occupied Building Location, 140
Operating/Maintenance Procedures, 385
    Consequences of Deviations, 387
    Considerations for Batch Procedures, 387
    Developing Procedures, 386
    Maintaining Procedures, 389
    Need for Procedures, 386
    Upper and Lower Safe Operating Limits, 387
Operating Conditions, 142
Overpressure Hazards, 74
Overpressure Impact, 91
Passive Design Solutions, 125

Physical Explosions, 74
Physical Location, 71
Pilot Operated Relief Valves, 335
Piping and Piping Components, 265
    Design Considerations, 280
    Design Solutions, 269
    Failure Scenarios, 269
    Past Incidents, 268
Piping Specifications and Layout, 280, 302
Pitfalls in Material Selection, 143
Pitting, 144
Plant Emergency Alarm and Surveillance, 351
Plant Siting and Layout, 137
Pressure Effects on Structures, 92
Pressure Relief Systems, 337
    Flashing Liquids, 337
    Liquid Service, 337
    Relief Valve Inlet and Outlet Sizing, 338
    Vapor Service, 337
Pressure-Vacuum Relief Valves, 336
Pressure/Vacuum Relief Systems, 332
    Pressure Relief Devices, 335
    Relief Design Scenarios, 333
Pressurized Storage Tanks, 181
Probit Analysis, 79
Probit Functions, 80
Procedural Design Solutions, 126
Process Chemicals, 142
Process Control Instrumentation, 265
Process Design/Process Chemistry, 135
Process Equipment Safe Operating Limits, 135
Process Hazard Analyses, 34, 96
Process Hazards, 64, 65
Process Knowledge Management, 379
Process Safety Information, 379
Process Variations, 142
Process Vents and Drains, 316
Process Vessels, 177
Properties of Materials, 141
Protection Layers, 315
Purging and Pressurized Enclosures, 319
Qualitative Risk Criteria, 113

INDEX

Quantitative Risk Analysis, 103, 109, 110, 117
Quantitative Risk Criteria, 114
Quench System, 371
Rapid Phase Transition Explosions, 74
Reactive Hazards, 72
Reactivity Evaluation Screening Tool (REST), 54
Reactivity, 15, 16
Reactors, 183
    Design Considerations, 191
    Design Solutions, 186
    Failure Scenarios, 185
    Past Incidents, 183
Registry For Toxic Effects of Chemical Substances (RTECS), 52
Reliability/Maintainability Analysis, 118
Risk Analysis, 23, 63, 108
Risk Based Process Safety, 5, 109, 381
Risk Criteria, 113
Risk Management, 108
Risk Tolerance, 117
Risk, 22, 26
Risk-Based Design, 21
Runaway Reactions, 31, 49, 192
Rupture Disks, 335
Safe Operating Limit (SOL), 136
Safeguarding Strategies, 123
    Active, 123, 125
    Characteristics of Design Solution Strategies, 126
    Inherent, 123, 124
    Passive, 123, 125
    Procedural, 123, 126
Safeguard Stewardship, 127
Safety Factor, 127
Safety Instrumented Systems, 128, 325
    Safety Integrity Level, 326
    SIS Life Cycle Overview, 326
Safety Relief Valves, 335
Selection of Insulation Materials, 152
Semi-Quantitative Hazard Evaluations, 100
Shelter-in-Place, 352
Site Layout, 138

Site Security Issues, 159
    Cyber/Electronic Security, 161
    Delay, 160
    Detect, 160
    Deter, 160
    Layers of Protection, 160
    Physical Security, 160
    Security Layers of Protection, 160
Sizing of Rupture Disks, 338
Societal Risk, 116
Solid-Fluid Separators, 237
    Design Considerations, 244
    Design Solutions, 239
    Failure Scenarios, 239
    Past Incidents, 237
Solids-Gas Mixtures, 287
Solids Handling and Processing Equipment, 245
    Design Considerations, 254
    Design Solutions, 249
    Failure Scenarios, 248
    Past Incidents, 246
Solids-Liquid Mixtures (Slurries), 286
Stability, 15
Static Electricity, 322
Steam Snuffing, 370
Steam/Condensate, 309
Storage and Warehousing, 304
Storage Layout, 140
Storage Tanks and Vessels, 178
Storage, 255
Stress Corrosion Cracking, 144
Stress-Related Corrosion, 144
Surface Drainage, 147
Temperature and Pressure, 70
Temperature Impact, 90
Test Strategies, 58
Testing Methods, 56
Thermal Expansion, 282

Thermal Insulation, 150
  Absorption of Liquids, 151
  Durability, 152
  Fabrication, 152
  Fire Safety, 151
  Thermal Performance, 151
Thermal Radiation, 73, 83, 84
Thermal Stability, 57
Thermoplastic, Plastic-Lined and FRP Piping, 284
Toxicity Hazards, 74
Toxicity, 15, 17
Trapped Liquids, 145
Troubleshooting Zone, 135
Tube Rupture, 266
Types of Explosions, 92
Types of Ignition Source, 76
Unacceptable or Unknown Operation Zone, 136
Underground Piping, 148
Unit Layout, 139
Utility Systems, 72, 143

Utility Systems, 306
  Design Considerations, 308
  Past Incidents, 307
Vacuum Equipment Considerations, 235
Vacuum Relief Devices, 337
Valves, 281
Vapor Cloud Explosions, 74
Vapor Control Systems, 349
Vapor Density, 41
Velocity Criteria, 280
Ventilation/Exhaust, 321
Vessels, 167
  Design Considerations, 177
  Design Solutions, 169
  Failure Scenarios, 168
  Past Incidents, 167
Vibration, 283
Water/Steam Curtain, 370
What-If Analysis, 98
Workplace Hazardous Materials Information System (WHMIS), 52
Zones of Operation, 136